W9-AOB-506

SANCTITY OF LIFE AND HUMAN DIGNITY

Philosophy and Medicine

VOLUME 52

The titles published in this series are listed at the end of this volume.

SANCTITY OF LIFE AND HUMAN DIGNITY

Edited by

KURT BAYERTZ

KLUWER ACADEMIC PUBLISHERS
DORDRECHT / BOSTON / LONDON

Library of Congress Cataloging-in-Publication Data

Sanctity of life and human dignity / edited by Kurt Bayertz.
        p.   cm. -- (Philosophy and medicine ; v. 52)
    Includes bibliographical references and index.
    ISBN 0-7923-3739-5 (hardbound : alk. paper)
    1. Medical ethics. 2. Dignity. 3. Life--Religious aspects.
 I. Bayertz, Kurt. II. Series.
 R724.S352 1996
 174'.2--dc20   95-36283

ISBN 0-7923-3739-5

Published by Kluwer Academic Publishers,
P.O. Box 17, 3300 AA Dordrecht, The Netherlands.

Kluwer Academic Publishers incorporates
the publishing programmes of
D. Reidel, Martinus Nijhoff, Dr W. Junk and MTP Press.

Sold and distributed in the U.S.A. and Canada
by Kluwer Academic Publishers,
101 Philip Drive, Norwell, MA 02061, U.S.A.

In all other countries, sold and distributed
by Kluwer Academic Publishers Group,
P.O. Box 322, 3300 AH Dordrecht, The Netherlands.

Printed on acid-free paper

Printed in the Netherlands

TABLE OF CONTENTS

KURT BAYERTZ and H. TRISTRAM ENGELHARDT, JR. / Preface   vii

KURT BAYERTZ / Introduction: Sanctity of Life and Human Dignity   xi

PART ONE / SANCTITY OF LIFE

JAMES F. KEENAN, S.J. / The Concept of Sanctity of Life and Its Use in Contemporary Bioethical Discussion   1

HELGA KUHSE / Sanctity of Life, Voluntary Euthanasia and the Dutch Experience: Some Implications for Public Policy   19

WOLFGANG LENZEN / Value of Life vs. Sanctity of Life – Outlines of a Bioethics that Does without the Concept of *Menschenwürde*   39

STEPHEN WEAR / Sanctity of Life and Human Dignity at the Bedside   57

PART TWO / HUMAN DIGNITY

KURT BAYERTZ / Human Dignity: Philosophical Origin and Scientific Erosion of an Idea   73

MARTIN HAILER and DIETRICH RITSCHL / The General Notion of Human Dignity and the Specific Arguments in Medical Ethics   91

DIETER BIRNBACHER / Ambiguities in the Concept of Menschenwürde   107

THOMAS PETERMANN / Human Dignity and Genetic Tests   123

PART THREE / THE CONCEPT OF A PERSON

LUDGER HONNEFELDER / The Concept of a Person in Moral
Philosophy   139
MARY C. RAWLINSON / Alterity and Judgment – Some Moral
Implications of Hegel's Concept of Life   161
ANTON LEIST / Persons as "Self-Originating Sources of Value"   177

PART FOUR / PROBLEMS OF CRITICAL CARE

H. TRISTRAM ENGELHARDT, JR. / Sanctity of Life and Menschen-
würde: Can These Concepts Help Direct the Use of Resources
in Critical Care?   201
JOHN C. MOSKOP / Not Sanctity or Dignity, but Justice and
Autonomy: Key Moral Concepts in the Allocation of Critical
Care   221
VOLKER VON LOEWENICH / Sanctity of Life and the Neonatalo-
gist's Dilemma   229

PART FIVE / THE ROLE OF THE STATE

KEVIN WM. WILDES, S.J. / The Sanctity of Human Life: Secular
Moral Authority, Biomedicine, and the Role of the State   241
MARTIN HONECKER / On the Appeal for the Recognition of
Human Dignity in Law and Morality   257

APPENDIX

BJÖRN HAFERKAMP / The Concept of Human Dignity: An
Annotated Bibliography   275
GEORGE KHUSHF / The Sanctity of Life: A Literature Review   293

Notes on Contributors   311

Index   313

PREFACE

The conference that gave birth to this volume was held in Bielefeld, Germany, October 1–3, 1992, at the Zentrum für interdisziplinäre Forschung (ZiF). It bore the rather pedantic title, "Die Begriffe 'Menschenwürde' und 'Sanctity of Life' und ihre Tragweite für ethische Konfliktlagen in der modernen Medizin". As the participants made their plans for the meeting, the expectation was of an intense three days of scholarly explorations of the historic and philosophic roots of "sanctity of life" and "human dignity" (*Menschenwürde*). In the English-speaking world and in Germany these two concepts have played rich and varied roles in reflections on how appropriately to protect persons and restrain improper technological interventions. The focus and concerns of the conference were primarily theoretical. An intense scholarly discussion in a closed conference setting was to be the basis for reassessing the significance and force of these concepts.

In inviting participants, there was much discussion of whom to select. Because of her prior publications and scholarship in the area, an obvious invitee was Helga Kuhse. However, the work of Helga Kuhse was associated with Peter Singer, who had been under intense informal ban in Germany and Austria. His invitations had been met with protests leading to the cancellation of meetings. Because she was an ideal contributor, and given the intellectual focus of the conference, Prof. Kuhse was invited. There was the hope that a closed small meeting could be the basis for academic exploration without a riot.

The result was a convulsion: An angry crowd gathered, protesting Prof. Kuhse's participation, as well as the conference as a whole. Attempts

vii

K. Bayertz (ed.), Sanctity of Life and Human Dignity, vii–ix.

to determine the points at issue and discuss them with the crowd were met with loud whistles and screams to prevent reasoned public exploration of the various allegations (e.g., that the conference was one on genetic engineering). One of the co-organizers, Prof. Kurt Bayertz, was doused with paint. The Rev. Prof. Kevin Wm. Wildes, S.J., sustained bodily injury. The glasses of Prof. Hans-Konrat Wellmer were broken. In general, threatening behavior was displayed by many. Under other circumstances, this may have led to the closing of a conference. Surely, similar conferences in Germany and Austria had been closed by similar threats.

Such did not occur at Bielefeld. The rector of the University of Bielefeld, Prof. Dr. Helmut Skowronek (who had just taken office), and Prof. Dr. Peter Weingart, member of the directorium of ZiF, stood by the conference organizers and the new goal that had been given to the conference: resolutely to stand by the principle that philosophical issues should be explored philosophically, not shouted down or silenced by threat of disruption. The decision was to stand by the commitment to a peaceable and uninterrupted scholarly discussion of the questions at issue. The conference was a success against all odds. Most importantly, against considerable odds, the principle of academic freedom was protected. We cannot thank those who stood by us enough. Our debt is not merely personal. Our debt is for their role in preserving the academy's commitment to the reasoned exploration of scholarly issues despite threats of disruption or of the imposition of secular orthodoxies or notions of political correctness. Whatever might be appropriate in religious or theological debates that might appropriately exclude some as heretics, secular academic discussions should be free to explore all issues peaceably with the confidence that open, reasoned discussion is superior to coerced silence.

We wish to underscore our deep debt to the *Zentrum für interdisziplinäre Forschung* and the *University of Bielefeld* that put their facilities at our disposal. They did this at considerable risk and with real damage sustained. We wholeheartedly appreciate all the members of the staff of the ZiF who stood by us and supported us, at times at some personal risk. They showed courage and good humor; they provided protection and hospitality. We are very grateful for the financial support from the *Fritz Thyssen Stiftung* and the *Akademie für Ethik in der Medizin* for their support of the conference and from the *Fritz Thyssen Stiftung* for the publication of this volume. Finally, we would like to thank Björn Haferkamp who managed to compose a respectable text from the various manuscripts received.

April 21, 1995

Kurt Bayertz
H. Tristram Engelhardt, Jr.

KURT BAYERTZ

INTRODUCTION: SANCTITY OF LIFE
AND HUMAN DIGNITY*

I. TWO ROCKS

The contributions within this book analyse two concepts which play a
key role in bioethical discussions: the terms "sanctity of life" and "human
dignity". Although these two terms have differing histories, and although
they are at home within differing cultures ("sanctity of life" mainly in
the Anglo-Saxon, "human dignity" mainly in the European and especially
German debates), they have similar "strategical" functions within their
own cultural contexts. In many cases, they are used to bring a difficult
or controversial debate to an end. This might be in the form of a gentle
caress, appealing to (supposed) common convictions: "Of course,
everyone agrees that human life and human dignity are inviolable! This
means, however and without doubt, that . . .". And because hardly
anybody doubts that human life is worthy of protection or human dignity
of respect, it is difficult to contradict what this "means without doubt".
In order to put forward one's own views, one would have to withdraw
from this caress and put paid to these supposed common convictions.
Anybody attempting just this is then frequently confronted with a second,
rather less gentle employment of these two terms, amounting to dis-
pensing with indispensible and elementary moral consensus, and to
doubting the cornerstones of morality itself. Appealing to "sanctity of
life" and "human dignity" thus functions as a kind of "stop" sign, putting
an end to all further discussion.

This attempt to identify cornerstones of morality and to protect them
from scepticism and criticism should come as no surprise. Ever rapidly
increasing technological innovation in the field of biomedicine has

xi

K. Bayertz (ed.), Sanctity of Life and Human Dignity, xi–xix.
© 1996 Kluwer Academic Publishers. Printed in the Netherlands.

created a multitude of new options which are viewed by many contemporaries as a source of uneasiness, if not alarm. Much of what used to be taken for granted has become doubtful and the subject of controversial debate. This includes the descriptive as well as the normative dimension. More and more frequently, as well as rapidly, many are reaching the point where they see doubt cast upon their deepest and most essential moral convictions; quite understandably, they would neither like to question these themselves, nor to have them questioned by others. In these circumstances, it seems obviously sensible to look around for stable rocks able to weather the storm of technological progress and moral pluralism. At the very least, a few normative stakes should be driven into the immense field of biological activity. "Sanctity of life" and "human dignity" are just such rocks or stakes.

There are three problem constellations which give rise to this uneasiness, and within which the two concepts thus play an important role. The principle that human life is holy and inviolable, and that physicians especially should refrain from destroying it, is an argument which runs through medical ethical debate from Ancient Times to the present day. This principle entered the realm of everyday U.S. politics in 1973, when the Supreme Court ruled that *abortion* be legalised. Since then, the "holiness of human life" has become the central argument against the acceptance of abortion to be focussed on by the ethical debate on abortion. Secondly, this principle plays an important role in the debates about modern *intensive care* and the connected problems of *terminating treatment* and *euthanasia*. Intensive care is based on attempts to maintain the vital functions of the human body in a crisis situation, thus giving the organism a little more time, necessary for the healing process: it involves saving life itself. Anybody doubting the current practice of intensive care, the – in many cases senseless – prolonging of life with technology, and who chooses to emphasise the enormous costs involved, is quickly confronted with the "sanctity of life" argument. This is even more true with regard to the ethical legitimacy of euthanasia. Preserving human life is regarded as the highest duty of the medical profession; human life is the last thing which may be calculated in terms of money.

Thirdly and finally, the ethical controversies surrounding technological intervention in human reproduction should be mentioned, as the term "human dignity" here too plays a key role, especially in the German-speaking countries. This came to our attention in the 1960s with the debate on legislating heterologous insemination. Since this form of reproduction technology removes the sperm donor from the "natural context of responsibility", truly "reifying" him, many authors have viewed it as an obvious violation of "human dignity". Similar arguments were put forward in the 1980s, when *in vitro* fertilisation and the progress of gene technology opened up a multitude of new options in the field of human reproduction. In Germany in 1991, as a result of this debate, interventions in the human germ-line, human cloning and the creation of human-animal hybrids were decreed worthy of criminal punishment: inflictions of this law carry a punishment of five years' imprisonment or a fine. Following a national ballot in 1992, Switzerland even outlawed these options as part of its Constitution. In both countries, the most important reason given for these restrictions was the State duty to protect "human dignity". In several European Documents concerning bioethical issues, this concept plays a key role, too. Back in 1982 the *European Parliament* declared that the rights to life and human dignity protected in Articles 2 and 3 of the *European Convention for Human Rights* include the right to a genome which has not been artificially manipulated. And the more recent "Bioethics Convention" of the *Council of Europe* defines in Article 1 its purpose and object as to "protect the dignity and integrity of all human beings". Already in the title of the convention the concept of dignity is prominent: *Convention for the Protection of Human Rights and Dignity of the Human Being with Regard to the Application of Biology and Medicine: Bioethics Convention.*

II. SOURCES OF AUTHORITY

Both concepts are able to play their "strategic role" more than anything because of a specific authority lending them far more importance than comparable ethical terms such as "autonomy" or "justice". This specific authority springs from two different sources. Firstly, both terms have *religious roots*. With regard to the term "sanctity of life", this is obvious from the vocabulary alone. This does not mean, however, that this term may only be used within a religious framework. There have been many attempts to separate it from its religious roots and to incorporate it into secular ethical debates. "Sanctity of life" then means no more than *life's freedom from injury* as an absolute moral principle. The situation is similar for the term "human dignity", the origins of which lie at least partly in the Christian idea of *imago dei*. As part of the human species, which has an extraordinary position within creation, each human being is attributed with an inherent dignity which – like its immortal soul – should ultimately be exempt from human interference. From this religious context, the idea of an inherent human dignity became part of Modern Philosophy and was reformulated in categories of secular reason. Kant, for example, connected human dignity with the reasonable nature of the human being: the individual possesses dignity because of its participation in the reason of the species and its autonomy.

The secularisation of these two terms has not led to a loss in authority, which may be attributed to the key fact that they have been absorbed by the Law, especially modern Constitutional Law. This is particularly true of the term "human dignity", which has gained radically in Constitutional importance during the 20th Century. Today, it is an important part of the Constitutions of many countries, including Canada, Ireland, Italy, Sweden, Portugal, Spain, Greece and Germany. In the U.S., the concept of human dignity has been used by the Supreme Court in connection with the First, Fourth, Fifth, Sixth, Eighth and Fourteenth ammendments. In addition, it has attained a firm position within numerous international documents, not least the Charter of the United Nations, which proclaims the "inherent dignity" and the "equal and inalienable rights of all members of the human family" to be "the foundation of freedom, justice and peace in the world". The Universal Declaration of Human Rights of 1948 took up and enforced this idea that human rights are anchored in the "inherent dignity" of the human being. The term "sanctity of life" has not been party to a comparable legal institutionalisation – at least not in these words. And yet there can be no doubt that the protection of (human) life is one of the highest priorities of a modern State. This is not only one of the basic insights of modern political philosophy, but is also beyond legal contradiction. The second source of authority for both these terms is therefore their anchored position in current national and international Law.

III. CONCEPTUAL DIFFICULTIES AND AMBIGUITIES

However accepted their factual authority may be, their rational authority is just as controversial. Within philosophical literature, both the term "sanctity of life" and the term "human dignity" are controversial. This is partly due to the fact that neither of these terms has been exactly defined, and partly due to the fact that their relative importance is unclear. Let us look first at the term "human dignity", which many have characterised as an "empty formula" to be filled with varying contents. A coherent interpretation of this term may not even be extrapolated from the legislation of the Federal Constitutional Court: on the one hand, decisions have been made which interpret "human dignity" in the sense of a positive, substantial image of man (rooted in Western Christian tradition), on the other hand, "human dignity" is bound to human subjectivity and self-determination. These diverging interpretations exactly mark the area of tension in which the debates on gene and reproduction technology are taking place, abortion and intensive care. Even in the Constitutionally binding interpretation of "human dignity", the term seems rather to give rise to a controversial range of meaning than to offer a neutral platform where all the different positions may meet.

The position with "sanctity of life" is similar, its secular interpretation also being confronted with various difficulties. The first of these has to do with terminology: does "sanctity of life" refer to *all* forms of life, or just to human life? Both possibilities are riddled with difficulties. If life *per se* is holy, then the human right to live is no higher than that of any animal or even a plant. If, on the other hand, the holiness is restricted to human life (and this is at least the case with regard to the binding "right to life" in the Constitution), then we have to ask how this sharp division between human and non-human life may be ethically justified, and how it is to be put into operation without resorting to biological characteristics. On top of all this is the question, in many cases not just theoretical but eminently practical, of whether *every* human life is to be considered inviolable, under *all* conceivable circumstances, and whether no price (not just financial) is too high in order to preserve a human life.

This brings us to the problem of relative importance, which surfaces wherever the postulate "sanctity of life" clashes with other principles, norms or vlaues. With regard to distributive justice, for example, there may be tension between the duty to protect life and other individual or collective interests. Resources in short supply (money, personnel, etc.) being used for intensive care cannot be used for preventive or educational purposes simultaneously. Especially interesting in our chosen context are, of course, the conflicts arising in certain cases between the principle of "sanctity of life" and the principle of "human dignity".

Two different cases are at stake here. On the one hand, "sanctity of life" and "human dignity" may clash with each other within the framework of an *interpersonal* conflict of values. This is the case with abortion, for example. There is a conflict between the embryo's right to live and the woman's right to self-determination, a fundamental element of human dignity. This kind of conflict leads to the formation of hierarchies, i.e. which of the various principles and rights involved is to outweigh the other(s). And yet just such hierarchical formations are controversial. The present debate on the "sanctity of life" principle is not so much concerned with its validity as with its relative importance. Is life an *absolute* value or can it be relativised through other values? Besides interpersonal collisions of principle, there are also *intrapersonal* conflicts of value. Intensive care, termination of treatment and – in extreme cases – euthanasia are probably the most difficult cases involving such collisions of principle. In many cases, the relatives of an irreversibly comatised ("brain dead") patient insist that treatment be terminated because they view it as undignified. The physician in charge, however, often counters with an appeal to the "sanctity of life" argument: he refuses to terminate treatment with the argument of being simply unable to "play God".

IV. BLOCKED DISCUSSIONS AND DECISIONS

These two terms which, *prima vista*, seem to amount to much the same thing are then, in such cases, suddenly and irreversibly opposed. They are used to justify as morally indispensable types of action which are totally incompatible. Argumentatively, hardly anything more can be done at this stage, since as soon as somebody appeals to "human dignity" or "sanctity of life", each quarrel is abruptly brought to an end. Both terms are brought forward as "winning" arguments. Both are intended to refer to highest values which cannot be weighed up against other, less significant values. This is also intended to prevent moral relativity on the theoretical level and individual arbitrariness on the practical level. And yet, however welcome such dependable orientations might be, we have to doubt whether they are to be gained in this manner. As yet, the actual consequences have taken another direction. On a theoretical level, the discussion is cancelled, or burdened with personal reproaches; necessary values are not ensured: moral convictions are immunised. Where, originally, life was supposed to be free from injury and dignity inviolable, particular arguments are suddenly helping themselves to the concepts in question. On the practical level, an appeal to "heavy concepts" easily changes from being a way of measuring the moral correctness of decisions to a vehicle for avoiding decisions. Anyone appealing to the holiness of life does not, in a concrete case, have to make difficult and painful decisions concerning life-prolonging measures vs. termination of treatment. The term "sanctity of life" has put paid to this decision; other points of view no longer have to be taken into consideration.

An instructive example of this was the discussion about the "Erlangen baby", which threw the German public into consternation for weeks. On 5th October, 1992, at 2 p.m., 18 year-old Marion Ploch had an accident on her way home from working as a doctor's receptionist in Southern Germany. Her car crashed into a tree. Twenty to thirty minutes went by before the emergency services arrived at the site, intubated the critically injured patient and drove her to hospital. Because of her serious brain damage she was transferred to the neurological intensive care unit of the University Hospital in Erlangen. Over the next few days, the patient's condition became increasingly critical. On 8th October, 1992, neurologists pronounced the patient to be brain dead. Normally the story would end here. When brain death is established, all the machines are switched off, as there is no point in artificially respirating and feeding the dead. Marion Ploch was, however, pregnant. A foetus was alive inside her body, and it would die if all the machines were to be turned off. Since the foetus seemed to be unimpaired and the dead woman's remaining bodily functions were stable, artificial respiration and feeding were continued. An attempt was made to save the life of the foetus. On November 16th, however, the "Erlangen baby" was spontaneously aborted, dead on arrival. Artificial respiration of the mother ceased. It remained unclear exactly what caused the foetus' death. The parents refused permission for an autopsy to be carried out on their daughter or their grandchild. Several other things also remained unclear. The parents told the media that they were put under pressure by the physicians: that they were blackmailed into agreeing with the physicians' decision. The responsible public prosecutors did not, however, start preliminary proceedings.

In the meantime an article featured in the German tabloid newspaper "Bild" had provoked a huge public discussion about the physicians' decision. A great many opinions were expressed throughout the media, ranging from acceptance to rejection. Different professional circles also expressed differing opinions. Even those who condoned the physicians' behaviour were uneasy and apprehensive. Most of the public, however, seemed to be repulsed and outraged. The occurrence was said to be "totally perverse", "macabre" and "a scandal"; it was condemned as being "outrageous, inhuman and an insult to women". Rejection was particularly widespread and extreme amongst women. On November 10th, the *Ärztinnenbund* (Association of Women Physicians), the *Grünen* (Green Party) and the *Evangelische Frauenarbeit* (Protestant Women's Group) submitted a petition containing 7000 signatures from all over Germany to the Minister for Women's Affairs, Angela Merkel, in Bonn, demanding that the "experiment" in Erlangen be stopped. They appealed to the Minister to make sure that then, and in the future, this kind of "medical experiment" be prohibited. They called for "official powers" to become involved; the Court responsible did not, however, take legal proceedings against the physicians in Erlangen. But criticism was not restricted to the woman's movement. One Member of Parliament challenged the German Government to give its opinion on whether artificially maintaining the bodily functions of a dead person for months on end in order to complete a term of pregnancy represents a violation of the *Menschenwürde* (human dignity) guaranteed in Article 1 of German Basic Law. The objection – like many others – was centered around the instrumentalisation argument: a human being is used exclusively as the means to an end, reduced to an "incubator" or "child-bearing machine".

On the other hand, the physicians in charge, but also much of the medical profession emphasised the duty to preserve human life: in this case the life of the foetus. They did not refer specifically to the "Heiligkeit des Lebens", but certainly to a "right to life". Important here is not so much the terminological difference but, on the one hand, the fact that this right is also and without limitation attributed to a 14-week old embryo and, on the other, the unquestioning nature with which the right to live of the "child" – as, interestingly enough, this foetus kept being called – was ranked supreme to all other considerations. If we follow the argumentation of the physicians in question, they had no other choice: they could only act in the manner in which they actually did.

V.

The contributions within this book are all concerned with the terms "sanctity of life" and "human dignity", from different philosophical, professional and cultural points of view. They do not reach agreement concerning the significance and scope of these terms, but, to a certain extent, they provide a reflection of the public controversies connected with both terms. Of course, there are some points of convergence. Even those deeming the concept of "human dignity" indispensable do, for example, admit the relative uncertainty of this term's content. "Human dignity" is therefore unable to provide a sufficiently clear normative orientation, instead having to be translated into more precisely defined norms and values; "blurred edges" necessarily remain.

Even more significant than such points of convergence is the fact that the contributions within this book contain a dialogue demonstrating how blocking each other with the aid of "heavy concepts" is not necessarily a consequence of employing such concepts. To put it another way: from the significance of concepts like "sanctity of life" and "human dignity" and from their ability to solve practical problems raised by the biomedical sciences and their application, there is little to be seen *a priori*. Retreating to ideologically prominent positions is of no help here: only working with and on these concepts.

Department of Philosophy
University of Münster
Germany

NOTE

* Translated into English by Sarah L. Kirkby.

JAMES F. KEENAN, S.J.

THE CONCEPT OF SANCTITY OF LIFE AND ITS USE IN CONTEMPORARY BIOETHICAL DISCUSSION

I. INTRODUCTION

I am a Roman Catholic moral theologian and have often seen the term "sanctity of life", but in preparing the topic, I am surprised at the fact that it receives very little attention in places where I would have expected some. For instance, in the fifteen volume collection of the *New Catholic Encyclopedia*, it has no entry. It appears as a modest after-thought in the later supplement. It is not found in new theological dictionaries from the United States, England, or Germany: *The New Dictionary of Theology*, *The Oxford Dictionary of the Christian Church*, *The Theological Dictionary*. It did not appear in the German *The Concise Dictionary of Christian Ethics*; in Palazzini's Italian *Dictionary of Moral Theology* there was only "Life, Respect for: see Murder, Suicide." Only the Anglican, John MacQuarrie who edited two dictionaries, ran entries in both, *A Dictionary of Christian Ethics* and *The Westminster Dictionary of Christian Ethics*. Thus, Roman Catholics, who are believed to be the most apt to use it, rarely explicitly analyze it. The oft-quoted words of Edward Shils ([61], p. 2) seem appropriate: "To persons who are not murderers, concentration camp administrators, or dreamers of sadistic fantasies, the inviolability of human life seems to be so self-evident that it might appear pointless to inquire into it."

Surprising, too, is the fact that the concept's origin has not been recorded nor its development narrated. Though often noted to be used without any exactitude (e.g., [39], p. 3), more surprising is the fact that we do not seem to know where the term came from, what its roots are, and why it has been appropriated elsewhere. Worse, this lack of

1

K. Bayertz (ed.), Sanctity of Life and Human Dignity, 1–18.
© 1996 *Kluwer Academic Publishers. Printed in the Netherlands.*

knowledge becomes a particular asset for those who oppose the values that the concept purports to protect. A reading of Glanville Williams's *The Sanctity of Life and the Criminal Law*, for example, critiques the term, by presenting a number of advocates of a particular interpretation of the term. The usual suspects who are "rounded-up" are those who handily make the very mistakes that the critic seeks to attack. This lack of scholarship on the roots of the phrase leads critics, like Williams, to create their own straw man. Their dubious method of selectivity cannot be countered, however, until the concept's actual background is established. I intend to do just that in these pages.

I begin by examining the phrase etymologically. I then turn to its use as it appears first as a *restrictive, absolute injunction* against the direct killing of human life and second as a *general, universal presumption* to revere human and other life forms. I evaluate each use as I present each.

As a Roman Catholic theologian my comments and bibliographical references tend considerably, though not at all exclusively, to Roman Catholic teaching. Besides exercising my competence I choose this for a number of reasons. First, Roman Catholic ethics has, since the high scholasticism of the 12th century, boasted a natural law based ethics that is arguably accessible to all persons of good will, independent of religious faith. Unlike the ethics of modern or contemporary Protestant theologians, notably Karl Barth and Paul Ramsey, its ethics backs away from any claims of being revealed or confessional (see [28]). On the contrary, as in Saint Paul's *Letter to the Romans* (1:18ff.), Roman Catholic ethics argues that one does not have to be a member of its community in order to talk about its ethics. Second, practically speaking, choosing Catholic ethics gives parameters to a discussion that could be unwieldy. Third, Catholic moral theologians have debated the topic with concerns parallel to those raised by philosophers, including those at the conference that addressed this topic in Bielefeld, Germany. Thus, hopefully, my paper will engage a broader constituency than hitherto achieved. Fourth, interestingly the term is gaining greater currency among Bishops and Catholic lawyers in the States. Let me begin.

II. ETYMOLOGICAL ROOTS

What does it mean to say *"sanctity* of life?" The term is distinguishable from equality of life, purity of life, and gift of life precisely by the quality of "sanctity". The *Oxford Latin Dictionary* notes, "*sanctitas*: The state of being protected by religious sanction, sacro-sanctity." The *Oxford English Dictionary* writes, "Sanctity: 1. Holiness of life, saintliness; 2. The quality of being sacred or hallowed; sacredness, claim to (religious) reverence; inviolability." *Random House Dictionary* writes, "Sanctity 1. holiness, saintliness, or godliness; 2. sacred or hallowed character; *inviolable sanctity of the temple*. 3. a sacred thing." These definitions suggest that sanctity is a quality which is reverenced as somehow touched by divinity and therefore untouchable for humans: sanctity is that which the divinity protects from violability. Forcellini's Latin *Lexicon* in its first entry on *sanctitas* underlines this forcefully with a quote from Sallust: "sanctitas, qualitas illa, qua res venerabiles et inviolabiles sunt." The overriding effect of sanctity is veneration and inviolability; the overriding reason for both is its sacral quality, a point found in the Scriptures, e.g., in *2 Maccabees* 3.12 and *1 Corinthians* 3.17. This quote from Sallust is central then for this paper: sanctity is used as a concept to uphold both reverence and inviolability. We will look at the second value first.

III. SANCTITY AS ABSOLUTE INVIOLABILITY

According to Sallust, sanctity can be used two ways. Later, however, the term more frequently emphasized the inviolable quality as opposed to a more general venerable quality. According to this singular meaning, sanctity does not enjoin a duty, but rather a limit. The term is not about what we ought to do, but, what we cannot do. Sanctity of the temple, sanctity of marriage and sanctity of life all function by informing us humans that we cannot violate a temple, a marriage bond, or a life. For instance, the term, "sanctity of marriage", in its historical context, does not mean primarily that one ought to love, care for, support, and/or cherish the spouse; the "sacrament" of marriage requires those duties. The sanctity of marriage, however, specifically functions to remind us that a marriage blessed receives a sacred quality and, therefore, cannot be violated: "What God has joined, let no one put asunder." (*Mark* 10.9)

This negative significance ought not to be underestimated. Palazzini's reference to respect for life under the entries of murder or suicide makes perfect sense here. Because human life is sacred, the sanctity of life means that we cannot murder or commit suicide: the sacred quality establishes a line that the human cannot transgress. Prerogatives here are only for the one who bestowed the sacred quality in the first place. Thus, not uncommonly we find, "No human being possesses the right to dispose of his life on his own authority. God alone, the author of life, has absolute dominion over it" [18].

This absolute prohibition upholding the inviolability of human life is found in the four most important modern documents from the Roman Catholic Church's teachings on matters dealing with killing. However, it is important to note that the actual term "sanctity of life" does not appear in these texts; for instance, in 1965, when abortion and infanticide are condemned at the Second Vatican Council, the term is not used. "God, the Lord of life" grounds the argument for protecting life from these "abominable crimes," but there is no use of the term ([19], pp. 954–55). In fact, I cannot find the phrase in any document from the Second Vatican Council. In the "Declaration on Procured Abortion" (1974), the Congregation for the Doctrine of the Faith presents arguments from faith and from reason. In their arguments from faith, they cite a series of statements from recent popes. Curiously none refers to the "sanctity of life", though all point in the same general direction, e.g., "As long as a man is not guilty, his life is untouchable, and therefore any act directly tending to destroy it is illicit . . ." ([20], pp. 443–44). In the Congregation's "Declaration on Euthanasia," (1980) we find, "most people regard life as something sacred and hold that no one may dispose of it at will" and later: "Intentionally causing one's own death or suicide is therefore equally as wrong as murder; such an action on the part of a person is to be considered as a rejection of God's sovereignty and loving plan." Likewise in summarizing the four arguments against euthanasia, the first is that it is the "violation of the divine law" ([20], pp. 510–517). In these three documents, then inviolability is associated with divine protection, but the specific term, "sanctity of life," is not invoked to name it as such.

Finally, in the document on reproductive technologies, *Donum vitae* [16], we find:

The inviolability of the innocent human being's right to life from the moment of conception until death is a sign and requirement of the very inviolability of the person to whom the Creator has given the gift of Life ([16], p. 10).

Human life is sacred . . . God alone is Lord of life from its beginning until its end: no one can, in any circumstance, claim for himself the right to destroy directly an innocent human being ([16], p. 11).

On the topic of destroying spare human embryos, the document states:

> By acting in this way the researcher usurps the place of God; and, even though he may be unaware of this, he sets himself up as the master of the destiny of others inasmuch as he arbitrarily chooses whom he will allow to live and whom he will send to death and kills defenseless human beings ([16], p. 18).

This last document is particularly important. It conveys the restrictive absolute prohibition, though again it does not invoke the actual expression. It argues that the inviolability of life restricts human prerogatives from usurping ones that solely belong to the divine; the theologian Gerry Coleman ([9], p. 108) explains it simply: "Human persons, then, have only a right to the use of human life. What makes killing forbidden is that it usurps a divine prerogative and violates divine rights."

How are we to judge this use of the concept? Before answering this, we need to recognize two fundamental insights. First, criticizing the sanctity of life as not being satisfactory for grounding an absolute prohibition against, say, active euthanasia or assisted suicide, does not mean that there are no grounds for arguing for a blanket social prohibition. Many advocate against assisted suicide, while still critical of sanctity of life. These same advocates are concerned that the concept will not achieve a successful consensus. That failure may lead others, who think that sanctity of life is the only argument against euthanasia, to believe in the moral liceity and social beneficence of euthanasia. Thus, criticism of the sanctity of life principle does not imply an advocacy for euthanasia. In many cases of criticism (e.g., [8], pp. 397–404; [32], pp. 42–45; [33], pp. 308–317; [34], pp. 825–837; [35], pp. 5–9), in fact, the opposite is true.

Secondly, whether any principle can enforce a prohibition as "absolute" is precisely what has animated a debate among Catholic moral theologians in the last decade. In particular, Josef Fuchs ([22], pp. 94–137) questioned whether we could ever first adequately anticipate all the circumstances required for moral estimation in any given situation and second then attempt to exhaust all possible situations so that one absolute rule would rightly apply in all these cases. Today, many Catholic moral theologians (see [7], pp. 601–29; [31]; [65], pp. 286–314) would argue that a prohibition against torture or suicide, for example, may be "virtually" always wrong, but not necessarily absolutely always wrong. Preserved here is the scholastic insight that general rules apply generally.

Thus, many moral theologians are reluctant to consider any rule as absolutely prohibitive. This being said, they do not argue against social laws for the common good which hold, for instance, that torture or assisted suicide ought never to be permitted. Rather, they distinguish whether we can judge a certain way of acting as absolutely and always wrong from whether we ought to maintain a law for the common good of contemporary society.

With these concerns stated, let us see three instances where the concept as defined here bears too much weight and cannot stand its ground. First, some theologians, bishops, philosophers and lawyers take the teaching further than the Church's teaching does. They do this particularly in applying these teachings to questions not about direct killing but about letting die. These writers exclude any evaluation of the patient's condition when making decisions about the uselessness of certain means of life support. Thus, the lawyers representing the New Jersey Catholic Conference argued when they filed an amicus curiae brief in the case of Nancy Jobes; they were countered by others [13] who argued that they were unjustifiably confusing and restricting the tradition. Similarly before the United States Senate, Warren Reich arguing for the dignity and equality of every human person contended that only the medical means and not the condition of the patient enters into the calculus of whether life-support is required by the principle of sanctity of life. Again, others (see [62], pp. 85–154) countered that such a restriction is unwarranted and added, again invoking an "authority argument", that the tradition does not support restricting sanctity of life to the exclusion of considering a patient's condition in decisions about foregoing medical treatment. Indeed, a brief survey ([10], p. 87; [37], pp. 203–220; [38], p. 132) of traditional manuals of the Catholic tradition shows that commonly the patient's condition is weighed in moral decisions about foregoing medical life-supports.

Still, it seemed that recently American Catholic bishops were moving in the direction of Reich and others. In taking an early position on the question of artificial hydration and nutrition for those in a persistent vegetative state, the United States Catholic Conference argued, as the lawyer Kevin Quinn ([58], p. 290) notes, for the "preservation of life without any regard to the quality of that life." Later, the New Jersey Catholic Conference argued similarly as did, to some extent, the bishops from Pennsylvania. But bishops from Oregon, Texas, and Washington argued that sanctity of life prohibits us from any direct taking of human

life, both active euthanasia and assisted suicide, but added that withholding or withdrawing a person from extraordinary means of life support
ought not to be confused with the prohibition against the taking of life.
Finally, the United States Bishops's Pro-Life Committee argued that
though they find arguments against withdrawing artificial hydration and
nutrition more cogent than those that permit it, still the latter remains
a valid option and ought not to be considered among those life-taking
prohibitions governed by the sanctity of life principle (see [36]). Thus
bishops do not now use sanctity of life arguments against policies for
the withdrawal of hydration and nutrition in the case of persons in a
persistent vegetative state.

Writers like Leonard Weber and Eugene Diamond join Reich and hold
that one cannot use a "condition of life" criterion without undermining
that facet of the sanctity of life principle that contends that every life
is of equal value regardless of condition. Richard McCormick responds
to the charge by saying that the equality of persons and the equality of
lives are distinguishable and that in every decision about foregoing life
support one must weigh a condition of life factor. Quality of life, argue
McCormick and Quinn (respectively, [50], pp. 339–351; [57], pp. 897–
937), is intrinsic to the sanctity of life principle. Similarly, Bishop Bullock
of Des Moines argues like McCormick: every person has the right to life,
but decisions regarding the suitability of treatment are dependent upon
patient's conditions. Logical distinctions must be observed. Thus, the
Roman Catholic tradition and contemporary bishops and theologians
contend against those who exclude the patient's condition from moral
decisions regarding the use of medical treatment.

Helga Kuhse in her critique of the sanctity of life principle characterizes it as Weber and Reich defined it, but targets the late Protestant
theologian Paul Ramsey. Ramsey ([59], pp. 191–92) extends the sanctity
of life principle to questions of foregoing treatment and excludes condition or quality of life issues, arguing that "all our days and years are
of equal worth whatever the consequences." She [42] demonstrates aptly
that his use of the principle is inconsistent, when he argues that for one
dying of Tay Sachs, no obligation exists to prolong the dying: "If all
our days are of equal worth whatever the consequence then why are
the days of life gained by treatments which prolong dying not also of
equal worth?"

Any attempt to extend the principle beyond inviolability in the form
of direct killing of the innocent has been rebuffed on three grounds:

that it contradicts the tradition that long espoused the principle; that it confuses the equality of persons with people's physical conditions; and that it is inconsistent since few are willing to argue that there is an absolute mandate to prolong dying.

Second, when the sanctity of life principle in traditions like Roman Catholicism claims that no person has the right ever to directly kill an innocent person, it tries to cover euthanasia, assisted suicide, abortion, and infanticide. In doing this it stretches itself too far. This is particularly true when it is used for abortion. That use is often critiqued and an effect of the critique is that many begin to believe that as the principle does not fully prove the absolute prohibition against abortion, likewise it applies insufficiently elsewhere. For instance, John Noonan, [54] the famous historian of moral argument, writes, "If I was popular in academic circles to point out that the rule on contraception could change, it was unpopular to note that the value underlying the abortion rule, the sanctity of innocent life, did not allow much flexibility in the rule protecting it." This is a curious statement. Though abortion was never permitted in the Catholic tradition (with the exception of a three hundred year discussion on the liceity of therapeutic abortion), still since the sixth till the twentieth century there has been one consistent, at times majority, position in every period of the Church's history arguing for a distinction between an unformed and a formed fetus. This position held that while abortions in both cases were always wrong, only the latter constituted homicide, i.e., the killing of a human person ([11, 53]). Thus Joseph Donceel ([15], p. 49) notes that for centuries the Church's "law forbade the faithful to baptize any premature birth which did not show at least some human shape or outline." In fact, when the above noted Vatican document, *Donum vitae*, asks, "How could a human individual not be a person?", it admits, "The Magisterium has not expressly committed itself to an affirmation of a philosophical nature . . .". ([16], p. 13) The problematic application of the principle of sanctity of life to abortion undermines its application elsewhere when used as an absolute prohibition.

In invoking the sanctity of life argument on abortion the question of person arises. Some argue from retortion that the burden of answering the personhood question belongs to those doing harm: they have to prove, since they are destroying life, that the life is not a human person's. Others accept the challenge and make the claim, extending non-personhood even to infants (besides Kuhse and Singer, see [17]). Applying sanctity of life as an absolute inviolability principle applicable to so many different medico-moral dilemmas only leaves the concept open to being problematic in this, that, or another application. It cannot and should not supersede other more specific and cogent arguments addressing the various dilemmas. Its failure as a useful argument in abortion leads many to think that the concept is equally weak elsewhere.

Third, the most problematic issue facing sanctity of life as the absolute inviolability of innocent life is that, as quotes above demonstrate, it is used often as a divine injunction. The concept becomes an argument from "God's dominion". A look at Aquinas's treatment of killing illustrates particularly how the argument works.

In his famous question in the *Summa Theologiae* on murder, Aquinas asks whether suicide is lawful. He answers negatively, giving three responses: it is against natural inclinations and the common good, and "because life is God's gift to man, and is subject to His power, Who kills and makes to live. Hence whoever takes his own life, sins against God." (II-II 64, 5c; see I-II 73, 9, ad2) He asks, however, whether Samson sinned in causing his own death. Instead of invoking the principle of double effect,[1] he quotes Augustine who argued that Samson was excused because God "had secretly commanded him to do this." Aquinas even adds that certain holy women, "who at the time of persecution took their own lives, and who are commemorated by the Church" were so addressed. (II-II 64, 5, ad4)

Later, Aquinas inquires into the liceity of Abraham's intention to kill his son and answers, "God is Lord of death and life, for by His decree both the sinful and the righteous die. Hence he who at God's command kills an innocent man does not sin, as neither does God whose behest he executes." (II–II 64, 6, ad1)[2]

Elsewhere, Aquinas underlines the positivistic nature of "sanctity". In distinguishing one meaning of sanctity as purity, he writes of the other, "it denotes firmness, wherefore in older times the term *sancta* was applied to such things as were upheld by law and were not to be violated. Hence a thing is said to be sacred when it is ratified by law." (II-II 81, 8c)

It ought to be noted, that Aquinas generally does not appeal to divine law language to validate his moral argument and that, indeed, in the section on killing it only appears in the suicide argument itself and in the case of Abraham's inspiration.[3] In fact, he only invokes specific divine commands as exceptions to the divine injunction when faced with cases of divine inspiration.

Divine injunction language is problematic on three fronts. First, obviously, secular society sees no compelling force in such language. Second, in Roman Catholic ethics the abdication of reasoned argument for voluntaristic ones marks a significant shift in a long standing tradition. Third, were anomalies accepted, still, pertinent questions must be asked regarding how we can have access to the mind, better, explicit will of God. For instance, in September 1992, the American bishops' fourth draft of the pastoral on women declared that the restriction of priestly ordination to men is expressive of the will of the Lord. This assertion is riddled with questions regarding hermeneutics, epistemology, moral reasoning, anthropology, and doctrine of God (see [22], pp. 28–49, 62–82). Such assertions only theomorphize a voluntaristic argument. In theological circles, we call them "conversations stoppers": they require obedience of the will and a cessation of reasoned reflection (see [60], pp. 81–98).

When the sanctity of life principle is translated into a restrictive claim of life being absolutely inviolable it actually weakens itself. Its voluntarism and its attempt to enjoin more than it can has been a considerable stumbling block but not for those who believe in the moral liceity of euthanasia, suicide, abortion or infanticide. For these proponents, sanctity of life has been a perfect target. Actually, it has been a stumbling block to those who argue against the moral liceity of these actions.

Like Aquinas, those of us who believe in opposing infanticide or euthanasia look elsewhere than sanctity of life when it expresses an absolute divine injunction. But those other arguments are not the topic of this paper. Rather, in considering the role of the sanctity of life in contemporary argument, I turn now to the second and newer use which, curiously, can be found in Sallust.

IV. SANCTITY AS REVERENCING LIFE

Noted above was the fact that for all the use of sanctity of life argumentation, Church documents do not invoke the term per se. An added curiosity is that recently with frequency the term is invoked in the United States. Though we find it used in the absolute sense in the legal brief from the New Jersey State Catholic Conference [52] in objecting to the removal of artificial and hydration, in another legal brief [49] we find it, not as an absolute principle but as a competing one, defining how the interests of the state embrace "two separate but related concerns: an interest in preserving the life of the particular patient and preserving the sanctity of all life." This same general and not absolutely specific use appears in an episcopal letter [51] on advance directives, subsequently recommended by many American bishops, and in the U. S. Bishops Pro-life Committee's pastoral on nutrition and hydration.

The frequent use of the term suggests its emerging role in social and political circles as a particular foundational, even contextual, device that conveys a basic disposition or attitude, rather than a specific absolute injunction. Rather than emphasize its absoluteness, it emphasizes its universality.

This disposition is appropriated by secular institutions. Without the emphasis on a divine and absolute injunction the term brings with it a utility that the *Hastings Center*, for instance, values:

Due regard for the sanctity of life can guard against the erosion of respect for life in our society. By creating a presumption in favor of continued treatment, the sanctity of life can also help to protect gravely ill patients who are vulnerable. This presumption can reassure society that the termination of treatment decisions are being made by individuals, institutions, or by society only after careful scrutiny and justification, and not out of ethically illicit motives. Sanctity of life, however, is a presumption; it does not by itself determine whether a particular treatment is appropriate for a patient [27].

When the term is freed of its restrictive, absolute sense, it gains a positive orientation. Rather than emphasizing what we cannot do or what we may not transgress, sanctity of life in this fuller sense brings obligations for what we are required to do, that is, to uphold life. This is what distinguishes the Hastings Center observation and, for instance, the two entries in the dictionaries edited by MacQuarrie, cited above.

When sanctity of life is used in the restrictive sense, it underlines the dominion of God. When it is used in the more universal and general sense, it emphasizes our stewardship. Though the more restricted definition uses this distinction between dominion and stewardship, it does so precisely to underline that we do not enjoy dominion. Consider, for example:

What is sacred, therefore, has been removed from all independent use by creatures . . . What has been consecrated to the Holy One is inviolate, not to be touched by man. Definitively removed from human authority and control, it is not subject to our will or to our initiatives. What is sacred is owned by God and by Him alone. Only the All-Holy can have the sacred as His Possession [56].

His dominion over his body and his faculties is one of stewardship only. Hence no human being possesses the right to dispose of his life on his own authority [18].

Human persons, then, have only a right to the use of human life, not dominion over human life ([9], p. 108).

When not used in the restrictive sense, stewardship becomes a task, rather than a boundary line. Stewardship can be applied to the disposition that informs us of obligations as we face questions for proceeding on issues regarding health care, genetic manipulation, environmental issues and animal rights. "Man's dominion is not a license to exploit creation but rather a charge to rule as God would rule. All life, since it comes from God, has a sacredness about it and demands respect for it . . . The special dignity and sanctity of human life comes from bearing the image of God and the responsibility to rule like God." ([3], p. 80) The applicability of the principle extends considerably, nearly universally, when its highly absolute and restrictive interpretation is abandoned. Daniel Callahan states, "On the basis of this principle, moral rules have been framed, human rights claimed and defended, and cultural, political, and social priorities established." (quoted in [29])

This positive, general, yet obliging disposition also makes itself more adaptable to secular audiences. By emphasizing stewardship, the term appeals to human experience. Without positing an absolute divine injunction, the term is "user-friendly" to non-believers as well. Edward Shils ([61], p. 12) recognizes this and writes,

The chief feature of the protoreligious, "natural metaphysic" is the affirmation that life *is* sacred. It is believed to be sacred not because it is a manifestation of a transcendent creator from whom life comes: It is believed to be sacred because it is life. The idea of sacredness is generated by the primordial experience of being alive, of experiencing the elemental sensation of vitality and the elemental fear of its extinction.

He ([61], pp. 18–19) adds, "Its sacredness is the most primordial of experiences."

Rather than rooting sanctity in the activity of a creator, in the assertion of God as Lord of Life, or in a call to obey the divine will, Shils looks not beyond human experience but within it and appeals to the common human experience of awe before "life". The argument then appeals not to injunctions or prerogatives being exercised elsewhere, but rather to reflection on what it means to belong to the species and, further, to the world of the living. The argument, though intuitive, makes its appeal basically on rational, reflective grounds, rather than the voluntaristic ones upon which the divine injunction is based.

Shils' assertion about this "sense" of life as possessing the quality of sanctity is not far from the insight that Thomas Long discusses in an on-going debate with Peter Singer and Helga Kuhse. Long's point ([45], p. 96) is that behind every argument in debates on euthanasia and other related matters is a "metaphysical sensibility". "Its acceptance or rejection is as much (if not more) a matter of how people feel about themselves and the world as it is about philosopher's arguments. Sensibilities emerge from and express varying degrees of thought, hope and desire." Unfortunately rather than demonstrate what "metaphysical sensibilities" could animate a "secular utilitarian" the author appears to reserve the one that embraces "sanctity of life" for religious believers. Still, the insight can be appropriated by a secular society, precisely as Shils suggests.

Finally, the concept so defined provides a ground that secular ethics in America lacks. Thus, Joseph Cardinal Bernardin gave a series of addresses throughout the country arguing for a consistent life ethic. This argument held that the sanctity of life argument could not be applied to abortion alone but to other life and death issues, such as war, capital punishment and euthanasia. That stance, however, became interpreted as emphasizing the absolute specific prohibition against direct killing while upholding God's dominion. As a result, many came to see the claims for a consistent life ethic riddled with inconsistencies. Pacifists found that the ethic did not extend as much absolute protection against war as against abortion. Similarly, opponents to capital punishment found that the application of the prohibition was more absolute than universal (see [12, 26, 55]). When the consistent ethic was narrowly and negatively defined, it lost its universal force.

On the other hand, others translated Bernardin's proposals in more general ways, emphasizing as Margaret Steinfels ([63], p. 270) did, such words as, "A consistent ethic of life should honor the complexity of the issues it must address . . . A systematic vision of life seeks to expand the moral imagination of a society, not partition it into airtight categories." When Bernardin's ethic promotes a moral imagination, rather than an absolute norm, it makes greater impact on American society lacking considerable moral imagination.

The need that Americans have for some foundational shared and collective point of view has been recently pointed out by Mary Ann Glendon. As noted on the cover of her recent work, *Rights Talk*,

"American rights talk is set apart from rights discourse in other liberal democracies by its exaggerated absoluteness, its hyper-individualism, its insularity, and its silence with respect to personal, civic and collective responsibilities." Her description of the moral and political terrain in the States expresses well our need for appreciating a context where the self-understanding of each individual is shared (see [4]). Instead, rooting ourselves individually into soil of entitlements without responsibilities, Americans have no context for moral discourse, particularly in setting an agenda of who we ought to become. Moral discourse here takes in large measure the form of a declaration of individual rights and prerogatives rather than an articulation of any shared destiny or vision.

Glendon ([24], p. 35) contrasts, for instance, the way West German (1975, 1976) and American (1973) decisions on abortion reflected respective long term concerns.

Although both courts substituted a set of values for those promoted by the respective legislatures, the content of those values differ markedly. Roe v. Wade, like its predecessor, Eisenstadt v. Baird, embodies a view of society as a collection of separate autonomous individuals. The West German decision emphasizes the connections among the woman, developing life, and the larger community. Donald Kommers has therefore characterized the difference between the two decisions as one between individualistic and communitarian values. He points to *Roe*'s emphasis on the individual woman, her privacy and autonomy, and contrasts it with the West German court's emphasis on the interest that society as a whole has, not only in the abortion decision itself, but in the long-range formation of beliefs and attitudes about human life.

Commenting further on privacy and its definition in the two different countries, she ([24], p. 37) adds,

When privacy was taken into West German constitutional law in 1949, it was given a somewhat different meaning from the one its American counterpart came to have. Privacy rights in West German constitutional law are based on Article 2(1) of the Basic Law ("Everyone shall have the right to the free development of his personality, insofar as he does not violate the rights of others or offend against the constitutional order or the moral code") in conjunction with Article 1(1) ("The dignity of man shall be inviolable"). The affirmative German right to the free development of one's personality contrasts with the negative American formulation of privacy as the right to be left alone. Whereas the American conceptualization of privacy emphasizes freedom *from* various kinds of restraints, the German version stresses and makes clear what this freedom is *for*.

Just as American law and its underlying ethos are in need of a positive unifying force, setting goals, however vague, about what our tasks are as a people, similarly the "sanctity of life" principle enhances moral argument when it engages shared self-understanding. When sanctity of life works not as a voluntaristic injunction but as an experienced insight that feeds our "moral imagination" and opens our dispositions to appreciate not simply our rights (or God's dominion) but our stewardship, then it brings something to American moral argument that is sorely lacking. In that stewardship, that will only be genuine when not exclusive of other life forms, we may learn even better reasons for reverencing life. Thus sanctity of life often translated into an absolute and negative sense finds a new function in a universal and positive interpretation. For the moment, here in this collection where another concept with different roots provides a similar function, we may find value, once again, in that quality Sallust first described.

Weston School of Theology
Cambridge, Massachusetts
USA

NOTES

[1] A principle he did not establish. See [23].

[2] See in this question Aquinas's interesting remark that a judge, who knows that a person is innocent of a capital crime but who cannot prove it, is excused of guilt if the judge assigns a penalty coinciding with the conviction, II–II 64, 6, ad3.

[3] I have the suspicion that Suarez, de Vitoria and Cardinal de Lugo may have developed the concept of killing as a violation of divine dominion. Certainly, it is in Aquinas's writings, but only, I think, in seminal form. The 16th and 17th century theologians, however, developed considerable arguments for the divine, positive law. See ([46], pp. 224–258 and [2], pp. 95, 101).

BIBLIOGRAPHY

1. Aquinas, Th.: 1948, *Summa Theologica*, Christian Classics, Westminster, Maryland.
2. Ashley, B.: 1992, 'Dominion or stewardship?: theological reflections', in K. Wildes (ed.), *Birth, Suffering and Death*, Kluwer Academic Publishers, Dordrecht, pp. 85–106.
3. Autiero, A.: 1992, 'Dignity, solidarity, and the sanctity of human life', in K. Wildes (ed.), *Birth, Suffering, and Death*, Kluwer Academic Publishers, Dordrecht, pp. 79–84.
4. Bellah, R.: 1985, *Habits of the Hearts*, Harper and Row, New York.
5. Bernardin, J.: 1988, *Consistent Ethic of Life*, Sheed and Ward, New York.
6. Bullock, W.: 1992, 'Assessing burdens and benefits of medical care', *Origins* 21, 553–555.

7. Cahill, L.S.: 1981, 'Teleology, utilitarianism, and Christian ethics', *Theological Studies* **42**, 601–29.
8. Callahan, D.: 1988, 'Vital distinctions, mortal questions', *Commonweal* **115**, 397–404.
9. Coleman, G.: 1989, 'Assisted suicide: an ethical perspective', in R. Baird *et al.* (eds.), *Euthanasia*, Prometheus Books, Buffalo, pp. 103–110.
10. Connell, F.: 1953, *Outlines of Moral Theology*, Bruce Publishing Company, Milwaukee.
11. Connery, J.: 1977, *Abortion: The Development of the Roman Catholic Perspective*, Loyola University Press, Chicago.
12. Connery, J.: 1988, 'A seamless garment in a sinful world', in P. Jung *et al.* (eds.), *Abortion and Catholicism*, Crossroad, New York, pp. 272–278.
13. Devine, R.: 1989, 'The Amicus Curiae brief', *America* **160**, 323–326, 334.
14. Diamond, E.: 1976, 'Quality vs. sanctity of life in the nursery', *America* **135**, 396–398.
15. Donceel, J.: 1988, 'A liberal Catholic's view', in P. Jung *et al.* (eds.), *Abortion and Catholicism*, Crossroad, New York, pp. 48–53
16. *Donum vitae*: 1987, Libreria Editrice Vaticana, Vatican City.
17. Engelhardt, H.: 1989, 'Ethical issues in aiding the death of young children', in R. Baird *et al.* (eds.), *Euthanasia*, Prometheus Books, Buffalo, pp. 141–154.
18. "Euthanasia": 1967, *New Catholic Encyclopedia*, McGraw-Hill, New York, vol. 5, p. 639.
19. Flannery, A.: 1975, *Vatican II: Conciliar and Postconciliar Documents*, Fowler Wright Books LTD., Leominster, Herefords.
20. Flannery, A.: 1982, *Vatican Council II: More Postconciliar Documents*, Fowler Wright Books LTD., Leominster, Herefords.
21. Fuchs, J.: 1979, 'The absoluteness of moral terms', in C. Curran *et al.* (eds.), *Readings in Moral Theology*, No. 1, Paulist Press, Mahwah, New Jersey, pp. 94–137.
22. Fuchs, J.: 1987, *Christian Morality: The Word Becomes Flesh*, Georgetown University Press, Washington, D.C.
23. Ghoos, J.: 1951, 'L'Acte a double effet, etude de theologie positive', *Ephemerides Theologicae Lovanienses* **27**, 30–52.
24. Glendon, M.A.: 1987, *Abortion and Divorce in Western Law*, Harvard University Press, Cambridge.
25. Glendon, M.A.: 1991, *Rights Talk*, Free Press, New York.
26. Gudorf, C.: 1988, 'To make a seamless garment, use a single piece of cloth', in P. Jung *et al.* (eds.), *Abortion and Catholicism*, Crossroad, New York, pp. 279–296.
27. 'Guidelines on the termination of life-sustaining treatment and the care of the dying': 1987, *Hastings Center* **131**, 9.
28. Gustafson, J.: 1977, 'A Protestant ethical approach', in J. Noonan (ed.), *The Morality of Abortion*, Harvard University Press, Cambridge, pp. 101–122.
29. Healy, J.: 1991, 'Sanctity of life', *Linacre Quarterly* **58**, 69.
30. Helm, P. (ed.): 1981, *Divine Commands and Morality*, Oxford University Press, Oxford.
31. Hoose, B.: 1987, *Proportionalism: The American Debate and Its European Roots*, Georgetown University Press, Washington, D.C.

32. Hughes, G.: 1975, 'Killing and letting die', *The Month* **236**, 42–45.
33. Kamisar, Y.: 1978, 'Some nonreligious views against proposed mercy killing legislation', in T. Beauchamp *et al.* (eds.), *Contemporary Issues in Bioethics*, Wadsworth Publishing Co., Belmont, CA., pp. 308–317.
34. Keenan, J.: 1983, 'Toeten oder Sterbenlassen?', *Stimmen der Zeit* **201**, 825–837.
35. Keenan, J.: 1992, 'Assisted suicide and the distinction between killing and letting die', *Catholic Medical Quarterly* **42**, 5–9.
36. Keenan, J. and Sheehan, M.: 1992, 'Life supports', *Church* **8**, 10–17.
37. Kelly, G.: 1950, 'The duty of using artificial means of preserving life', *Theological Studies* **11**, 203–220.
38. Kelly, G.: 1957, *Medico-Moral Problems*, Catholic Hospital Association, St. Louis.
39. Kuhse, H.: 1987, *The Sanctity-of-Life Doctrine in Medicine*, Clarendon Press, Oxford.
40. Kuhse, H. and Singer, P.: 1985, *Should the Baby Live?*, Oxford University Press, Oxford.
41. Kuhse, H. and Singer, P.: 1986, 'For sometimes letting- and helping-die', *Law, Medicine, and Health Care* **3**, 149–153.
42. Kuhse, H. and Singer, P.: 1988, 'Resolving arguments about the sanctity of life', *Journal of Medical Ethics* **14**, 198–199.
43. Kuhse, H. and Singer, P.: 1991, 'Prolonging dying is the same as prolonging living', *Journal of Medical Ethics* **17**, 205–206.
44. Long, T.: 1988, 'Infanticide for handicapped infants', *Journal of Medical Ethics* **14**, 79–81.
45. Long, T.: 1990, 'Two philosophers in search of a contradiction', *Journal of Medical Ethics* **16**, 95–96.
46. Mahoney, J.: 1987, *The Making of Moral Theology*, Clarendon Press, Oxford.
47. Mangan, J.: 1949, 'An historical analysis of the principle of double effect', *Theological Studies* **10**, 41–61.
48. Martini, C.M.: 1989, 'Some basic considerations on moral teaching in the Church', in E. Pellegrino *et al.* (eds.), *Catholic Perspectives on Medical Morals*, Kluwer Academic Publishers, Dordrecht, pp. 9–21.
49. *Matter of Conroy*, 98 NJ 321, 349 (1985).
50. McCormick, R.: 1981, *How Brave a New World?* SCM Press Ltd., London.
51. Myers, J.: 1991, 'Advance directives and the Catholic health facility', *Origins* **21**, 276, 280.
52. New Jersey State Catholic Conference: 1987, 'Providing food and fluids to severely brain damaged patients', *Origins* **16**, 582–584.
53. Noonan, J.: 1970, 'An almost absolute value in history', in J. Noonan (ed.), *The Morality of Abortion*, Harvard UP, Cambridge, pp. 1–59.
54. Noonan, J.: 1992, 'A backward look', *Religious Studies Review* **18**, 111–112.
55. Overberg, K.: 1980, *An Inconsistent Ethic: Teachings of the American Catholic Bishops*, University Press of America, Lanham.
56. Quay, P.: 1992, 'The sacredness of the human person', *Linacre Quarterly* **59**, 77.
57. Quinn, K.: 1988, 'The best interests of incompetent patients', *California Law Review* **76**, 897–937.
58. Quinn, K.: 1990, 'The bishops' misstep', *Commonweal* **117**, 288–292.
59. Ramsey, P.: 1978, *Ethics at the Edges of Life*, Yale University Press, New Haven.

60. Schueller, B.: 1990, 'Paranesis and moral argument in *Donum vitae*', in E. Pellegrino (ed.), *Gift of Life*, Georgetown University Press, Washington, D.C., pp. 81–98.
61. Shils, E. (ed.): 1968, *Life or Death: Ethics and Options*, University of Washington Press, Seattle.
62. Sparks, R.: 1988, *To Treat or Not to Treat?* Paulist Press, Mahwah.
63. Steinfels, M.: 1988, 'Consider the seamless garment', in P. Jung *et al.* (eds.), *Abortion and Catholicism*, Crossroad, New York, pp. 268–271.
64. Bishops' Pro–Life Committee: 1992, 'Nutrition and hydration: moral and pastoral reflections', *Origins* **21**, 705–710.
65. Vacek, E.: 1985, 'Proportionalism: one view of the debate', *Theological Studies* **46**, 286–314.
66. Weber, L.: 1976, *Who Shall Live?* Paulist, New York.
67. Williams, G.: 1957, *The Sanctity of Life and the Criminal Law*, Alfred A. Knopf, New York.

HELGA KUHSE

SANCTITY OF LIFE, VOLUNTARY EUTHANASIA AND THE DUTCH EXPERIENCE: SOME IMPLICATIONS FOR PUBLIC POLICY[1]

I. INTRODUCTION

In its 1980 *Declaration on Euthanasia,* the Roman Catholic Church re-confirmed its firm opposition to euthanasia of any kind: ". . . nothing and no one can in any way permit the killing of an innocent human being, whether a foetus or an embryo, an infant or an adult, an old person or one suffering from an incurable disease, or a person who is dying" ([20], p. 7). The *Declaration* captures what I want to call the "sanctity-of-life view". This is the view that all innocent human lives are absolutely inviolable and equally valuable, and that the intentional termination of such lives is always morally wrong [11].

The sanctity-of-life view has its source in the Judaeo-Christian tradition and has shaped Western thinking for many hundreds of years [21].[2] It is now enshrined in professional codes of conduct, in public policies and the law. As one writer in the field recently put it, both the Hippocratic tradition and the English criminal law "subscribe to the principle of the sanctity of human life which holds that, because all lives are intrinsically valuable, it is always wrong intentionally to kill an innocent human being" [9].[3]

Voluntary euthanasia offends against the sanctity-of-life view because it is an instance of the intentional termination of life. One person deliberately and purposefully terminates the life of an incurably ill person, at that person's request. But are all cases of the intentional termination of life intrinsically or, as the sanctity-of-life view would hold, absolutely wrong? People who approach ethics from different moral, cultural or religious perspectives will often arrive at different answers. Because these

19

K. Bayertz (ed.), Sanctity of Life and Human Dignity, 19–37.
© 1996 *Kluwer Academic Publishers. Printed in the Netherlands.*

different answers have their source in particular value systems, they cannot be shown to be true or false, in the ordinary sense of those terms [3].

This raises the question of an appropriate social response. Given that there is fundamental disagreement about the morality of a practice, how should modern pluralist societies such as our own respond to it? Should they allow or prohibit the practice, and on what grounds?

It is now widely accepted that personal autonomy or liberty is a very important value and that it is inappropriate for the state to either adopt a paternalistic stance towards its mature citizens, or to restrict their freedom through the enforcement of a particular moral point of view. Only if one person's actions cause harm to others is it legitimate for the state to step in, and to bring in laws that restrict individual liberty. As John Stuart Mill put it in his famous essay *On Liberty:* "The only purpose for which power can be rightfully exercised over any member of a civilised community, against his will, is to prevent harm to others. . . . Over himself, over his own body and mind, the individual is sovereign" ([17], p. 135).

The argument from liberty or autonomy suggests that people should, under the appropriate circumstances, be free to commit suicide, and that those who are terminally or incurably ill should be able to enlist the help of willing doctors to end their lives [1]. While there is considerable public and professional support for a liberal approach towards voluntary euthanasia, it has, however, been claimed that a public policy that allows the practice will have harmful consequences for others and that attempts to reform the law must therefore be resisted.

Until the early 1980's, this type of argument against voluntary euthanasia remained fairly weak. There was simply no evidence to back up the claim that a liberal approach would have harmful consequences for others. By that time, however, a significant development had occurred. The Dutch Courts had developed guidelines which allowed doctors to practise voluntary euthanasia [12]. Today, more than 10 years later, the Netherlands is still the only country in the world where doctors are permitted to directly and purposefully end the life of a patient, at the patient's request.[4]

The world at large has taken a keen interest in what is sometimes called "the Dutch experiment". Some articles claimed that there was evidence of abuse, of harm to others. There was, however, little more than anecdotal evidence and conjecture to substantiate these claims.

But then, in 1990, the Dutch government commissioned a large nation-wide study on voluntary euthanasia and other medical end-of-life decisions, to add some facts to the national and international speculations. The results of the study were published in 1991 and I shall refer to the study as "the Dutch survey" [25].[5] This study, opponents of the liberalization of voluntary euthanasia laws claim, clearly demonstrates that harm to others will result if a society allows its medical practitioners to directly cut short the lives of their patients ([22]; [6]; [23]; [19]).

But does it? Before we can answer that question, we must take a closer look at the survey and draw some important distinctions.

A word of warning is in order. The discussion, like its subject matter, is complex. Past discussions have often been marred by serious confusions – the result of the writers' inability or unwillingness to draw conceptually and practically important distinctions ([24]; [13]). A central aim of this paper is to draw these distinctions, and to highlight some of their implications for public policy.[6]

II. THE NATURE OF MEDICAL END-OF-LIFE DECISIONS

The Dutch survey details the findings of three separate studies. All three studies had the aim of quantifying how Dutch people die. This was done by asking doctors, who had treated a recently deceased patient, to provide a range of data on the patient's death. Had the patient died because nothing could be done to save or prolong her life, or had she died because the doctor had engaged in a deliberate life-shortening action or omission? If the answer was "yes", what kind of an action or omission was this, and had the patient given her consent?[7]

The study is probably the largest and most sophisticated survey of medical end-of-life decisions ever undertaken anywhere in the world. It categorized end-of-life decisions according to three distinctions which have traditionally been imbued with moral significance: whether the doctor caused the patient's death, or merely allowed it to happen; whether the doctor directly intended the death (or merely foresaw it), and whether the patient did or did not consent[8] to the end-of-life decision. The first two distinctions – whether the doctor caused the patient's death, or merely allowed it to happen, and whether the doctor directly intended the death (or merely foresaw it) – are central to the moral and legal discussions surrounding the sanctity-of-life view. They determine whether an action or an omission that leads to death is a case of the intentional termination of life ([11], Chapters II and III). The third distinction – whether the patient did or did not consent to the end-of-life decision – is relevant to moral and legal discussions that take as their starting point the principle of personal autonomy or liberty.[9]

Causation

To determine whether the death was caused by life-shortening palliative care, by a lethal drug, or whether the patient died from an untreated medical condition, the study distinguished between three types of end-of-life decisions:

(A) death resulting from the withdrawal or withholding of treatment;
(B) death resulting from the intensifying of pain and/or symptom control;
(C) death resulting from assisted suicide or euthanasia, that is, from the administration, supplying or prescribing of a drug, with the explicit purpose of hastening the death of the patient.[10]

Intention

Next, a distinction was drawn in terms of the doctors' intentions. Did doctors forego treatment or administer drugs with the purpose of hastening the death of the patient, or did they merely take into account the probability that death would be hastened by the decision? [This additional distinction is relevant to (A) and (B) only. In the case of (C), it is presupposed that all actions in this group were performed with the explicit intention of bringing the patient's death about.] This, then, gives us the following five categories of end-of-life decisions:

(a) death resulting from the withholding or withdrawing of treatment, having taken into account the probability that this decision would hasten the death of the patient;
(b) death resulting from the withholding or withdrawing of treatment, with the explicit purpose of hastening the death of the patient;
(c) death resulting from the intensifying of the treatment of pain and/or symptoms, having taken into account the probability that this would hasten the death of the patient;
(d) death resulting from the intensifying of the treatment of pain and/or

symptoms, with the explicit purpose of hastening the death of the patient;

(e) death resulting from euthanasia or assisted suicide, that is, the prescribing, supplying or administering of drugs with the explicit purpose of hastening death.[11]

Consent

Finally, there is the question of consent. Was it a consented-to death, that is, did the doctor consult with the patient, obtaining her consent, before implementing either one of the above end-of-life decisions?

The answer to this question will give us ten distinct categories of end-of-life decisions, where each of the above five categories is divided into deaths that were consented to and those that were not.

(1) death resulting from the consented-to withholding or withdrawing of treatment, having taken into account the probability that this would hasten the death of the patient;

(2) death resulting from the unconsented-to withholding or withdrawing of treatment, having taken into account the probability that this would hasten the death of the patient;

(3) death resulting from the consented-to withholding or withdrawing of treatment with the explicit purpose of hastening the death of the patient;

(4) death resulting from the unconsented-to withholding or withdrawing of treatment with the explicit purpose of hastening the death of the patient;

(5) death resulting from the consented-to intensifying of the treatment of pain and/or symptoms, having taken into account the probability that this would hasten the patient's death;

(6) death resulting from the unconsented-to intensifying of the treatment of pain and/or symptoms, having taken into account the probability that this would hasten the patient's death;

(7) death resulting from the consented-to intensifying of the treatment of pain and/or symptom control with the explicit purpose of hastening the death of the patient;

(8) death resulting from the unconsented-to intensifying of the treatment of pain and/or symptom control with the explicit purpose of hastening the death of the patient;

(9)   death resulting from consented-to euthanasia and assisted suicide*),
        that is, the prescribing, supplying or administering of drugs with
        the explicit purpose of hastening the end of life;
(10) death resulting from unconsented-to euthanasia and assisted
        suicide*), that is, the prescribing, supplying or administering of
        drugs with the explicit purpose of hastening the end of life.

[*) In all cases of assisted suicide, where the patient herself performs
the final act, consent must be presupposed. This means that category (10)
applies to cases of euthanasia only, where the doctor, rather than the
patient, administers the lethal drug.]

III. WHAT IS EUTHANASIA?

In 1990, there were 128,786 deaths in The Netherlands. The Dutch survey
estimates that 38% of all deaths (and 54% of all non-acute deaths) were
the result of a medical decision. In other words, approx. 48,700 patients
had died because a decision had been taken not to prolong their lives,
or to engage in life-shortening actions. Let us take a closer look at the
composition of these deaths.[12]

Table 1. Incidence of end-of-life decisions

Category	Cause of Death	% of Total Deaths	Nos. (rounded)
(A)	withholding or withdrawing a treatment	17.5	22,500
(B)	pain and symptom control	17.5	22,500
(C)	euthanasia[13]	2.9	3,700

The above break-up suggests that cases of euthanasia were relatively
infrequent. There is, however, disagreement and confusion when it
comes to defining the term "euthanasia". Some understand euthanasia
in a narrow sense, where a doctor directly causes death by, for example,
injecting a lethal dose of a muscle relaxant or a drug, such as potassium chloride, into a patient's vein. Some will regard all active interventions, including withdrawal of treatment, as cases of euthanasia; others
will call only consented-to life-terminating acts "euthanasia". Some will
regard the withdrawal or withholding of "ordinary means" (but not of

"extraordinary means") as an instance of euthanasia, and yet others will focus on the doctor's intentions, or on whether the doctor knew, or should have know, that death would result as a consequence of her action or omission. Depending, then, on one's particular understanding of "euthanasia", there will be more or less cases of the practice – and those who have written about the Dutch survey have arrived at figures ranging from some 2,300 to more than 26,000 cases of euthanasia per year ([5]; [6]; [8]; [18]; [22]; [26]).

Traditionally, euthanasia has been defined in terms of the agent's intention. The Vatican's *Declaration on Euthanasia* thus understands "euthanasia" as "an action or an omission which of itself or by intention causes death" ([20], p. 6). On this traditional understanding, a doctor can practise euthanasia by an action or an omission, or by killing a patient or by allowing her to die. Such actions or omissions are understood to be cases of euthanasia if the doctor *directly intends* the patient's death. If the doctor merely *foresees* that death will result as a consequence of what she does or does not do, the life-shortening action or omission is not regarded as a case of euthanasia.[14]

Euthanasia, then, involves the intentional or purposeful termination of life of one person by another, by either an action or an omission. On this traditional and widely accepted understanding of the term, there would have been many more cases of euthanasia in the Netherlands than category (C) above suggests. Doctors would have practised euthanasia not only when they performed actions that fitted the description in that category (I shall from now on refer to these actions as cases of "direct euthanasia"), but also with regard to deaths intentionally and purposefully brought about by non-treatment or the administration of life-shortening palliative care. Thus, if we now refer to the list of end-of-life decisions headed "Intention", euthanasia was practised not only in the case of category (e), but also in the case of categories (b) and (d).

We already know that 17.5% of all deaths were the result of non-treatment decisions, and that 17.5% of doctors administered what they thought was potentially life-shortening palliative care. But how frequently did doctors intend to cut short the life of the patient by these means?

In the case of non-treatment decisions, life-sustaining treatment was withheld in approximately half the cases (49%) with the explicit intention of not prolonging the patient's life. This means that some 11,000 deaths resulted from euthanasia by non-treatment.

In the case of palliative care, pain and symptom control was increased in 11% of cases with the primary purpose of shortening life. This means that approx. 2,500 deaths in this category were the result of euthanasia. Moreover, in 17% of cases, the respondents had also indicated that hastening the patient's death was *part* of the purpose. If one were to classify these deaths as cases of euthanasia as well, another 4,000 deaths would have to be added, bringing the total number of euthanasia by way of palliative care to 6,500.[15] This will give us the following break-up of euthanasia cases.

Table 2. Incidence of intended deaths (euthanasia)

Category	Cause of Death	Intended Deaths	Nos.(rounded)
(b)	withholding or withdrawing a treatment	49%	11,000
(d)	pain and symptom control	28%	6,500
(e)	"direct" euthanasia	100%	3,700

This means that approx. 21,200 deaths were the result of euthanasia, understood as an instance of the intentional termination of life of one human being by another.

Those who take the view that it is always intrinsically wrong for a doctor to intentionally cut short the life of a patient would be worried by these results. But intrinsic wrongness is not our concern. Our question is whether there has been harm to others. More specifically, our question is whether harm, understood as the unconsented-to termination of life, has resulted from doctors being permitted to practise voluntary euthanasia.

IV. HARM TO OTHERS

How, then, might others be harmed through doctors being permitted to practise voluntary euthanasia? A typical argument – and the one I shall focus on – is this one: if *voluntary* euthanasia is permitted, this will soon lead to non-voluntary or involuntary euthanasia, that is, to the killing of patients who have not given their consent.

This is a so-called "slippery-slope" argument. It asserts that the introduction of a certain practice (which may be morally good or morally neutral) will lead to other unjustified consequences. While it might thus be possible to justify voluntary euthanasia by appeal to the principle of autonomy, this principle cannot justify non-voluntary or involuntary euthanasia, and it is here – in the area of unconsented-to killings – that harm is said to arise.[16]

It is correct that unconsented-to terminations of life cannot be justified by appeal to the principle of autonomy or liberty, although it might of course be possible to justify at least some of these life-shortening actions in some other way (for example, by appealing – especially in the case of incompetent patients – to the principle of beneficence, that is, by claiming that the unconsented-to termination of life is in the patient's best interests). But let us, for the purposes of our present discussion, grant that in-voluntary and non-voluntary euthanasia always involve harm for the patient whose life is cut short. To remind ourselves that this premise is merely granted, rather than accepted, I shall henceforth write "Harm".

We must, however, make one other (by no means obvious assumption): that "Harm" to these others would not have resulted, that is, that they would not have had their lives cut short, if doctors had been prohibited, by law, to practise *voluntary* euthanasia.[17] In other words, we are assuming for the purposes of our discussion, that it is the liberalisation of voluntary euthanasia which has led to non-voluntary or involuntary euthanasia.[18]

Having explicitly stated these assumptions, we must establish how many patients did or did not give their consent to the relevant life-shortening action or omission.

V. UNCONSENTED-TO EUTHANASIA

To do this, we now turn to the list of end-of-life decisions headed "Consent". Given our assumptions about harm to others, we should be worried about categories (4), (8) and (10), where consent to euthanasia was lacking.

In the case of Category (4) – euthanasia by withholding or withdrawing a treatment – consent appears not to have been obtained in 62% of cases. This suggests that some 6,800 deaths in this category were the result of non-voluntary or involuntary euthanasia.

In the case of Category (8) – euthanasia by way of pain and symptom control – doctors had not discussed this decision with 40% of their patients. This means that 2,600 deaths were the result of unconsented-to euthanasia.

Finally, there is direct euthanasia, as per category (e). In 28% of cases, consent had not been obtained, which means that 1,030 deaths had been the result of unconsented-to direct euthanasia.[19]

Table 3. Incidence of unconsented-to euthanasia

Category	Cause of Death	Unconsented-to Euthanasia	Nos.
(4)	withholding or withdrawing a treatment	62%	6,800
(8)	pain and symptom control	40%	2,600
(10)	"direct" euthanasia	28%	1,030

Altogether, then, we have approximately 10,430 cases of unconsented-to euthanasia, and, given our assumptions, a very large number of patients has been "Harmed" as a consequence of doctors being permitted to practise *voluntary* euthanasia.

We should also note, and this is a point to which we shall return in the next Section, that *many more* patients have apparently been "Harmed" through euthanasia being practised in a palliative care or non-treatment context, than have been "Harmed" by the administration of "direct" euthanasia.

VI. CONCLUSION: SOME IMPLICATIONS FOR PUBLIC POLICY

Discussions about voluntary euthanasia in The Netherlands have often centered on "direct" voluntary euthanasia, that is, on "medical killings", where doctors terminate the lives of their patient's through the direct and purposeful injection of a lethal substance. Those opposed to the idea, will generally express outrage at the suggestion that doctors should be able to engage in so heinous a practice, and then point to the alleged harm that has resulted from relaxing the prohibition on voluntary euthanasia. The underlying argument is that "direct" voluntary euthanasia cannot be regulated, and once a country allows its doctors to practise it, unconsenting others will suffer as a consequence.

As we have seen, "direct" euthanasia is, however, just one method by way of which euthanasia may be practised. Doctors can also intentionally and purposefully end the lives of patients by two other methods – through increased pain and symptom control, and through the withholding or withdrawing of treatment. (I shall, for the remainder of my discussion, refer to these two other methods as "indirect" euthanasia, or "indirect" life-shortening decisions.) "Indirect" euthanasia is, as we noted above, resorted to much more frequently than "direct" euthanasia. 17,500 deaths were the result of "indirect" euthanasia, and 3,700 were the result of "direct" euthanasia, and – given our assumptions – many more people have been "Harmed" by the first practice (9,400) than by the second practice (1,030). These results suggest that it is unreasonable for those intent on prohibiting voluntary euthanasia on the grounds that it causes "Harm" to others to lavish most of their attention on "direct" euthanasia, to the relative neglect of "indirect" euthanasia.

In the case of "direct" euthanasia, it is always assumed that the doctor's direct intention or purpose is the ending of life. When it comes to "indirect" euthanasia, on the other hand, things are far less clear. The withdrawal or non-employment of life-sustaining treatment has long been an accepted part of standard medical practice, as has the adequate administration of pain and other symptom control, even when there is little or no doubt that the particular action or omission will lead to death.

Now, on the view we have been discussing – namely the view that understands euthanasia as an instance of the intentional or purposeful termination of life – these "indirect" life-shortening actions are described as euthanasia only if the doctor undertakes them with the specific intention or purpose of bringing the patient's death about. If the very same action or omission is undertaken without that intention (although the expected outcome is still the death of the patient), the doctor is not deemed to have practised euthanasia, but to have engaged in practices that lie, at least *prima facie*, within the range of traditionally accepted medical and moral options.

The definition of "euthanasia" we have so far been working with thus rests on the previously mentioned Principle of Double Effect, which has long been an important component of the sanctity-of-life view. This principle focuses on the directly intended consequences of an action or an omission, to the relative neglect of merely foreseen consequences. It will, under the appropriate circumstances, allow doctors to bring a patient's death about, provided the death is merely foreseen and not directly intended. The Principle of Double Effect is, however, a *moral* principle, and not accepted by everyone. Some people think it has deep moral significance, others dismiss it as a hypocritical device that allows people, who ostensibly subscribe to a strict sanctity-of-life view, to nonetheless bring a patient's death about without taking full responsibility for it. Be this as it may, we have, for the purposes of our discussion, set such contested moral questions aside to focus on the issue of whether voluntary euthanasia leads to "Harm" to others.

In light of the above, we must, however, broaden our focus. We should no longer be concerned with euthanasia only, that is, with unconsented deaths that are directly intended by doctors, but also with all those unconsented-to deaths that are a merely foreseen consequence of a deliberate and socially sanctioned medical end-of-life decision. The reason should be clear. We are concerned with exploring the issue of "Harm" to others, where harm-to-others is understood as an instance of the unconsented-to termination of life. This means that we are interested in all deliberate actions which, through their being deemed permissible in some contexts, will lead to "Harm" to others. From this perspective, the subjective mental states of doctors, that is, whether they directly do or do not intend a death, become quite irrelevant. In short, then, we should be worried not only about unconsented-to cases of euthanasia, but also, and equally, about all unconsented-to foreseen deaths that are the outcome of a deliberate medical decision.

Looking at all non-treatment decisions, and (a) ignoring the distinction between directly intended and merely foreseen consequences and (b) assuming that consent was not obtained in 62% of all the cases, we find that 13,950 deaths were unconsented-to deaths; in the case of potentially life-shortening pain and symptom control, the relevant figures are 40% (9,000 deaths); and in the case of "direct" euthanasia they are, as we saw above, 28% (1,030 deaths). This means that in 23,980 cases doctors decided on a a course of action which they thought would lead to death, without obtaining the patient's consent. In 1,030 cases, the mode was unconsented-to "direct" euthanasia, and in 22,950 cases it was "indirect" euthanasia.

These are truly staggering figures. What conclusions should we draw from them? We may safely, it seems to me, draw at least one conclusion: If the relatively small number of unconsented-to deaths in the context of a public policy which allows voluntary or consented-to euthanasia are thought to be an adequate reason for prohibiting the practice, then one would find even more convincing reasons for a restrictive public policy approach in the much larger number of unconsented-to deaths that result from the socially condoned non-employment of life-sustaining treatment and the administration of life-shortening palliative care. Consistency would demand, then, that a society adopt a strict sanctity-of-life approach, where doctors would not only be prohibited from practising "direct" voluntary euthanasia, but where they would also be prohibited from engaging in any other life-shortening action or omission – with or without the patient's consent.

Such a policy would not have many supporters. If it could be enforced, it would certainly not be cost-free. Sanctity-of-life would always trump patient autonomy and suffering. Life would, to turn Thomas Hobbes on his head, be brutish, nasty and long.

The alternative is to change our focus. When it comes to end-of-life decisions for competent patients, we should develop frameworks which focus on consent, rather than on the traditional notions of intention and causation. Here is why.

In countries other than the Netherlands, doctors are not permitted to openly practice "direct" voluntary euthanasia. They are, however, permitted (and indeed legally required) to engage in other life-shortening actions or omissions. In Britain and Australia, for example, competent patients have a common law right to refuse medical treatment, and doctors must not treat patients against their will. There is also an emerging legal consensus that doctors may, under the appropriate circumstances, administer potentially life-shortening palliative care. What doctors must not do, under existing laws, is to intentionally bring about the death of a patient. But when is that the case, and why should intention matter?

Our laws are none too clear on the first part of the question. What can be said with some confidence is, however, this: doctors are not taken to intend all the foreseen and probable "natural" consequences of their life-shortening actions or omissions. If the notion of intention were understood in this wide "objective" sense, then the death resulting from each and every deliberate medical end-of-life decision (to the extent that the death is not unintended or accidental) would be taken to be the intended consequence. Since this does not seem to be the current legal view, this means that our regulatory frameworks and laws (like some traditional moralities) are at least to some extent based on a subjective notion of intention ([7], n.d.a.).

Regulatory frameworks based on subjective notions of intention and causation are, however, not likely to work well. Very often, only the doctor herself is able to say whether a death was her primary or her secondary intention, or whether it was not intended at all and merely the result of some other morally or legally defensible action. Bringing in the notion of "direct" causation may occasionally allow us to bring a Dr. Cox ([2], p. 1311) before the courts, who uses potassium chloride, when most of his colleagues would have used some other "indirect" method to end life, but it is not very helpful in ensuring that patients have given their consent to a life-shortening action or omission.

The Dutch study clearly shows that the contemporary focus on "direct" voluntary euthanasia as a possible source of "Harm" to others is misplaced. If patients are "Harmed" by unconsented-to life-short-ening actions or omissions, then much greater "Harm" is caused by our social acceptance of "indirect" end-of-life decisions – the foregoing of life-sustaining treatment, and the administration of life-shortening pal-liative care.

The time has come to briefly return to the substantive question of harm, already touched on above. I do not think that all unconsented-to life-shortening actions or omissions involve harm for the patient. On the contrary, I take the view that incompetent patients can sometimes be harmed by doctors *not* engaging in these end-of-life decisions ([14], pp. 140ff). Nor am I alone in this assumption. Even supporters of the sanctity-of-life view believe that incompetent patients can be harmed by having their lives prolonged by so-called disproportionate means of treatment, or by not receiving adequate (but life-shortening) pain and symptom control [11]. While there is thus agreement that not all life-shortening actions or omissions involve harm to the patient, there is, however, no substantive agreement as to when, and why, harm has occurred. Nor are such arguments generally provided by those who point to the Dutch survey, in an attempt to substantiate the claim that voluntary euthanasia leads to harm or abuse, understood as the uncon-sented-to termination of life ([5]; [8]; [6]; [18]).

Two conclusions follow from this. The first conclusion is that it is ille-gitimate to speak of harm until and unless substantive criteria have been provided that allow us to distinguish between cases where an uncon-sented-to medical end-of-life decision harms a patient and when it does not. The second conclusion is that it is illegitimate to point to the Netherlands and to claim that the fact that some 24,000 patients did not consent to an end-of-life decision (or that doctors directly intended the death of some 10,500 unconsenting patients) shows that there has been "abuse", or that there has been more "abuse" in the Netherlands than in any other country or state.[20]

If we want to ensure that decision-making in the practice of medicine is soundly based, we must firstly ensure that it respects the important value of personal autonomy or liberty. This means, as the Dutch survey has shown, that we should focus neither on the doctor's intentions, nor on the method by which death is brought about. Rather we should take as our focal point the notion of consent, and attempt to find ways and means by which it can be ensured. The reason is simple. A consented-to death respects the patient's autonomy, an unconsented-to death does not. And when it comes to the treatment of incompetent patients, we need to provide substantive criteria that will allow us to distinguish between those end-of-life decisions that harm an incompetent patient and those that do not. The mere fact that an end-of-life decision is of one type rather than of another, or that the doctor does or does not directly intend the death in question, does not answer the question of whether that decision has harmed or benefited an incompetent patient.

Our continuing pre-occupation with the doctor's intentions and, to some extent, causation is, I suspect, a legacy of the traditional sanctity-of-life view which dictates that doctors must never intentionally cause death. But that view should not – for all the reasons outlined – continue to shape the policies of liberal pluralist societies.

Monash University
Clayton, Victoria
Australia

NOTES

[1]   A shorter version of this paper was presented at a conference organized by the Centre for Human Bioethics at Monash University on "Active voluntary euthanasia: The current issues", Royal Australasian College of Surgeons, November 15, 1993. *Active Voluntary Euthanasia: The Current Issues*, John McKie (ed.), Melbourne, Centre for Human Bioethics, 1994, pp. 9–22.

[2]   Even before the rise of Christianity some Greek philosophical and religious schools expressed strong respect for human life, independently of Judaeo-Christian influence. The Hippocratic Oath, attributed to the fourth century BC, thus already strongly disapproves of abortion, euthanasia, and suicide. There are, however, good reasons to believe that this oath was not representative of Greek opinion. See, for example, [21], p. 6.

[3]   Keown appears to be wrong when he says that the taking of human life is, on the

sanctity-of-life view, wrong because all human lives are intrinsically valuable. On the traditional sanctity-of-life, the wrongness of taking human life lies in acting contrary to the will of God. See [11], pp. 18–19.

[4]   In the Netherlands voluntary euthanasia has not been legalized but decriminalized. Physicians who perform voluntary euthanasia in accordance with the relevant procedural rules will generally not be prosecuted.

[5]   A Summary of the findings was published by P.J. van der Maas, J.J.M. van Delden, L. Pijnenborg and C.W.N. Looman under the title "Euthanasia and other medical decisions concerning the end of life" in *The Lancet* [26]. It should be noted that the Dutch survey is detailed, complex and sometimes ambiguous. My brief paper cannot take account of all the complexities, and the theoretical and practical implications of the study. The arguments presented here must therefore not be taken out of context.

[6]   I would like to thank Loes Pijnenborg for her helpful comments on an earlier version of this paper.

[7]   The focus of the Dutch survey was on medical end-of-life *decisions*, not outcomes. Medical end-of-life decisions were defined as "decisions taken by physicians concerning actions and performed with the purpose of hastening the end of life of the patient or decisions for which the physician has taken into account the probability that the end of life of the patient will be hastened." ([25], p. xv.) Such actions or omissions will not always lead to the intended or expected death of the patient. The difference between a decision-perspective and an outcome perspective raises interesting philosophical and practical problems, but I will ignore these for the purposes of the present paper and will write as if there were congruence between an end-of-life decision and the expected outcome. This approach seems apt since supporters of the sanctity-of-life view focus primarily on the agent's intentions, not on the actual outcomes of decisions.

[8]   The Dutch survey did not rely on the narrow notion of "consent", where the patient typically agrees to a course of action proposed by the doctor. The survey speaks of "discussions" between doctor and patient, and of end-of-life decisions performed "at the request of" the patient. I shall use the term "consent", but understand it in a wider sense, that is, I shall regard an end-of-life decision as a consented-to decision if it has been discussed with the patient.

[9]   The study also looked at competency, that is, at the question of whether the patient was competent and could have been consulted (but was not), or whether the patient was incompetent and could not be consulted. For the purposes of this brief investigation, I shall ignore the important ethical issues that arise out of this distinction.

[10]   In the Dutch study, the term "euthanasia" is used in the sense of "voluntary euthanasia", that is, an intentional life terminating act, performed at the request of the patient. I shall use the term "euthanasia" in the wider sense commonly used in the Anglo-American literature, namely as an instance of the intentional termination of life.

[11]   The Dutch survey is somewhat ambiguous as to what is or is not "[voluntary] euthanasia", that is, the intentional termination of life, with or without request. This, the authors explain, is due to the fact that the boundaries between the different categories of end-of-life decisions (for example, between life-shortening palliative care and direct euthanasia) are not always clear. This prompted the authors to speak of a boundary area around euthanasia. ([24], pp. 326–27)

[12]   These and all subsequent figures are based on the Dutch survey [25], and on personal communications with the the authors.

[13] This group also contains suicides – 0.3% of all deaths. For the purposes of this discussion, I will not distinguish between suicide and euthanasia. Suicides will later on in my discussion be grouped together with cases of voluntary euthanasia because of the common element of consent.

[14] This is, of course, the so-called principle of Double Effect. For a discussion see [11], Chapter III.

[15] To the extent that the hastening of the patients' deaths was part of the purpose, they ought probably to be classified as intended deaths, in accordance with the parameters set by the Principle of Double Effect, and I have done so here. See [11], Chapter III.

[16] "Slippery slope" arguments can take many forms, and those who appeal to them often claim that different mechanisms are at work. Slippery slope arguments do, however, generally have one thing in common: the claim that the introduction of voluntary euthanasia will lead to non-voluntary and involuntary euthanasia, and to harm to others. With regard to the Dutch situation, there is, of course, also the specific claim that the Dutch survey demonstrates the existence of a slippery slope. Those who make the latter claim seem unaware of the fact that it is not possible to demonstrate a "slippery slope", unless one has two sets of empirical data – one set of data from the time before the relevant practice was introduced, and one set of data from the time after the practice was introduced. Only then would one have some evidence to show that there has been an increase in the harmful consequences, following the introduction of the practice. There is, however, only one set of data from the Netherlands – the data found in the survey we are discussing. There is thus no reference point against which the figures in the survey can be compared. Those wanting to base their "slippery slope" arguments on the figures found in the Dutch survey are therefore committing a serious error. On this point, see [13].

[17] Surveys of doctors and nurses have shown that even in societies where voluntary euthanasia is unlawful, it is relatively widely practised by doctors. This indicates that unlawfulness as such may not be a decisive deterrent to actions judged appropriate by doctors ([10]; [15]; [16]; [27]).

[18] A lot of other things would, strictly, also have to be assumed before one could say that the the harm caused by the liberalization of voluntary euthanasia would be a sufficient reason for banning the practice on the grounds of harm to others. In the public policy area we are concerned with maximizing outcomes. Hence, one would not only have to assume that the harm resulting from non-voluntary or involuntary euthanasia would have been prevented, but also that the prevention of the harm in that area would be larger than, for example, the harm caused by the introduction of restrictive laws, such as the thwarting of autonomous peoples' desire to die quickly and painlessly at a time of their choosing, increased suffering, anxiety (at not being able to have voluntary euthanasia if and when one needs it), and so on.

[19] In some 40% of cases, there had been earlier discussions between the doctor and the patient about euthanasia, but there had not been an explicit request at the time. If one were to take these earlier discussions as implied consent, then the number of unconsented-to cases of direct euthanasia would be 600.

[20] While few quantitative studies on unconsented-to end-of-life decision have been undertaken outside the Netherlands, there is no reason to think that such decisions occur less frequently in other countries that have a similar level of medical technology at their disposal. A recent study by the Centre for Biomediocal Ethics at the University of Minnesota involved 287 patients who were approaching the end of their lives. It found

that 84% (240 patients) had died after a decision had been taken to forego at least one type of life-sustaining treatment. Only 35% of these patients (84) had been able to participate in the decision. This means that 65% of all end-of-life all decisions were taken without the patient's consent. ([4], p. 1)

BIBLIOGRAPHY

1. Charlesworth, M.: 1993, *Bioethics in a Liberal Society*, Cambridge University Press, Cambridge.
2. Dyer, C.: 1992, 'GMC tempers justice with mercy in Cox case', *British Medical Journal* **305**, 28 November 1992, 1311–1312.
3. Engelhardt, H.T.: 1986, *The Foundations of Bioethics*, Oxford University Press, New York.
4. Faber-Langendoen, K.: 1994, 'Center project continues its focus on care of the dying', *Newsletter – The Center for Biomedical Ethics*, Spring, 1–2.
5. Feningsen, R.: 1991, 'The report of the Dutch governmental committee on euthanasia', *Issues in Law and Medicine* **7**, 339–44.
6. Fleming, J.I.: 1992, 'Euthanasia, the Netherlands and slippery slopes', *Bioethics Research Notes*, Supplement, Vol. 4, No. 2, 1–4.
7. Griffiths, J.: 1992, 'The regulation of medical procedures which shorten life', unpublished paper.
8. Gunning, K.F.: 1991, 'Euthanasia', *The Lancet* **338**, 1010.
9. Keown, J.: 1993, 'Courting euthanasia?: Tony Bland and the law Lords', *Ethics and Medicine* **9**(3), 34–37.
10. Kinsella, T.D. and Verhoef, M.J.: 1993, 'Alberta euthanasia survey: 1. physician's opinions about the morality and legalization of active euthanasia', *Canadian Medical Association* **148**(11), 1921–1932.
11. Kuhse, H.: 1987a, *The Sanctity-of-Life Doctrine in Medicine – A Critique*, Oxford University Press, Oxford. (German Edition: 1994, *Die Heiligkeit des Lebens in der Medizin – Eine philosophische Kritik*, Harald Fischer Verlag, Erlangen.)
12. Kuhse, H.: 1987b, 'Voluntary euthanasia in the Netherlands', *Medical Journal of Australia* **147**, 394–396.
13. Kuhse, H.: 1992, 'Voluntary euthanasia in the Netherlands and slippery slopes', *Bioethics News* **11**(4), 1–7.
14. Kuhse, H. and Singer, P.: 1985, *Should the Baby Live? – The Problem of Handicapped Infants*, Oxford University Press, Oxford.
15. Kuhse, H. and Singer P.: 1988, 'Doctors' practices and attitudes regarding voluntary euthanasia', *Medical Journal of Australia* **148**, 623–627.
16. Kuhse, H. and Singer, P.: 1993, 'Voluntary euthanasia and the nurse: an Australian survey', *International Journal of Nursing Studies* **30**(4), 311–322.
17. Mill, J.St.: 1968, 'On liberty', in M. Warnock (ed.), *Utilitarianism – John Stuart Mill*, Collins, London, pp. 126–225.
18. Pollard, B.: 1992, 'Euthanasia in Holland', *Quadrant*, November, 42–6.
19. Pollnitz, R.: 1992, 'The case against euthanasia: a response to Dr. Syme', *Australian Medicine*, June 1, 15.

20. Sacred Congregation for the Doctrine of the Faith: 1980, *Declaration on Euthanasia*, Vatican City.

21. Temkin, O. and Temkin C.L. (eds.): 1967, *Ancient Medicine: Selected Papers of Ludwig Edelstein*, John Hopkins University Press, Baltimore.

22. ten Have, H.A.M.J. and Welie, J.V.M.: 1992, 'Euthanasia: normal practice?', *Hastings Center Report* **22**(2), 34–38.

23. Tracy, G.D.: 1992, 'Medical aspects of euthanasia', *Medical Journal of Australia* **156**, 579–80.

24. van Delden, J.J.M., Pijnenborg, L. and van der Maas, P.J.: 1993, 'Dances with data', *Bioethics* **7**(4), 323 –329.

25. van der Maas, P.J., van Delden, J.J.M., and Pijnenborg, L.: 1992, *Euthanasia and Other Medical Decisions Concerning the End of Life*, Elsevier, Amsterdam.

26. van der Maas, P.J., van Delden, J.J.M., Pijnenborg, L. and Looman, C.W.N.: 1991, 'Euthanasia and other medical decisions concerning the end of life', *The Lancet* **338**, 669 –674.

27. Ward, B.J. and Tate, P.A.: 1994, 'Attitudes among NHS doctors to requests for euthanasia', *British Medical Journal* **308**, 1332–1334.

WOLFGANG LENZEN

VALUE OF LIFE VS. SANCTITY OF LIFE – OUTLINES OF A BIOETHICS THAT DOES WITHOUT THE CONCEPT OF *MENSCHENWÜRDE*

I. INTRODUCTION AND SUMMARY

The aim of the Bielefeld-conference was to determine the role played by the concepts of *Menschenwürde* and of "sanctity of life" within ethical considerations of modern medicine and biosciences. The lively discussion has shown that, on the one hand, there is a variety of morally wrong actions which violate the "Menschenwürde" of certain individuals; on the other hand, despite its frequent use in philosophical, theological, and legal argumentations the concept of *Menschenwürde* turned out to be largely dispensable in the sense that the immorality of most, if not all, of these actions may as well be explained without resorting to it. At least it was my personal impression that the papers presented at the conference failed to provide convincing examples of medical or bioscientific practices which (1) are generally considered as morally illicit; (2) violate the principle of "Menschenwürde"; and (3) are such that their moral wrongness cannot be established on independent grounds. The aim of the present paper is to reinforce the claim of the moral redundancy of the concept of *Menschenwürde* by outlining a system of bioethics which does without it. The key idea of this system is to replace the notion of "sanctity" of life by a more fruitful conception of the value of life to be developed in section III. First, however, let me present my ethical theory.

39

K. Bayertz (ed.), Sanctity of Life and Human Dignity, 39–55.
© 1996 *Kluwer Academic Publishers. Printed in the Netherlands.*

II. ETHICAL THEORY

The following investigations will be based on a blend and refinement of three well-known principles of moral philosophy, namely the Golden Rule, the maxim *Neminem laedere*, and the principle of Utilitarianism. In its traditional version, the Golden Rule forbids us to "do to others what we do not want to be done to ourselves". The idea behind this maxim is perhaps better expressed by saying that we should never do to others what *they* do not want to be done to *them*. Furthermore, in order to avoid problems resulting from unreasonable or capricious wishes, one may add the further qualification that we should never do to others what they *reasonably* want not to be done to them. The notion of someone reasonably wanting something shall be interpreted in such a way that individual X has a reason for wanting action A not to be done if and only if (iff, for short) A conflicts with X's legitimate interests. Although the latter concept of "legitimacy" is not without problems, it shall be presupposed as a primitive concept here.

The related maxim *Neminem laedere* simply says "Do no harm". Accordingly action A of an agent X is to be considered as morally wrong iff A does harm to at least one other individual Y. The concept of harm is meant to cover not only physical but also mental or psychic injuries. Here it will be understood in a very broad sense so that action A does harm to an individual Y whenever A violates the legitimate interests of Y. This interpretation of *Neminem laedere* basically expresses the same ethical principle as the Golden Rule. In what follows, however, it will be taken only as a *necessary* but not as a sufficient condition for an action to be morally wrong[1] and it will be referred to as *Minimal Principle of Ethics*:

(ETH$_{min}$) Action A of a person X is morally wrong *only if* A violates the legitimate interests of at least one other individual Y.

This weak principle is strong enough to solve certain bioethical problems. As will be shown below, ETH$_{min}$ allows a clear evaluation of the morality of (at least some instances of) suicide and also of (at least some instances of) euthanasia. However, ETH$_{min}$ is evidently too weak to judge on those actions which favour the legitimate interests of some individual Y and at the same time violate the legitimate interests of some other individual Z. Here we need the conceptual means to compare the harm or the good done to the one with the harm or the good done to the other individual. This is where Utilitarianism enters the scene.

Utilitarian ethics is based on the assumption that for every relevant action A and for every individual Y_i (from a certain set $\{Y_1, \ldots, Y_n\}$) we are given a numerical value $u(A,Y_i)$ – the utility of A for Y_i – which somehow corresponds to the degree to which the individual Y_i wants A to happen or wants A not to happen. With the help of these values the *total utility* of an action A is defined as the sum of the utilities of A for all individuals involved, i.e.

$$U(A) = \sum_{i \leq n} u(A, Y_i).$$

The total utility of A represents or indicates the moral quality of A in the sense that action A is morally better than some other action B iff the total utility of A is greater than the total utility of B. The problem, however, is to find an absolute or classificatory criterium telling us which actions are morally right and which are morally wrong.

In its usual formulation, utilitarianism entails the following super-erogatory principle which requires us to *maximize* the interests of all individuals concerned:

($UTIL_{sup}$) Action A_1 is morally right if *and only if* it maximizes the interests of all individuals concerned in the sense that, for every alternative action A_j (j = 2, . . . , m), $U(A_1) \geq U(A_j)$.

This principle is much too demanding, however. According to $UTIL_{sup}$ every action A which is not optimal, i.e. for which there exists an A' such that $U(A') > U(A)$, would have to be classified as morally wrong. But most of our everyday activities, such as drinking a cup of coffee, smoking a cigarette, or walking along the river, fail to be optimal; it is easy to construct alternative actions with a higher total utility, e.g., drinking no coffee, smoking no cigarettes and donating the money saved to charity. Nevertheless we would be reluctant to consider these activities as immoral.

In order to develop a more realistic utilitarian ethics, one has to introduce the notion of a *normal* utility function u* which represents the preferences of the individual Y in the following "natural" way: u*(A,Y) is > 0 whenever action A favours Y's interests and whenever therefore Y wants A to happen, while u*(B,Y) is < 0 whenever B violates the interests of Y and Y therefore wants B not to happen.[2] In particular, a positive value u*(A,Y) may be interpreted as the amount of money which Y is willing to pay for making A happen while a negative value u*(B,Y) corresponds to the amount of money which Y is willing to pay for preventing B from happening. In what follows we presuppose that every utility function u* is normal so that for every individual Y and for every action A, a positive value u*(A,Y) expresses the degree to which A favours the interests of Y, i.e. the degree to which Y wants A to be done; analogously, a negative value u*(B,Y) characterizes the degree to which B conflicts with the interests of Y, i.e. the degree to which Y wants B not to be done.

Let us further presuppose that the utility-functions u** are *intersubjectively comparable* in the following sense. Whenever an action A is positive for two individuals Y_1 and Y_2, then $u^{**}(A,Y_1) > u^{**}(A,Y_2)$ iff A favours the interests of Y_1 more than it favours the interests of Y_2; and if action B is negative for Y_1 and for Y_2, then $u^{**}(B,Y_1) < u^{**}(B,Y_2)$ iff B violates the interests of Y_1 to a larger extent than the interests of Y_2. This can be guaranteed, e.g., by interpreting $u^{**}(A,Y_i)$ specifically as the amount of money that Y_i would be willing to pay for making A happen (or unhappen), if Y_i has the same fixed amount of money at his disposal as every other Y_j.

Now if we restrict our considerations to normal and intersubjectively comparable utility functions u**, then the *total utility* of an action A, U**(A), is positive iff the good done by A to some of the individuals concerned outweighs the harm done to the others. Therefore we can state the following utilitarian principle which avoids the problem of supererogation and, at the same time, generalizes the idea of *Neminem laedere* and thus supplements ETH_{min}:

(UTIL) Action A is morally wrong iff its normal total utility, $U^{**}(A) = \sum_{i \leq n} u^{**}(A,Y_i)$, is negative.

III. SANCTITY OF LIFE VS. VALUE OF LIFE

In its strictest form, the Christian principle of the "sanctity" of life forbids the killing of any living being because each life is supposed to be created by God and therefore sacred or untouchable. This view, although it appears to have been held by Albert Schweitzer, cannot seriously be accepted because otherwise also the lives of viruses, wood-worms, tsetse-flies, cock-roaches, grasshoppers, etc. would count as sacred. In recent discussions the principle of the "sanctity" of life is therefore usually taken in a much narrower sense as restricted to *human* life only. A theological argument that might be given in support of this restriction derives from the Old Testament's story according to which the Creator of all things and beings has formed only *man* after his own image. If one does not want to refer to the authority of the Bible, one may interpret the principle of the "sanctity" of life more mundanely as the following principle of the *absolute value of human life*:

(AVHL)   Human life has an absolute value; therefore one may not kill any human being.

According to this principle, not only common murder but also every other kind of bringing human life to a premature end, in particular suicide, euthanasia, and killing in war or in self-defence, would count as morally wrong. This problem is discussed at length in [11]. Moreover, as has been criticised by Singer [21, 22], AVHL represents a "speciesistic" ethics insofar as it remains entirely silent on the issue of killing animals. Third, AVHL fails to provide an adequate basis for bioethical issues which deal with the coming into existence rather than the ending of (human and non-human) life. In view of these shortcomings, it seems favorable to develop an alternative approach which (1) does not rest on theological premises; (2) explains under which circumstances killing a human being might be justified; and (3) can in principle be generalized so as to take into account also the value of non-human life. The basic idea is to assume that the life of each individual X – whether human or not – has a value that depends on the sum of values of everything that X ever experiences during his life. Accordingly AVHL shall be replaced by the following more realistic claim:

(VL)   Human life has a very high value; but the life of a non-human sentient being also has a certain value which will in general differ from one species to another.[3]

For reasons of space, however, the issue of animals must be skipped here.

In order to render the first sentence of VL somewhat more precise, let us consider the completed life of a person X. If X dies, say, at the age of 75, her past life may be split up either into temporal units or into a sequence of states and events $A_1, \ldots, A_n$. This sequence might begin with some prenatal events or with the first sucking at the mother's breast. It will contain some typical events as hurting a knee as a baby, celebrating Christmas, writing maths homeworks, lying in the sun, reading newspapers, meeting lovers and friends, etc. Typically it will end with illness, pain, and the fear of death. Each of these states or events has a value for X which can in principle be measured by a normal and intersubjectively comparable utility function u. Since most of the things we experience in life are positive for us, the value of X's total life,

$$V(L, X) = \sum_{j \leq n} u(A_j, X),$$

will normally be much greater than 0. Moreover, at almost every instance t of her life, not only has the *past* life of X normally had a positive value

$$V(L_{\leq t}, X) = \sum_{j \leq k(t)} u(A_j, X),$$

but the value of her *future* life,

$$V(L_{>t}, X) = \sum_{j > k(t)} u(A_j, X),$$

will also, in general, contain more pleasure than pain and thus be positive. Accordingly, X wants to stay alive, because (she believes that) $V(L_{>t}, X) > 0$. So if X had died earlier at a certain time t, she would have suffered a loss the worth of which can in first approximation be identified with the value of the rest of her life, $V(L_{>t}, X)$. In this sense the harm done by killing a person X is *prima facie* as great as the value of the span of life that X would otherwise have lived.

Two qualifications may be in order. First, the entire value of the *completed* life of an individual X was *retrospectively* divided in two parts; the value of life that X lived up to a certain time t, $V(L_{\leq t}, X)$, plus the value of life that X lived from t onwards, $V(L_{>t}, X)$. The latter value was then equated with the loss that X *would have* suffered if she *had* been killed at t. However, in order to evaluate the harm done to someone who is *going to* be killed at some time t, one has to determine the value of X's future life *prospectively*. This so-called *expected* value of X's future life might be calculated in roughly the same way as the expected utility of an action A is calculated within the theory of decision and action. E.g., starting from the class of all possible lives $\{L_1, \ldots, L_m\}$ open to X at t, one might (1) determine their respective values for X,

$V(L_j,X)$; (2) try to fix the probabilities, $P(L_j)$, that X will live the life L_j; and (3) calculate the expected value of X's future life as the weighted sum

$$V_{exp}(L_{>t},X) = \sum_{j \leq m} V(L_j,X) \cdot P(L_j).$$

Although in practice it may be very difficult, if not impossible, to figure out $V_{exp}(L_{>t},X)$ in a precise and reliable way, the basic idea seems to be clear enough to be used for judging the morality of most instances of killing.[4]

Second, sometimes it may even be necessary to distinguish different expected values of the future life of one and the same individual X. So far it has been tacitly presupposed that both the values $V(L_j,X)$ and the probabilities $P(L_j)$ are determined by X's own subjective valuations. But there is empirical evidence that in certain situations certain individuals X just have "wrong" probabilities. In particular, many young people trying to commit suicide falsely believe that there is no chance for their leading a happy life L_h, and thus they mistakenly consider their future life as worthless. As will be argued in the subsequent section, in such a situation it is better to base the moral evaluations on a modified expected value of X's future life where X's "wrong" subjective probabilities are replaced by more correct or more "objective" probabilities $P'(L_j)$.

IV. SUICIDE; ASSISTED SUICIDE

Christian tradition considers life as a gift of God and forbids suicide because God alone is held to be entitled to take what he has given to man. Thus, e.g., Eibach ([3], 122) maintains: "From a theological point of view it is part of [. . .] the essence of the finite creature man that [. . .] his life is already disposed of. Man does not give himself life, [. . .] life is not man's property but a donation, a gift [of which God alone] may dispose". However, as has been argued, e.g., by Birnbacher [2], such a view is far from convincing because gifts normally may be rejected or given back.

According to Eth_{min} one would hold in contrast that there are prima facie no intrinsic moral objections to suicide because primarily or even exclusively only the interests of the suicide himself are concerned. In a way it does not even matter whether the suicide has sufficient reasons for killing himself, i.e. whether his life is in fact as worthless as he thinks it to be. If he is right, then the suicide will be in his own interest; if he is wrong, then the suicide will be directed against his "true" interest; he will then harm himself and the suicide will be a fatal and irreversible folly. Nevertheless, according to $ETH_{min,}$ the suicide is not *immoral* because – or, more exactly: as long as – it does not harm anybody else.

Now the latter qualification indicates under which circumstances suicide might be immoral after all. Suicide will often touch upon the interests of others individuals. In the most extreme case a suicide may endanger the life or health of other people; or it may create serious financial problems for the family or for business partners. Perhaps more often than involving such material detriments, suicide will do psychic harm to friends and near relatives who may feel guilty for the death or who simply feel sad about the loss of the deceased. In such cases suicide of a person X is morally justified, according to UTIL, only if the (supposed) good for X outweighs the harm done to her friends and relatives. If and when this is the case is a matter of delicate individual considerations which clearly cannot be settled by abstract philosophical analysis.

Some philosophers have argued that the moral acceptability of suicide entails that assisted suicide would have to be morally acceptable, too. Thus, according to Lamb ([12], p. 49), "if a right to suicide is established then it follows that one has a right to assisted suicide [. . .] The right to suicide – if such a right were recognised – must also entail the right to any assistance in fulfilling that right". This, however, is a clear *non sequitur*. The moral situation of someone, A, to assist (or even to encourage) S in committing suicide differs quite considerably from S's own moral situation. For A it is crucial to know whether the suicide would be in S's true interest or not, i.e. whether the expected value of S's future life really is negative. Experience shows that many would-be-suicides, especially when attempted by younger people, represent panic or folly actions which are *afterwards* recognised as such by the suicides themselves. If therefore A is asked to help S in committing suicide, or if A is at least able to prevent S from committing suicide, he may often take it for granted that the death would be a great loss for S. And if A knows or has reason to believe that this is the case, then A acts immoral if he does nothing to prevent the suicide. Even if S does not harm anybody else and thus does not act immoral when killing himself, A would act immoral because *he* would do harm, or at least allow that harm is done, to *someone else*, namely to S!

V. EUTHANASIA

The concept of euthanasia as used in philosophical discussions must be carefully distinguished from the notion of so-called "euthanasia" with which the Nazi's tried to conceal the mass extermination of lives allegedly not worth living. We adopt the philosophically relevant "definition of euthanasia which brings under this heading only cases of opting for death *for the sake of* the one who dies" ([4], 86; my emphasis). This definition suggests that any proper instance of euthanasia is at the same time a case of *voluntary* euthanasia. The issue of nonvoluntary[5] euthanasia especially of gravely deformed and severely retarded newborns and infants will be discussed below. First, however, let us consider voluntary euthanasia in either passive or active form where, according to Rachels ([18], p. 148), active euthanasia is to be understood as a "positive action designed to kill the patient; for example giving him a lethal injection of potassium chloride. 'Passive euthanasia', on the other hand, means simply refraining from doing anything to keep the patient alive. In passive euthanasia we [. . .] let the patient die 'naturally' of whatever ills already afflict him."

Voluntary euthanasia of a seriously ill and suffering patient may be regarded as a special form of assisted suicide of someone who considers his life no longer worth living and accordingly wants to die. The two main differences between this particular situation and assisted suicide in general consist in the fact (1) that the patient is physically unable to set an end to his life by himself, and (2) that his wish to die is "objectively" understandable. He needs the help of somebody else, preferably a doctor, to set an end to his suffering. Now if the doctor comes to the conclusion that the value of the patient's life really is negative; if, according to the doctor's medical experience, this life contains nothing but pain and suffering, or at least more pain than pleasure; if, in other words, death really would be for the sake of the patient; and if the patient unmistakenly declares that he wants to die, then, in accordance with the principle *Neminem laedere*, the doctor is morally entitled to set an end to the patient's life, no matter whether in active or in passive form.

This view has come under attack especially from theologians. Thus, according to Koch ([10], pp. 108/9), "killing a person can never be an act of love". In particular, euthanasia should be strictly forbidden because the "patient must be able to trust the doctor in that [. . .] he always wants the best for [the patient's] welfare but never intentionally wants [his] death".[6] This objection fails to realise, however, that in every genuine case of euthanasia both the doctor and the patient are convinced that death *is* "the best for the patient's welfare".

Another objection points to the potential danger of misuse of active euthanasia. There is supposed to be a "slippery slope" which sooner or later might lead to the killing of patients who do not really want to die. A detailed discussion of this argument may be found in [12] and in some of the papers collected in [8]. Suffice it to mention here that recent experiences in countries, such as the Netherlands, where active euthanasia has been legalised, do not confirm the increased misuse as predicted by proponents of a "slippery slope".

The real problem of euthanasia rather seems to consist in cases of nonvoluntary euthanasia, where on the one hand death is (supposed to be) "for the sake of the one who dies" although the one who is going to die did not explicitly express his wish to be killed. This situation mainly applies to the abortion or killing of seriously handicapped fetuses or newborns and to active and passive euthanasia of incurably ill and suffering patients which are permanently unconscious or otherwise unable to articulate a wish to be killed.

One may, of course, give an easy, abstract judgment on the morality of such actions; to wit: Nonvoluntary euthanasia – just as voluntary euthanasia – is morally all right if and only if death *is in the true interest of the one who dies*. However, the big problem is to find out with sufficient certainty whether the crucial condition really is satisfied, i.e. whether the patient would have asked the doctor for being killed if he had been able to articulate wishes at all. One may perhaps assume this to be the case if, at some earlier time, the patient has unmistakenly declared that he wanted to be killed if his illness would become worse and if he would then be unable to repeat this wish. In such a situation the German Society for Humanitarian Death recommend the following procedure: "If the patient is almost or entirely unable to express his mind before [the act of euthanasia], his wish to be released from his sufferings may be laid down in a patient's declaration designed for this particular purpose. The declaration must not be testified, however, by a possible beneficiary."

Although the latter proviso offers some protection against misuse, this sort of nonvoluntary euthanasia is not without problems. After all the preferences of the patient concerning life or death may have changed in the meantime. Therefore the doctor should try everything he can to find out whether the patient still subcribes to his earlier declaration. As long as the patient regains consciousness then and when, he will be able to answer the question "Do you still want to die?" either by talking or by shaking his head or in some other way. But then the situation turns out to be a voluntary rather than an nonvoluntary euthanasia which has been discussed before.

If the patient, V, is permanently vegetating without consciousness, however, the doctor has no means to assure himself of the validity of V's former declaration. In this case there is no compelling reason for euthanasia either, because, being permanently unconscious, V no longer feels any pain and thus needs not to be released from his sufferings. Here active or passive "euthanasia" would *not be in V's (true) interest* but rather be motivated by other factors, e.g. by financial considerations or perhaps by the fact that the life-supporting machinery is urgently needed for the care of other patients with better prospects of life. On the other hand one has to note that "euthanasia" of a permanently vegetating patient would *not* be directed *against V's (true) interest* either. If V remains in a state of unconsciousness, then his future "life" lacks any (positive or negative) experiences and thus the value of V's life is simply 0. In this deadlock it may seem advisable to err on the safe side and abstain from "euthanasia"; after all no doctor can predict with absolute certainty that an unconscious patient will never "wake up" from his vegetative state.

To conclude let us consider the nonvoluntary "euthanasia" of severely deformed or handicapped infants which are unable to tell the doctor whether they want to be released from their illness or not (the special issue of *abortion* will be discussed in the subsequent section). Some diseases such as *spina bifida* may allow to infer the amount of the infant's suffering from his behaviour, but in the end every attempt of an external evaluation remains dubious. Singer ([21], p. 133) mentions that "some doctors closely connected with children suffering from severe spina bifida believe that the lives of some of these children are so miserable that it is wrong to resort to surgery to keep them alive. This implies that their lives are [according to the doctors' opinions!] not worth living". The moral justification of euthanasia requires, however, that the patient himself rather than the doctor considers his life as not worth living.

Summing up one can say that the moral problem of euthanasia does not consist – as the Roman Congregation of Faith pathetically maintains – in that "nobody can ever bestow the right to kill an [. . .] incurably ill or dying person [. . .]. For this would amount to a violation of a divine law, an offence of the dignity of man, and an attack against the human race."[7] The problem of euthanasia rather consists in the fact that neither relatives nor doctors know with absolute certainty whether death really is for the sake, or in the true interest, of the one who dies.

VI. ABORTION AND CONTRACEPTION

The fact, explained in section III, that the future life of an individual normally has a very high value immediately entails the *prima facie* wrongness of abortion. If a fetus, F, is killed even before making his first experiences, he is thereby deprived of the whole life he would otherwise live. Although we cannot know for certain the value of F's life, we are normally pretty sure that the fetus, if he is not aborted, will live a life that is worth living (unless F has a serious handicap that would drastically lower the value or quality of his life). Thus – according to Marquis [17] – the harm done to F by abortion is equivalent to the expected value of the whole life that he would otherwise live. Hence the principle of Utilitarianism can justify an abortion only if $V_{exp}(L,F)$ is smaller than the sum of the harm – or better: the sum of the disadvantages – which the mother and perhaps some other people would suffer if the fetus were allowed to live his life.

This position has been attacked by quite a number of philosophers who think that there is nothing wrong about abortion as long as it occurs before the fetus acquires at least a rudimentary form of consciousness.[8] The main rationale for this claim consists in the idea that a human fetus at an early stage of development or, more generally, an individual without even a rudimentary form of consciousness has no interests or preferences and hence has no *right* to live. This view seems to be mistaken in one decisive respect. It is, of course, true that a being like a stone or a tree which *qua* its nature will never gain consciousness cannot therefore be harmed in a morally relevant sense. But it is equally true that the morality of an action A depends not only on what the individuals in question prefer or wish *at the moment* when A is done; the *future* preferences, as far as they are affected by the action in question, must be taken into account as well. Let me illustrate this point by means of an example. Suppose that the victim of a traffic accident, T, is brought into the hospital in deep coma. After a careful examination of T's brain-injury the neurosurgeon concludes that T has suffered a complete amnesia and will remain 6 months in a state of unconsciousness. But afterwards T will slowly recover and one year later he will be able to live a normal happy life. Although during the first 6 months after the accident T has not wishes at all and thus killing him during this period would not violate his preferences, it seems evident that it would be morally wrong to kill him (or not to keep him alive).

Therefore a political or legal rule, such as the so-called "Fristenregelung" which permits any abortion up to the 13th week of pregnancy seems to be morally unacceptable. This does not mean, however, that abortion must be morally wrong in every case and under all circumstances. The moral principle UTIL instead supports the so-called "Indikationenlösung" in the sense that, on a utilitarian balance, abortion appears to be justifiable (1) in almost every instance of so-called medical indication; (2) in most instances of so-called eugenic indication; (3) in many instances of so-called criminal indication; and (4) perhaps also in some instances of so-called social indication.

A more detailed discussion of this issue and of the related issue of abortive or quasi-abortive measures of contraception may be found in [14] and ([16], ch. 4). Suffice it to mention here that – contrary to what Hare and some other philosophers have claimed[9] – "ordinary" contraception is morally absolutely right as long as it only aims at preventing an ovum from becoming fertilized. An unfertilized egg does not by itself develop into a being that lives a life worth while living. Its "life" normally ends with the menstruation, and as such has no more value than, say, the "life" of a finger-nail or the "life" of an appendix. Contraception does no harm to the ovum and doesn't violate the ovum's interests, because an unfertilized egg *qua* its nature is not capable of making experiences; therefore by itself it has no interests, and thus it cannot be harmed by any action in any conceivable way.

VII. IN-VITRO-FERTILISATION

In-vitro-fertilisation (IVF) ideally proceeds as follows. A certain number of eggs are taken from the woman's ovary and then inseminated in a test-tube (*vitrum*) with the sperm preferably taken from the husband; after a period of about two days all fertilised eggs are replanted into the woman's uterus. At present the success rate, i.e. the rate of pregnancies initiated by IVF, is still very low at around 10%. In view of these figures, Leist ([13], pp. 182/3) concluded: "Procreating by means of IVF means to bring embryos predictably and intentionally into a situation where they will die with a high degree of probability". According to Leist, IVF is comparable to a "Russian roulette" where a person is killed with a chance of 90%. This view, however, is untenable. IVF neither intends nor even entails the "killing" of an embryo which would otherwise, i.e. without IVF, have had any prospect of life. IVF just means to procreate new lives, but not to procreate one new life on the cost of nine others. *Prima facie* each ovum fertilised *in vitro* has an equal chance to survive, and if in the end only one or even none of the embryos actually survives, the attempt to procreate new life still must not be blamed for having "destroyed" or "killed" other lives. For if IVF had not been attempted at all, then the respective eggs would have remained unfertilised, and, to repeat, unfertilised eggs cannot be morally harmed in any conceivable way.

In order to increase the success rate of IVF, sometimes more eggs are taken from the woman's ovary than can be replanted after fertilisation. Accordingly some "spare" embryos are produced which are then either cryoconserved for possible future reimplantation or which are nourished for some time in order to serve as objects of medical studies (and to die afterwards). The Catholic Church categorically rejects such experiments by requiring that "all fertilised eggs have to be implanted. [. . .] Cryconservation [. . . and] experimentation with embryos are beyond the line of what can be morally accepted".[10]

The minimal principle of ethics developed in section II seems to suggest that using spare embryos for medical research is morally all right. As has been argued, e.g., in Warnock ([24], 6), such an embryo "could not be harmed by being used for research purposes and then destroyed". On the one hand, "it would die anyway if not implanted"; on the other hand, "embryos suffer, themselves, no pain, since at this stage they have no central nervous system, and no brain".

This argument is open to a serious objection, however. It is certainly true that (1) *if* an embryo E will not be implanted, E is bound to "die anyway"; and (2) any embryo E which is going to "die anyway" will not be harmed by being used for medical experimentations. But from this it does not follow (3) that E *could not* be implanted and hence *had to* "die anyway"! In other words, given that spare embryos do exist, medical research using them may be morally defensible; but the crucial form of IVF might be blamed for the fact of intentionally producing spare embryos. Legislation in Germany accordingly prohibits forms of IVF where more eggs are fertilised than are needed for the purpose of pregnancy, to wit, more than three eggs. And the "Embryonenschutzgesetz" also requires that each embryo procreated *in vitro* has to be implanted into the woman's uterus. The philosophical rationale behind this law has been expressed in a "minority report" of the "Committee of Inquiry into Human Fertilization and Embryology" (cf. [24], 10) as follows: "[T]he embryo has a special status *because of its potential for development* to a stage where everyone would accord it the status of a human person. It is, in our view, wrong to create something with the potential for becoming a human person and then deliberately to destroy it. We therefore recommend that nothing should be done that would reduce the chance of successful implantation of the embryo".

To be sure, this argument at best proves procreating spare embryos to be morally wrong *from the point of view of Neminem laedere*. Like in the case of abortion, however, *utilitarian* considerations might well speak in favour of the incriminated practices. According to Warnock ([24], 7), "the results of medical research using embryos is beneficial to the infertile, and to those whose interest is in reducing crippling genetic disease, or to those whose aim is simply to increase our knowledge of the earliest stages of embryonic development".

Department of Philosophy
University of Osnabrück
Germany

NOTES

[1] In particular, if an agent X faces a moral dilemma such that whatever he decides to do he cannot prevent doing harm to *some* individual Y, many people would be reluctant to say that X necessarily acts morally wrong.

[2] Alternatively, one may require that every normal utility function u* assigns the value 0 to the "tautological action" T to be interpreted intuitively as "A is either done or not done".

[3] Cf. ([12], p. 43): "A better course is to replace the sanctity of life principle [. . .] with a principle concerning the 'value of life', expressed in the statement that 'life is worth living'."

[4] From the *prima facie* point of view of the maxim *Neminem laedere*, killing X is morally permissible iff the expected value of X's future life is negative. So all one has to know

in the first instance is whether $V_{exp}(L_{>t}, X)$ is positive or not. If, on the other hand, from a utilitarian point of view it should become necessary to quantify the harm of X's death, the knowledge of certain basic facts of X's life such as age, health, etc. will often be sufficient to allow at least a rough estimation of $V_{exp}(L_{>t}, X)$.

[5] Singer ([21], pp. 129/30) distinguishes involuntary and nonvoluntary euthanasia as follows: "I shall regard euthanasia as involuntary when the person killed is capable of consenting to her own death, but does not do so [. . .] If a human being is not capable of understanding the choice between life and death, euthanasia would be neither voluntary nor involuntary, but nonvoluntary." Since it is very hard to imagine cases where killing X would be "for the sake of the one who dies" even though X himself refuses to be killed, the logically possible category of involuntary euthanasia is – as Singer himself came to admit – in reality empty.

[6] A similar critique is put forward in [3] and in [19].

[7] Quoted from the German newsmagazine *Der Spiegel*, issue 52, 1990.

[8] Cf. e.g. [1, 9, 23].

[9] Cf. in particular [5, 6, 7] and the critical discussion in [15]. Hare still seems to hold the view that "contraception and abortion really are on a par" ([6], 219) or that "Abortion is [only] a more tricky procedure medically than contraception" ([7], 12).

[10] ([19], 41). Reiter furthermore postulates that one "must not go beyond the biological structure of the family, i.e. the ovum and the sperm must originate from husband and wife". This requirement is based on a catholic view of sexuality and marriage which, among many other objections to IVF, is aptly refuted in [20].

BIBLIOGRAPHY

1. Bassen, P.: 1982, 'Present sakes and future prospects: the status of early abortion', *Philosophy & Public Affairs* **11**, 314–337.
2. Birnbacher, D.: 1990, 'Selbstmord und Selbstmordvorsorge aus ethischer Sicht', in A. Leist (ed.), *Um Leben und Tod*, Suhrkamp Verlag, Frankfurt , pp. 395–422.
3. Eibach, U.: 1988, 'Sterbehilfe – Tötung auf Verlangen? Theologische und ethische Gesichtspunkte', *Zeitschrift für Evangelische Ethik* **32**, 220–229.
4. Foot, Ph.: 1977, 'Euthanasia', *Philosophy & Public Affairs* **6**, 85–112.
5. Hare, R.: 1975, 'Abortion and the golden rule', *Philosophy & Public Affairs* **4**, 201–222.
6. Hare, R.: 1988, 'When does potentiality count? a comment on Lockwood', *Bioethics* **2**, 214–226.
7. Hare, R.: 1989, 'A Kantian approach to abortion', *Social Theory and Practice* **15**, 1–14.
8. Hegselmann, R. and Merkel, R. (eds): 1991, *Zur Debatte über Euthanasie*, Suhrkamp Verlag, Frankfurt.
9. Ketchum, S. A.: 1987, 'Medicine and the control of reproduction', in E. E. Shelp (ed.), *Sexuality and Medicine Vol. II Ethical Viewpoints in Transition*, Reidel, Dordrecht, pp. 17–37.
10. Koch, T.: 1987, ' "Sterbehilfe" oder "Euthanasie" als Thema der Ethik', *Zeitschrift für Theologie und Kirche* **84**, 86–117.

11. Kuhse, H.: 1987, *The Sanctity-of-Life Doctrine in Medicine. A Critique*, Oxford University Press, Oxford.
12. Lamb, D.: 1988, *Down the Slippery Slope – Arguing in Applied Ethics*, Croom Helm, London etc.
13. Leist, A.: 1990, *Eine Frage des Lebens – Ethik der Abtreibung und künstlichen Befruchtung*, Campus Verlag, Frankfurt.
14. Lenzen, W.: 1991, 'Wie schlimm ist es, tot zu sein – Moralphilosophische Reflexionen', in R. Ochsmann (ed.), *Lebens-Ende*, Asanger Verlag, Heidelberg, pp. 61–178.
15. Lenzen, W.: 1995, 'Hare über Abtreibung, Empfängnisverhütung und Zeugungspflicht', in Ch. Fehige and G. Meggle (eds), *Zum moralischen Denken*, Suhrkamp Verlag, Frankfurt, pp. 225–239.
16. Lenzen, W.: 1996, *Liebe, Leben, Tod. Ein moralphilosophischer Essay*, Suhrkamp Verlag, Frankfurt, forthcoming.
17. Marquis, D.: 1989, 'Why abortion is immoral', *Journal of Philosophy* 86, 183–202.
18. Rachels, J.: 1979, 'Euthanasia, killing and letting die', in J. Ladd (ed.), *Ethical Issues Relating to Life and Death*, Oxford University Press, Oxford, pp. 146–163.
19. Reiter, J.: 1987, 'Ethische Probleme um den Lebensende – Bausteine für eine Bioethik', *Theologie der Gegenwart* 30, 38–46.
20. Schöne-Seifert, B.: 1990, 'Philosophische Überlegungen zu "Menschenwürde" und Fortpflanzungs-Medizin', *Zeitschrift für philosophische Forschung* 44, 442–473.
21. Singer, P.: 1979, *Practical Ethics*, Cambridge University Press, Cambridge.
22. Singer, P.: 1979a, 'Unsanctifying human life', in J. Ladd (ed.), *Ethical Issues Relating to Life and Death*, Oxford University Press, Oxford, pp. 41–61.
23. Tooley, M.: 1972, 'Abortion and infanticide', *Philosophy & Public Affairs* 2, 37–65.
24. Warnock, M.: 1987, 'Do human cells have rights?', *Bioethics* 1, 1–14.

SANCTITY OF LIFE AND HUMAN DIGNITY AT THE BEDSIDE

However much the history of the relationship between physician and patient has been that of the "silent world" that Jay Katz describes [2], that world is increasingly being impacted by a cacophony of voices in the present day. Whether basic communication and mutual understanding is actually increasing remains debatable. But endless discussions and reflections, in the clinical and bioethics literature, in the popular media, in courses and seminars for students and practitioners of health care, in rooms where team members or ethics committees meet, and at the bedside, are occurring, particularly concerning the appropriate response to chronic, terminal and devastating disease. And such discussions are not just between physicians as in the past; all members of the health care team, ethics consultants and committees, as well as patients and their families, are increasingly attempting to have their voices heard.

Beyond such sources, the information and insights being injected into such discussions have also become increasingly varied and complex. Numerous studies of patient outcomes, specifically keyed to patients' states of health or disease, are available to be appealed to and evaluated for relevance and import. Speculations and inquiries regarding what "quality of life" the patient had beforehand, or might have after the fact, are common, as is the search for or solicitation of prior statements from patients as to what they might or might not have wanted in the situations at hand. And all parties are increasingly bringing their more or less articulate ethical views and distinctions to the bedside and waiting room discussions. It is as if the Socratic injunction to philosophize has finally been embraced by all concerned.

Within such a cacophony of voices and insights, the injection of

K. Bayertz (ed.), Sanctity of Life and Human Dignity, 57–71.
© 1996 Kluwer Academic Publishers. Printed in the Netherlands.

"concepts" such as "the sanctity of life" or "human dignity" often seem
to have a jarring effect. It is as if they are anachronistic terms that were
thought to be well left behind given everyone else's further thought
and experience. Simply speaking, they look like quite inadequate "tools"
to apply to the task at hand, and markedly inferior to others.

The term "sanctity of life" seems out of place to many given the widespread acceptance of "quality of life" thinking, the idea that medicine
can only do so much, and that limits should be placed on how much is
attempted with patients who, as the Hippocratic corpus put it, are "over-
mastered" by disease. Such language often seems to signify an
unreflective "vitalism" that refuses to acknowledge that a human life
can end up being deprived of all chance of real benefit, and continues
to insist on our obligations to an empty husk.

The term "human dignity" seems to have no more real content, being
also an anachronistic term. However useful once in opposing the "treat
at all costs and to the bitter end" mindset, initially a battle of slogan
against slogan, it can now seem like another vague term that lacks the
precision and level of insight that now characterizes current debate.
Though of opposite import to the "sanctity of life" appeal,[1] the injection of concern about the "patient's dignity" seems equally jarring as
other discussants seek to balance complex outcomes with the "cost",
e.g. suffering, to the patient of trying to secure them. Much that is
undignified, i.e. connection to all those machines and lines, severe edema
or disorientation, etc., may well be necessary as positive outcomes are
pursued. So the patient's abstract dignity, whatever that might be, is
hardly useful currency in the complex debate over possible benefits,
and the costs of pursuing them.

In this paper, I shall attempt to clarify the function of such terms,
particularly within discussions at the bedside, about the management
of terminally ill or otherwise devastated patients. I believe that much
more is going on with such terms than simply the outpourings of untutored or unreflective minds. The sources of such language are, in one
sense, clear enough: such "jarring" language can come from any of the
usual players in such discussions, i.e. clinicians, non-clinician consultants, whether ethicists or clergy, and patients and family members. What
they actually have in mind (conceptually and otherwise) is, however,
often quite unclear, and this is partially the reason for their jarring quality.
For the other parties to the "discussion", the task at hand is tough enough
without having to attempt to educate and counsel someone who does

not even have a basic sense of the currency and character of the issues at hand.

Getting clear about why, in what sense, and to what purpose people use such terms will be the specific task of this essay. Its reflections will be, in large part, quite conjectural. Utilizing my own experience as an ethics consultant, and a member of a number of ethics committees, I will attempt to array what seems to be behind such language, to the extent it actually has any content or import. But the sources of such insights will mainly *not* be self-revelations from those who use such terms, but rather from the speculations of those who are confronted with them. To anticipate, much of this usage, I will suggest, is just what it seems: unreflective and untutored. It also stems from conditions that call mainly for education and counseling, e.g. when the "recalcitrantly unrealistic" family member justifies a demand for continued aggressive treatment on "sanctity of life" grounds. But, equally, my concern here will be whether there are not also some grains of truth, or important concerns, that such language harbors or serves, and that the current "complex" discussion at the bedside may well need to be more mindful of.

To put this another way: from the point of view of ethics consultation, whatever the source and import of such language, it is always important to take it seriously, to attempt to elucidate whatever content and import it has. The following is not, then, an attempt simply to marginalize "concepts" like "sanctity of life" or "human dignity", and thus allow those of us who are "more sophisticated" in such matters to proceed by ignoring them. This is so, even if no "legitimate" content or agenda can be found for such terms in any given case, given that a primary goal of ethics consultation is to fashion a sense of the case at hand, and the ethical and spiritual values involved, so that all parties might come to an agreement. And such consensus will often occur only when the concerns and agendas implicit in such language are made explicit.

I. VARIOUS INTELLECTUAL MEANINGS OF AND AGENDAS BEHIND "SANCTITY OF LIFE" AND "HUMAN DIGNITY" EXPRESSIONS

Many of the meanings of such language have been described and discussed by others in this volume, so I will not rehash such definitions in any detail. "Sanctity of Life" as a concept is used to signal views about the "sacredness or holiness" of human life, that it has a certain "inviolability" or "infinite value".[2] Other meanings can include claims for its high moral purpose and significance, as well as the emphasis on its being a gift from God which no man should destroy, whether one is speaking of abortion, suicide or euthanasia. In some quarters, a functional synonym occurs in the assertion of the right to life. And the language here is as much negative, i.e., oppositional, as it is positive: it is used to oppose the proponents of abortion, euthanasia, "quality of life" thinking, and those who believe that our fiscal obligations to the sick can be somehow discussed and calculated as if human life was just one more quality to be weighed in the balance against others.

Given such meanings, it is easier to see why the insertion of the language of sanctity of life is jarring at the bedside. In brief, it just runs counter to the specific calculations and balancings that the other participants are attempting to produce. It is, in effect, much like being trumped, or being told: "Sorry, but we are not going to play that particular game". It is, in the end, the ultimate discussion stopper, the final way that a patient or family member can refuse to engage in the complex reflection on costs, benefits, and the probabilities and possibilities that others are attempting. Infinite value puts everything into a different league against which no combination of finite values or disvalues can ever amount to enough to lodge a creditable challenge.

The "concept" of "human dignity" has been advanced with much more positive, detailed content in certain contexts. It has been used as a primary ground for patients' rights, e.g. to be informed, to make decisions, to do last things, and so forth, and more generally as a source of the insistence that people receive proper respect and consideration within the often impersonal assembly line of modern health care. In sum, as a term, human dignity can be used to emphasize and protect any particular quality that we value in human beings, and particularly those which are threatened by illness and the nature of modern health care. In this sense, it also often moves into a negative, oppositional mode, as the effects of illness, or the nature of the care the patient is receiving, are identified for complaint. It can even be used to object to something as seemingly mundane as the fact the patient has been placed in a room with other patients, rather than in a private room.

As with sanctity of life language, generic statements (or complaints) regarding the patient's dignity can also have a jarring effect at the bedside. And, again, this is because such expressions seem to signal an untutored quality to the person expressing them. Part of it may simply be exasperation that one is faced with someone who doesn't realize what is going on or possible, as when human dignity is appealed to for the sake of an unavailable private room. Sticking to more poignant matters, the patient, or more often family member, who appeals to human dignity as being assaulted by the exigencies of intensive care also seems to be trumping the ongoing discussion. Whatever the real agenda, when expressed, the other parties tend to sense that the source is attempting to object, in a global, abstract fashion, to all they are attempting to do for the patient, as well as objecting to the "complex" discussion that they are attempting to carry out. Such language can function simply as a rejection of the legitimacy of both enterprises.

Whatever the abstract intellectual meanings, and other agendas behind these terms or their synonyms, it is their usage in jarring ways, as attempts to shortcircuit the "usual" manner of discussion, that I am particularly concerned with in this paper. Sometimes such language is readily unpacked by those who initially offer it, and the discussion can proceed in its own fashion. At other times, such language is used as a trump, as discussion stopper, with no attempt or ability to unpack it any further, and it is to such "usage" that I particularly want to refer. I am, in other words, much more concerned to speculate on the reasons that people have for using these terms than their abstract intellectual meanings. What are they trying to protect or provide for?

II. UNDERLYING MEANINGS AND AGENDAS OF "SANCTITY OF LIFE" AND "HUMAN DIGNITY" EXPRESSIONS AT THE BEDSIDE

The first usage I want to discuss is the seemingly simple one where, early on, patients or families use such terms to stop the discussion that the staff is attempting to have with them. Beyond such early usages, which often function as attempts to reassert some control in situations where patients and/or families feel powerless and overwhelmed, I will seek to discuss three underlying, more complex states that appear to be functioning when some "player" in the discussion continues, especially in the face of much opposition, to advance such terms. These underlying states will be described, rather loosely and arbitrarily, as (1) dissonance, (2) denial, and (3) discomfort.

A. *Patients and/or Families Who Throw out the Anchor Early on*

Not seldom, as staff attempts to respond to chronic, terminal or potentially devastating illness, their initial approach to the patient and/or family
is met by what amounts to a rebuff, an attempt to derail or shortcircuit
the sort of discussion that staff believes should ensue. Whether it is an
initial discussion of the possibility that some major catastrophe may have
occurred and a do-not resuscitate order be considered,[3] or where informed
consent for a more invasive procedure that signals the seriousness of
the patient's condition is attempted, the response may well be, respectively, of the form: "Oh No! Just do everything you can to save him",
or "I'm not sure: I do not want him to suffer". Toward our particular
focus here, such responses not seldom incorporate "sanctity of life", or
alternatively, "human dignity" language.
    Part of the problem is often caused by staff themselves by lack of
proper preparation or anticipation. Little or no sense may have previously
been given of the seriousness of the patient's condition, either in the sense
that major invasive procedures may be necessary, or that things may "not
work out". But even when such preparation is attempted, similar discussion stopping responses can occur. I submit that something more
fundamental than "untutored mentation" is often at work here. Staff, from
their point of view, may be quite comfortable with how things are going
for the patient. Diagnoses and an initial plan of management may have
emerged for a patient who presented mainly as a large question mark,
and the concurrent task is to "bring the patient or family up to speed".
Things are thus "under control" for the staff; uncertainty has given way
to some degree of clarity and direction.
    For patients and families, however, all may well be just a "booming,
buzzing confusion", and their primary sense of things may well be that
everything is quite out of control. Having recently had an experience
on the "other side" as the family member of an acutely ill loved one, I
will simply report what we all know, but often tend to forget when
approaching any such patient or family member. The experience on the
"other side" is a profoundly exhausting, alienating and alarming one,
characterized mainly by a sense of powerlessness, fear and, simply and
directly counter to the staffs' perception, one where everything seems
to be completely "out of control". Our life or that of our loved one is
in the hands of strangers who we need to trust but have no grounds for
doing so. We "sit in the dark" in our beds, or some distant waiting

room, time drips by, and even reassurance heightens the sense of gnawing ambiguity as we do not know if it is accurate, or just for our own temporary comfort, compassionately provided but perhaps groundless. We are told that someone will be back to discuss things shortly, or that a certain test is pending, but what is unfolding in a timely matter for staff occupies a debilitating eternity for us. Swept away from familiar homes and loved ones, we are completely at the mercy of strangers in the most foreign of environments.

That patients and/or families attempt to derail any such initial discussion in such situations should not thus be surprising. If anything is surprising, it is that they do not do it more often. However compassionate, available, and anticipatory the staff is, patients and families are often simply not prepared for what is happening. How could they be? And the usage of our key terms should thus be seen, not necessarily as any intellectual statement of conviction or bias, but perhaps simply as an attempt to beat back the threat and alienation being felt, to somehow destroy the gnawing ambiguity that, however accurate, is simply intolerable.

Sensitive staff can, relatively quickly, help patients and families overcome such debilitating feelings. The biggest danger, at such an early juncture, is that staff will react in exasperation, and not respond to what may well only be a cry for help, reassurance and support. But, at other times, more than just the natural bewilderment and fear of patients and families is operative, and sensitive education, counseling and support do not lead patients and families to an acceptance of what is going on, at least an acceptance of the way staff has learned to think about and approach such problems. To such deeper rooted factors, we now turn.

B. *Cognitive Dissonance at the Bedside*

Three forms or sources of dissonance for patients or families at the bedside may be usefully discussed here, viz. dissonance of role, reality and past experience. All three may be illustrated by a commonly received patient, in sum: "Father". And what *is* Father, to himself, and to his family? To himself, Father may well have always tried to be in control, the decision-maker in the family, the buttress and foundation of the family's viability, and an absolutely essential part of its reality. To his family, the sense may well have been the same, leavened by further realities such as having never made decisions without him (or *for* him), and having depended on his guidance and contribution for survival. He may well have been a "fighter", as families often put it, not the sort to give up, or conversely, not the sort that needs any particular help from anyone else to make his way in the world.

Whatever "pathology" women's liberation may see in all this, it is certainly, to some extent, descriptive of the reality, roles and past experience that not a few patients and families bring to the bedside. Similar analyses can be accurate in the case of "Mother" for that matter. But the point is that the sort of discussion that the staff legitimately wants to engage the family in, and the nature of what they have to say, is quite disconsonant with much of what has been an ongoing, basic reality for years. That families do not readily accept a radically new way of seeing all this is, again, not surprising. To ask the wife, for example, threatened in her own mortality and basic security by present circumstances, to make decisions for the "decision-maker", to consider that things may well not "work out" for him (and thus also for her), and to entertain the idea that things will not somehow be quickly and easily back to normal, is to ask a great deal. And, again, if the rhetoric that we are concerned with emerges, either explicitly, or in the "do whatever you can to get him back home to me" or "I don't want him to suffer" form, such statements should again be received not necessarily as intellectual declarations. It is as likely that they indicate that the person approached is unprepared or presently unable to engage in the sort of discussion or role that staff, in the abstract, is legitimately attempting to offer.

Another basic source of cognitive (or emotional) dissonance may be usefully noted here. The search for prior statements by the patient may yield vague senses that "he was always a fighter", or a specific statement, perhaps made early on in treatment when there was still hope, that the patient "wanted everything done". Such sentiments may well conflict with medical realities when it turns out that no benefit is possible, that the objective grounds for hope are gone. Conversely, in an offhand remark, or regarding the experience of another acquaintance or family member, the patient may have indicated that he never wanted to be put on one of those "breathing machines", or didn't want to be a "burden". But such statements may just result in more dissonance, as when ventilation for a potentially reversible infiltrate is "legitimately" being considered by staff, or the probability that things may "not work out" is leavened by the staffs' accurate insistence that there is still a significant chance to "turn things around".

Other dissonant factors can include: (1) being asked to discuss what the patient steadfastly refused to discuss when he or she had the chance, whether it be the reality of a cancer, or what he or she would or would not want if "things do not work out"; (2) being asked to consider aggressive treatment for a critically ill loved one when the last experience the family had of such a situation was the extended and horrifying treatment and death of another loved one, for whom there was also a chance (or perhaps the previous case involved caregivers who overindulged in optimism that proved baseless and, as a result, quite cruel – many families have had such unfortunate experiences and they profoundly effect their view of the next case, however disanalogous);[4] and (3) being asked to "sign off" on a document that removes support from someone that the party had always been in a relationship of support to, and for whom the last thing they want to consider is that that person is gone. Here parents being asked to agree to limitations of treatment on their children is particularly disconsonant. The difficulty of making the extra step of formally accepting the "reality" that staff has presented them, and which they have vaguely recognized and accepted verbally, has led one institution at which I consult to allow families to skip any formal signing, and even to just defer to what "staff thinks is best" without verbally giving explicit agreement to it. Institutional counsel, for obvious reasons, was not happy with this, but the alternative had been that families had been unable to explicitly agree even when they clearly agreed to staffs' appraisal, and either they were thus forced into doing it, or the treatment had to be continued against everyone's clear sense to the contrary.

C. *Denial at the Bedside*

Much of the preceding can present as (or evolve into) states where the legitimate diagnostic and prognostic findings of staff are simply rejected by patients or families. As in the preceding, such rejections can go either way. The patient may refuse to accept the terminal, non-treatable nature of his or her illness. Similarly, family members may refuse to accept such findings. Conversely, patients or families can reject invasive treatment options either because they refuse to believe things are that serious as to require them, or they cannot accept the diminished state of affairs that such invasions signify or lead to. The classic example of the latter situation is when a person, accustomed to a life of health and vitality, is informed that he or she has a dead, gangrenous leg that must be amputated or he or she will die. Not seldom the initial response is rejection of all this, including that there is any positive value to a life minus that leg. Again, "dignity" language may be used in support of such a rejection. Equally, the disconsonance of having "Father" in the dire straights previously mentioned can lead to flat, entrenched denial, catering as much to the threat to the family member's own life as to the patient. Not a few spouses have been at pains to make us understand how "they can not live without him", that everything they have had and worked for will be gone. It is as if, in a sense that is only abstractly symbolic, they are reacting to the specter of their own lives coming to an end. In important ways, this is exactly the case and denial responses by families can have similar roots to those of patients.

The interesting conceptual feature of such denial is often that the situation has moved from one of exhausting, gnawing ambiguity to situations where clarity and knowledge are now becoming available. But the clarity offered may be equally unpalatable, and the patient or family member attempts to reject it by a firm, intense statement of allegiance to its contrary, however unrealistic or diminished in possibility. Our terms can thus be used as a buttress against emerging knowledge of the patient's actual situation and prospects. Again, it is a confusion to simply accept such responses as intellectual statements of moral belief, or simply the results of the exercise of patient or family autonomy. They as well may signify conditions that patients and/or family members need considerable help and support to work through.

D. *Discomfort at the Bedside*

I have proceeded here as if the jarring, discussion stopping or rejecting usage of our terms mainly comes from patients and/or families. In my experience, this is more often the case these days, but in the past, staff were as likely to be the source of such language (and agendas). But though much improvement has occurred, some staff remain equally capable of such "behavior". Doctors who can not accept that they have failed to save or cure their patient are proverbial. An equivalent, although converse, problem arises when physicians are unable to see or appreciate the value of markedly diminished states, which patients, for whom nothing else is left, may nevertheless appreciate.

To my mind, staff are more likely to have such responses in the face of gnawing ambiguity, although I have seen staff who will still assert "sanctity of life" even for patients in clearly irreversible coma. In sum, the uncertainty of many medical situations can be as disconsonant for staff as it is to patients or families. I believe, in fact, that this is as much a reason for overtreatment of patients by staff as any other factor. And one response to such unpalatable ambiguity is, again, reliance on the rhetoric of sanctity of life whereby the possibility that one is needlessly causing the patient to suffer is held at bay. Conversely, human dignity can be appealed to in order to justify a non-aggressive approach to a patient who still retains some chance of being "salvaged". Again, as with patients and families, I think it is often more apt to see such statements as attempts to defend against the unpalatable and profoundly discomforting, rather than statements of some intellectual viewpoint. Either way, such language may find its main use in the attempt to avoid the "tough choices" that are ingredient in many patients' situations.

Speaking particularly of staff, another usage of our terms here merits identification. Although it is generally accepted that (1) no treatment is so ordinary as to always be required, and (2) withdrawal of treatment is not more problematic than withholding it, nevertheless some staff resist such conclusions. And again our terms are utilized in service of such opposition. Significant numbers of staff still seem to think that the withholding of life-support is preferable to its withdrawal, and some that only withholding is morally acceptable.[5] I tend to place such opinions, and their grounds, under the heading of "discomfort at the bedside" for the simple reason that often staff justify their reticence or refusal precisely on the ground that they are "uncomfortable" with what is being proposed, e.g. the probably fatal removal of ventilatory support. And, again, something about the "sanctity of life" or the patient's "dignity" may be offered in support of such reticence.

The usage of the term "discomfort" here is, in fact, apt, particularly in the sense that we are not dealing with a well worked out distinction between withholding and withdrawing, but just a declaration of discomfort with the latter. It is as if the intensity of being involved in the withdrawal of care is more than such staff can accept; such involvement is avoided in a way that is similar to the traditional avoidance many staff exercised when telling patients the "bad news". That such "convictions" lead to behaviors that are quite contrary to patients' interests and good medical practice is the case. Consider the patient with advanced chronic obstructive pulmonary disease who presents with acute shortness of breath. It is still often the approach of many staff to attempt to get a do-not resuscitate order and the rejection of intubation from such a patient. But what this amounts to is that a patient, who may only have an eminently reversible infiltrate for which short term intubation is reasonable management, is being approached as if, without sufficient reason, they are instead experiencing a final exacerbation of their lung disease and will probably end up "stuck on the ventilator" if placed on one. What is preferable: not intubating a patient who *may* end up stuck on a ventilator, or withdrawing ventilatory support from a patient who *is* stuck? But the discomfort of "pulling plugs" leads certain staff to engage in behaviors that they never would engage in otherwise, including trying to get hypoxic patients to make such crucial decisions. The suggestion, again, is that the occurrence of "sanctity of life" or "human dignity" language in support of such behavior does not merit reception as an intellectual view that will bear (or is grounded in) any analysis. It may well be, again, the attempt to disconnect or distance oneself from the unpalatable, to not face the distressingly ambiguous or clearly brutal facts.

E. *Simply Inappropriate Agendas*

Though quite rare (fortunately), there are occasionally much darker considerations to which our terms can be put that merit notice here, i.e. where they are enlisted in support of quite inappropriate agendas. By "inappropriate" here, I simply mean when such language is used in the service of personal agendas or interests of the person expressing them which are in direct conflict with (or at least unrelated to) the interests of the patient at issue. The focus here is on family members, and often of a particular sort, i.e., family member(s) who do not attend to the "complexities" of the discussion, seem to be tuning them out, and also seldom come to see the patient or assist with his or her care, but are quite clear about what they want or do not want done. Regarding "human dignity" language, this may be used by families who, at a minimum, do not seem really interested in any positive chances the patient may have, but are particularly intent on the idea that the patient is suffering "needless indignity" from further aggressive treatment, and are pushing for nature to be allowed to "take its course". In such cases, staff may have the impression that, if they are engaged at all, it is by their own exhaustion with the course of treatment, or the chronic disease process, and are seeking relief. One extreme example of such behavior occured when the sister of a younger man (his ranking legal surrogate), in the face of all appeals, demanded extubation and "death with dignity" for her brother who, critically ill from an as yet undiagnosed acute infiltrate, was still expected to be returned to a productive, fully sapient life. It was later determined that they had been alienated for years and were currently engaged in a battle over the estate of a parent who died without a will. Another example, this time regarding the "sanctity of life" usage, involved a spouse who demanded that a vegetative, unresponsive, ventilator dependent patient be aggressively managed because "all life is sacred." Later, in consultation with a social worker, it was discovered that the patient had a substantial pension that the family continues to receive and which will cease if the patient dies.

III. CONCLUDING REMARKS

For all of the preceding, the attempt to insist on the sanctity of life or human dignity can well have a positive, necessary place in discussions at the bedside, however much slogans must evolve into specific views and concerns to have any practical effect. To attempt to remind strangers that this is not, in fact, just another patient with a certain problem, or that the indignities ingredient in medical care must not simply be accepted without concern or further comment, can be a crucial service. Likewise to insist that human life is not just another quality to be thrown into the balance with fiscal concerns or the inconvenience to others of valuing or respecting it, will always be needed as a check against callous and disrespectful responses. I have mainly taken as my task in this essay to isolate those meanings, concerns and agendas that seem to often underlie the usage of such language at the bedside; uses that thwart rather than further legitimate discussions, that seek to hide, dispel or otherwise protect against realities that should be directly addressed. Colleagues in the rest of this volume have accomplished the more positive task.

Finally, to take the preceding as tending to discredit patients or families as decision-makers and participants in the "discussion" is neither intended nor indicated. In my experience, if given half a chance, both do quite well in both regards, not seldom bringing a good deal of wisdom and courage to the bedside, and their input is absolutely necessary however difficult it may be to get it into a useful form. My concern is that we do not take such pronouncements from such sources as more (or less) than they are; *that*, to my mind, is where the discrediting and disrespect come in, not in helping distraught strangers come to terms with the unfamiliar and distressing realities present at the bedside.

State University of New York
Buffalo
USA

NOTES

[1] I refer the reader here to [1] by H. Tristram Engelhardt, Jr. where he points out that the meaning of "human dignity" expressions on the continent are different than those used in America, and may as well have meanings that are consonant with "sanctity of life" expressions.
[2] I refer the reader here particularly to H. Tristram Engelhardt, Jr's. discussion of the meaning of this term in his companion article in this volume: ([1], 203–208).
[3] This may just sound inappropriate, but in some situations the devastating quality of what has occurred can be quite clear to staff early on, e.g. when a patient suffers a cardiac arrest in the community, and sustains major, irreversible, anoxic damage to the brain. In some cases, the profound nature of the devastation can be pretty clear in a day or so, at least to experienced staff. That families are quite unprepared for such a discussion does not mean it should not occur.
[4] The most stark example of this sort of disconsonance in my experience involved a woman who had been diagnosed with an eminently treatable, discrete bowel tumor. In response to all appeals, she continued to refuse surgical intervention, continually reminding staff that they were telling her the same things that other staff had told her sister, a woman who had recently suffered greatly from extended, probably futile treatment for metastatic pancreatic carcinoma, and still died. Beyond the specifics of what she said, her generic reason for refusing was that she "wanted to die with dignity". It took much effort and time to get this patient to understand that her situation was quite disanalogous to that of her sister.
[5] For an extended discussion of the reality of and possible responses to such reticent staff members, see [3] and [4].

BIBLIOGRAPHY

1. Engelhardt, H.: 1996, 'Sanctity of life and Menschenwürde: can these concepts help direct the use of resources in critical care?', this volume, pp. 201–219.
2. Katz, J.: 1984, *The Silent World of the Doctor and Patient*, Free Press, New York.

3. Wear, S., LaGaipa, S. and Logue, G.: 1994, 'Toleration of moral diversity and the conscientious refusal by clinicians to withdraw life-sustaining treatment', *Journal of Medicine and Philosophy* **19**, 147–159.
4. Wear, S. and Logue, G.: 1994, 'Policy options when an attending physician refuses to honor a legally enforceable demand to withdraw life-sustaining treatment', *VA National Center for Clinical Ethics News* **2**, 1–2, 9, 11.

KURT BAYERTZ

HUMAN DIGNITY: PHILOSOPHICAL ORIGIN AND SCIENTIFIC EROSION OF AN IDEA*

I. DIGNITY AND SUBJECTIVITY

The concept of a specifically human dignity is one of the fundamental philosophical innovations of the Renaissance. Parallel to the emergence of the portrait as an independent artistic genre within the field of painting and the autobiography as a new literary genre, a comprehensive series of writings on the *dignitas hominis* by Petrarca, Giannozzo Manetti and Pico della Mirandola grew to the dimensions of an independent literary genre. These works of art, literature and philosophy lent expression to that new, human self-understanding which was to become fundamental to the Modern Age. Of course, the concept of human dignity did not emerge as a *creatio ex nihilo*: it is rooted in Ancient philosophy, as well as in Christian theology. In Ancient times, the concept of dignity usually referred to respect for individuals with a high social status: a Greek king or a Roman senator, for example. It was the Stoics who first developed the idea of a dignity attributable to the human being *per se*, i.e. independently of individual characteristics. In Cicero's writings, we find both interpretations side by side. Christianity picked up on the second meaning and interpreted the dignity of all human beings theologically: the latter's origins may be attributed to the special position which the human being assumes within creation as *imago dei*. Human dignity is viewed here as reflecting the dignity of God.

The philosophy of the Renaissance picked up on these older views, then reinterpreting them, however. Most important was disassociation from the theological view of the world prevalent in the Middle Ages. The terrestrial life of the human being was no longer regarded as being merely

73

K. Bayertz (ed.), Sanctity of Life and Human Dignity, 73–90.
© 1996 *Kluwer Academic Publishers. Printed in the Netherlands.*

a test previous to eternal life, but was recognised as being valuable in itself. The *dignitas* literature sprang from the polemic against Mediaeval ideology of the Earth as a vale of tears. Human dignity acquired the role of a concept to counter the traditional Christian view of human existence as *miseria*. (Still in the mid-18th Century, in David Hume's essay "Of the Dignity or Meanness of Human Nature" [10] this confrontation of "dignitas" and "miseria" was prevalent.) Manetti dedicates the entire fourth book of his Treatise to a detailed description of not only Ancient – including "Heathen" – but also, and especially, Biblical and Christian demonstrations of a despisal of life. He is especially critical of Pope Innocence III's piece of writing entitled *On the Misery of Human Life* ([13], pp. 100, 114–116, 124–130). He programmatically views it as his task to

refute what, as we have seen, very many older and younger authors have written in praise of death and its advantageousness or about the misery of human life, since it contradicts, to a certain extent, what we have dealt with before. In our contradiction of the aforementioned, partly worthless and false views, we would like to adhere to a particular order, however, so that this matter is treated with more seriousness and severity. First of all, we shall answer the objections referring to the frailness of the human body, then those touching the lowness of the soul, and finally those referring to the human being as a whole ([13], p. 98).

Whereas the theology of the Middle Ages had essentially deduced human dignity from the human being's special relationship to God, this idea of the human being as *imago dei* is increasingly covered up (not necessarily replaced) over the centuries to follow by classifications decreeing the human being an independent being within the terrestrial world. Dignity is thus no longer comprehended as reflection, falling upon the human being from the transcendental world, but as the epitome of everything the human being represents within this world. The concept of dignity is used as a battering ram against a view of the world which aims to distract the human being from its existence within *this* world, presenting this world as a vale of tears, and every attempt to improve one's existence within it as vain. The concept of dignity is designed to give the human being a new self-awareness and the confidence necessary in order to improve this world and the human being's Fate within it.

There are three central elements constituting human dignity within Modern philosophy. The first of these is the human being's consciousness and ability to think. In a famous fragment of his *Thoughts*, Pascal wrote:

Man is but a reed, the most feeble thing in nature, but he is a thinking reed. The entire universe need not arm itself to crush him. A vapour, a drop of water suffices to kill him. But, if the universe were to crush him, man would still be more noble than that which killed him, because he knows that he dies and the advantage which the universe has over him; the universe knows nothing of this. All our dignity consists then in thought ([16], Fragment 347).

Human dignity is based on the fact that the human being does not just exist unconsciously, but is conscious of its existence; the human being thinks. This idea is fundamental to almost the entire spectrum of Modern philosophy. According to Mill, open discussion of the greatest questions which can occupy humanity raises "even persons of the most ordinary intellect to something of the dignity of thinking beings" ([15], p. 234). Modern philosophy picks up here on the conviction already prevalent in Ancient Times and the Middle Ages that reason is the most characteristic feature of the human race; it separates the human being from the animal. The Modern Age does, however, reinterpret the classical *topos* of the *animal rationale*, by restricting the concept of reason. Since Bacon and Descartes, the idea of rationality has become increasingly closely connected with decoding the secrets of Nature, in principle a never-ending process. Reason no longer implies the possession of sure knowledge about the goal and purpose of human existence, but refers, more than anything, to the examination of, and control over Nature.

This progressive examination of, and control over Nature is simultaneously the basis for the second element within the Modern concept of human dignity. It is the theory that the human being is different from other creatures because it is not fixed. Whereas all other living creatures are committed by Nature to a certain way of life, neither Nature nor God has provided the human being with a particular path. The human being is the only terrestrial creature which is free to decide its own way of life and to opt between various choices. This idea was already central to Pico della Mirandola's pioneering speech, in which Pico describes how God the Father assigned man "as a creature of indeterminate nature" a place in the middle of the world and addressed him thus:

Neither a fixed abode nor a form that is thine alone nor any function peculiar to thyself have we given thee, Adam, to the end that according to thy longing and according to thy judgment thou mayest have and possess what abode, what form, and what function thou thyself shalt desire. The nature of all other beings is limited and constrained within the bounds of laws prescribed by Us. Thou, constrained by no limits, in accordance with thine own free will, in whose hand We placed thee, shalt ordain for thyself the limits of thy nature ([17], pp. 224–225).

This concept of human unfixedness within God's plan of creation – or Nature – was to play a key role within philosophical human self-understanding over the centuries to follow. It rejects the idea of a human being constrained in the manner characteristic within Classical anthropology, and defines the essence of the human being as its permanent self-alteration. According to a famous definition by Rousseau, it is therefore "the faculty of improvement" ([18], p. 179) which distinguishes the human being from the animal. This laid the foundations for the idea of never-ending human progress, as programmatically formulated by Condorcet towards the end of the 18th Century. The intention behind his work is:

to show, from reasoning and from facts, that no bounds have been fixed to the improvement of the human faculties; that the perfectibility [sic!] of man is absolutely indefinite; that the progress of this perfectibility, henceforth above the control of every power that would impede it, has no other limit than the duration of the globe upon which nature has placed us ([6], p. 11).

One of the prerequisites for this idea of continual human self-creation is a farewell to teleology. The human being can only be truly free to create itself if Nature does not impose any goals and if Nature does not have any goals which could be rendered impossible to realise by human action. Free self-creation is only possible if all the natural prerequisites for human existence, as well as the peripheral circumstances, are reduced to the status of contingent facts. The metaphysical theory that Nature does not have any goals, nor any inherent value, that she is normatively neutral, corresponds on an epistemological level with Hume's principle, which states that an "ought" may not be deduced from an "is" ([9], p. 469), and on an *axiological* level with faith in the human being as the source of all normativity. But if Nature does not possess any general meaning, giving rise to normative, binding goals for human action, then only the human being itself can be the source of such a sense and the origin of such goals. The human being is thus not only its own creator, but also the creator of values and norms. This is the third component of human dignity. Immanuel Kant made this concept of moral self-legislation the centre of his philosophy, systematically developing it and declaring it to be the reason for human dignity: "*Autonomy* is therefore the ground of the dignity of human nature and of every rational nature" ([11], p. 103). In "Natural Law", an article within his *Encyclopédie*, Diderot had already made an explicit connection between autonomy and human dignity. Diderot discusses the question of who is to decide "the nature of the just and the unjust". According to the reply, it cannot be the individual, but only humanity itself:

Only humanity can decide, because its sole passion is the good of all. The particular will is suspicious, since it can be good or evil, but the general will is always good; it has never deceived and will never deceive. If animals belonged to a genus which was more or less the same as our own; if there was a sure method of communication between them and us; if they could tell us their feelings and thoughts very clearly, and could be party to our thoughts and feelings just as clearly; briefly, if they could all come together with us and vote, then we would have to include them, and then the matter of *natural law* would no longer be negotiated before *humanity* but before *animality*. But animals are separated from us by unchangeable and eternal barriers, and we are dealing here with an ordering of knowledge and ideas which are specific to the human genus, which result from the dignity of humanity and which are constitutive to it ([7], pp. 247–248).

Without being able to go into more detail about the concept of human dignity here, we may note that this concept has adopted a central position within Modern philosophy. It has undergone a radical change in meaning compared with its application in Ancient Times and the Middle Ages. The three fundamental components of the concept of human dignity – rationality, perfectibility and autonomy – lend it a previously inconceivable level of self-determination. To put it succinctly: with the concept of human dignity in its specifically Modern interpretation, the human being defines its own essence as *subjectivity*. Neither God, nor Fate, nor Nature tell the human being what to think or what to do. The human being is its own master. To put it pointedly, the human being is no longer just an image of God, but has become a kind of God itself, capable of thinking and deciding rationally, of shaping its environment and itself, and, ultimately, of creating its own values and norms. The only difference to the Gods is human mortality – and the fact that human greatness and dignity are not simply given, but set. They form a potential within the human race which it is capable of realising during an historical process of self-unfolding through personal activity, and which it *should* realise.

II. DIGNITY AND INDIVIDUALITY

During the historical process of self-unfolding, however, this inner-world God is frequently confronted by barriers: firstly, *natural* barriers, posed by external Nature and by human nature itself. And yet the more the human being learns to overcome these natural barriers with the help of science and technology, the more painfully it becomes aware of other barriers, posed by fellow human beings. Again and again, it is forced to undergo the bitter experience that it not only has to hold its own against adverse "Fate" and its enemy Nature, but also against other human beings, as well as social institutions growing up as consequences of human action. The fellow human being is always a potential competitor, and social institutions are always a source of threat. Relationships between human beings are not only characterised by love, solidarity and cooperation, but also by diverging interests and the conflicts which result. Fellow human beings can be the cause of insult, humiliation, exploitation and suppression – in extreme cases the cause of violent death. But if human self-determination and self-unfolding are dependent upon social and political circumstances, then it is only right that the concept of human dignity should extend to this sphere. At the end of the 18th Century, Georg Forster wrote:

The wealthy man who is able to enjoy the bounties of his fields and meadows, who is well-dressed, and who lives in a nice, clean, well-equipped house merits preference with respect to spirit, feeling, principles, knowledge – merits preference as *a human being*. He is doing well in all respects; in this comfortable situation, he looks around him, finds out who he is, where he comes from, and for what purpose, thus putting the best part of his being, his reason, which raises him above the rest of creation, to its intended purpose, and begins to become conscious of his *human dignity**. The emaciated slave of the Sarmantic nobleman, on the other hand, living in a ramschackle, smoky, bare hut, wearing a dirty sheepskin, half-eaten by insects, hard-working, eating little, often unhealthy food, is only familiar with animal sensations, likes to rest after his efforts, empty of thoughts, and ultimately dies without ever having enjoyed higher emotions, or his spiritual powers, without even having become aware of them, totally deprived of the purpose of his existence ([8], *my italics).

Forster sees in this unfolding of human physical and spiritual powers the goal which all social actions and political institutions should be serving. At the end of his considerations, he demands that the "false picture of *happiness*", for so long viewed as the goal of human development, be dashed from its pedestal and replaced by *human dignity* as the true "guide to life" [8]. In the course of the 19th Century, this socio-political application of the term human dignity became increasingly influential. In the 20th Century, it attained legal status within the Constitutions of various countries (Ireland, Italy, Sweden, Portugal, Spain, Greece, Canada and Germany) and within numerous international documents. In the U.S., the concept of human dignity is not part of the Constitution, but it has nevertheless been used by the Supreme Court in connection with the First, Fourth, Fifth, Sixth, Eighth and Fourteenth ammendments ([14], p. 3). Within the Preamble of the *Universal Declaration of Human Rights*, approved by the General Meeting of the United Nations on 10th December, 1948, recognition of the "inherent dignity" of all human beings is characterised as the basis of freedom, justice and World peace. Art. 1 of the Declaration states: "All human beings are born free, equal in dignity and human rights. They are endowed with reason and conscience and should act towards one another in a spirit of brotherhood."

The legal institutionalisation of the term human dignity is, on the one hand, a late effect of a profound social change which has been taking place since the beginning of the Modern Age. Hierarchically structured social relationships under the umbrella of a uniformly substantial system of values have been replaced in the Modern Age by a functionally complex society, opening up great scope for the behaviour of individuals and groups. The traditional concept of honour, which has an important bridging function between the self and society in hierarchically structured societies, has been replaced by the concept of a dignity attributable to all human beings, independent of their natural characteristics and social status.

Dignity, as against honor, always relates to the intrinsic humanity divested of all socially imposed roles or norms. It pertains to the self as such, to the individual regardless of his position in society. This becomes very clear in the classic formulations of human rights, from the Preamble to the Declaration of Independence to the Universal Declaration of Human Rights of the United Nations. These rights always pertain to the individual 'irrespective of race, color or creed' – or, indeed, of sex, age, physical condition or any conceivable social status. There is an implicit sociology and an implicit anthropology here. The implicit sociology views all biological and historical differentiations among men as either downright unreal or essentially irrelevant. The implicit anthropology locates the real self over and beyond all these differentiations ([5], p. 176).

Although the concept of human dignity may be comprehended as an expression of structural characteristics within modern society, its institutionalisation within national and international law is also a reaction to very concrete, 20th Century historical experiences, namely two World Wars and the industrial extermination of millions of human beings under the Fascist dictatorship in Germany. The elevated status of the term human dignity in numerous international documents, as well as the specific duty to protect human dignity imposed on the State of Germany by its *Basic Law*, are to be viewed as the consequence of a history which is not to repeat itself. The concentration camps made it blatantly clear that the modern human being no longer has, primarily, to hold its own against remorseless "Fate" or omnipotent Nature, but against itself.

In a certain sense, the socio-political application of the term human dignity and its legal institutionalisation may thus be viewed as an extension of the philosophical concept of human dignity. Both cases are concerned with the goal of securing human subjectivity. Originally conceived as a declaration of human independence from transcendent bodies and metaphysical restrictions, the term human dignity in its socio-political application is aimed at threats to humanity, human self-determination and self-unfolding, which originate from human action and its social consequences. However, behind this similarity there obviously lies a difference. The two applications of the term have different "inclinations". The *philosophical* concept of human dignity is directed against "vertical" limitations to human subjectivity: "upwards" against the influence of transcendent bodies, and "downwards" against the "apron-strings" of natural determinants. In its *socio-political* interpretation, the human dignity concept is supposed to repel limitations which are threatening on a "horizontal" level: limitations which originate from the actions of (other) human beings and from social circumstances and institutions created by human beings, especially within the State. Protection is thus no longer directed towards the human being as the *subject* of its own action, but as the *object* of the actions of other human beings. It is not doing away with barriers to human subjectivity *per se* which is now the goal, but erecting them, wherever free exercise of human subjectivity would limit the subjectivity of other individuals. Human dignity is no longer a term used to oppose a metaphysically irrepellable "miseria", but a bulwark against the leviathan.

A second difference becomes clear if we recall that the philosophical concept of human dignity is aimed at the human race in its entirety. The mortality of human beings has always limited their individual perfectibility. An unlimited self-unfolding can therefore only take place on the level of the genus. Moral autonomy is another thing which only exists for the human race as a whole. Diderot said that "the particular will is suspicious" and Kant also spoke specifically of the "dignity of man" ([11], p. 107), stemming from its ability to be universally legislative. The socio-political application of the term human dignity, on the other hand, is precisely not aimed at the entire human race. Its goal is not the self-unfolding of the *genus*, but of *individuals*. Securing human dignity also aims to prevent individual human rights from being sacrificed for the good of the genus – by calculating the sum of utility, for example.

This opposition between the individual and the genus could be avoided, however, if the dignity of the individual were to be interpreted as expression of its membership within the human genus. Kant had already adopted this path when he spoke of the duty: "practically to recognise the dignity of humanity in each other human being" ([12], p. 462). The famous "object formula" within the categorical imperative (which became an important basis for interpreting the term human dignity in German *Basic Law*) is based on deducing the dignity of the individual from the dignity of humanity: "Act in such a way that you always treat humanity, whether in your own person or in the person of any other, never simply as a means, but always at the same time as an end" ([11], p. 96). This interpretation does, however, hide the fact that there is latent tension between the philosophical and the socio-political interpretations of human dignity. Self-unfolding of the genus and individual self-unfolding do not simply coincide. Under certain circumstances, the fact that many or few individuals are hindered in their personal self-unfolding may be very useful for the self-unfolding of the genus. One example of this is the great cultural achievements in the past which were brought about by slavery.

Theoretically, this tension could also increase as a result of historical progress. This may be deduced from many connected factors which I shall proceed to outline briefly. (1) All the progress made by human subjectivity is based on cooperation and division of labour. This is also true of science and technology, which have become decisive vehicles in the growth of subjectivity. Science and technology are not only the product of (synchronous and diachronous) cooperation between many individuals; their implementation and practical application form a necessarily cooperative and labour-dividing enterprise. Each increase in subjectivity is thus primarily not an increase in the subjectivity of individuals, but in that of *the genus*. (2) Each increase in human power is, at the same time, directly or indirectly, an increase in human power over human beings. Individuals are thus not only subjects of the historical process of increasing subjectivity, but also, and always, its objects. To summarise: if each increase in the subjectivity of the genus implies an increase in power over individuals, then each increase in human subjectivity marks progress for human dignity in the philosophical sense of the term – as well as a (at least potential) threat to human dignity in its political sense. (3) This becomes especially clear when we realise that subjectivity and objectivity are never distributed equally. "The genus" is abstract, and not a subject capable of acting. Only individuals may act, and then not individuals as such, but individuals as concrete social beings under concrete historical circumtances. The means necessary to exercise subjectivity are usually available to those individuals who possess power due to their membership within particular social classes, their role within particular social institutions or their membership within particular professions. Briefly, these are the groups in political or economic power at the time. Sometimes it is even enough to belong to the group of the living in order to have power over other individuals – in this case those as yet unborn.

One could object that the idea of a dignity attributable to every human being independently of his or her characteristics is supposed to prevent just such a subordination of individuals to heteronomous purposes; it is supposed to prevent the genus from progressing alone at the cost of individuals. And yet this objection merely confirms the theory that tension exists between the philosophical concept of human dignity (as the rationality, autonomy, perfectibility of the genus) and the socio-political concept of individual dignity, tension which may lead to conflict at any time.

III. DIGNITY AND TECHNOLOGY

Such tensions, which can amount to contradiction, also occur within a field where the concept of human dignity is often brought into play, a field which has gained increasing attention over the past few decades: that of science and technology. On the one hand, there can be no doubt that science and technology in their specifically modern interpretations are manifestations of human dignity. They have become (theoretically) perhaps the most unambiguous expression of human rationality, and (practically) decisive means within the process of human self-unfolding.

The more the human being acknowledges the empirical conditions affecting its actions, and the more it is capable of controlling them, then the better it is doubtlessly capable of unfolding its subjectivity. Science and technology are the most important means of achieving this. With their help, the human being has been able to overcome numerous natural barriers. Today, each part of terrestrial Nature is either under the direct technological control of human beings, or is at least influenced by it. Even ventures beyond the Earth's surface into its atmosphere, or even beyond that into space, have become possible. Briefly: the magnificent expansion of human subjectivity which has taken place within just a few centuries has its foundations in science and technology.

There can be little doubt, however, that this success is profoundly ambivalent. Progress in science and technology, and the industrial practices which have thus been made possible, have begun to undermine the natural basis of human existence. If they were to extend beyond the presently relatively small number of developed countries to the entire Earth, the (as yet still) existing ecological balance would be irreversibly destroyed, and human life in its current form rendered impossible. Just like other biological species before it, the human race could meet its end as a result of its own (short-term) successes. Some authors have attributed this destructive practice to a demoralisation of Nature at the beginning of the Modern Age, the degradation of Nature to mere matter. It is not possible here to discuss the possible truth of this theory (cf. [2]). And yet one connection to the concept of human dignity, which Hannah Arendt has drawn attention to, cannot be overlooked:

The anthropocentric utilitarianism of *homo faber* has found its greatest expression in the Kantian formula that no man must ever become a means to an end, that every human being is an end in himself . . . For the same operation which establishes man as the 'supreme end' permits him 'if he can [to] subject the whole of nature to it', that is, to degrade nature and the world into mere means, robbing both of their independent dignity ([1], pp. 155–156).

It is not possible to discuss here whether this diagnosis is accurate (cf. [2]). But it is certainly conceivable that the idea of human dignity belongs within the context of a philosophical superparadigm, the practical realisation of which is a danger to the principles underlying the continued existence of the human genus. At this point, I would not like to go any deeper into the problematic relationship between humanity and Nature, turning instead to the no less problematic relationship humanity has to its own nature. The human being, in its continual efforts to unfold, not only comes up against barriers set by Nature herself, but also, and frequently, against barriers imposed by its own nature. For the human being as subject, its body is part of the outside world. Early on, Descartes declared the human body to be part of *res extensa*, ontologically on the same level as animals, plants or stones. This naturalisation of the human race has been increasingly substantiated, empirically and theoretically, in the further course of modern philosophy. The theory of evolution states that the human being derived from monkey-like animals and is thus a mere mammal. The empirical evidence available today regarding the close biological relationship between *homo sapiens* and the primates is almost suffocating. The difference in DNA between the human being and the chimpanzee is approximately one per cent.

Actually, the cold evolutionary fact is that the differences that really matter are even fewer than that. Most mutational changes in DNA have no overt effect on either the person or the chimpanzee who carries them. It is extraordinary, and rather humbling, to realize that the essential genetic differences between humans and chimpanzees probably amount to something between a tenth and a hundredth of one per cent of our respective genomes. And it is even more extraordinary to think that we will soon be able to track down just what these differences are. The trick will be to separate the differences that matter from those that don't. As we continue our exploration of the genome, we will find some rules that may help us to make that distinction ([20], 92).

For most contemporaries, however, such findings are no significant cause for provocation. At the end of the 20th Century, it is hard to comprehend the disturbance caused in the last Century by the theory that humans and apes are biologically related. With the exception of a few ignorami and religious fundamentalists who choose to shut out the overpowering scientific evidence, most contemporaries have got used to the idea of originating from the animal kingdom. And yet this familiarity should not lead us to forget that these findings infringe the empirical prerequisites upon which the concept of human dignity is based. This concept presupposes that there is a clear and unbridgable difference between the human race and the rest of Nature. Even Diderot, who was anything but a spiritualist, postulated "unalterable and eternal barriers" which separate us from animals. And yet modern biology maintains the reverse; and the further science progresses, and the more exact our knowledge about human nature becomes, the shorter the distance between the rest of Nature and "ours" becomes – the more the concept of human dignity melts to a postulate without empirical basis.

But it does not end here. The progression of knowledge renders it increasingly impossible to comfort ourselves with the impression that it is just the physical nature of the human being which science is increasingly in a position to explain and to break the spell of. Descartes declared the human body to be *res extensa* and placed it on the same level as animals, plants and stones, but *res cogitans* remained unaffected: the thinking ego remained to a certain extent "exterritorial". Thus attempts were made to limit the derivation of the human being from the animal kingdom to its physis, and to protect the human spirit from the harsh light of the evolution theory. This, however, was difficult from the very start. If the human being developed from the animal kingdom, then it is not plausible why these origins should only apply to its physical features. As early as the 19th Century, various theories arose attempting to explain in an evolutionary manner the rational and moral capabilities of the human being. These theories are just as attractive nowadays; in addition, their empirical content and the level of their experimental sophistication have steadily increased. Neurophysiology, behavioural genetics, experimental psychology and numerous other sciences aspire to explain human thought and action with the help of physiological and genetic determinants or environmental conditioning. B.F. Skinner insisted that the experimental analysis of the human being and its behaviour would lead to the disappearance of "autonomous man". All of the characteristics traditionally attributed to the "ego" are transferred within an analysis to the peripheral experimental circumstances, i.e. the test person's environment.

It is in the nature of an experimental analysis of human behavior that it should strip away the functions previously assigned to autonomous man and transfer them one by one to the controlling environment . . . In the scientific picture a person is a member of a species shaped by evolutionary contingencies of survival, displaying behavioral processess which bring him under the control of the environment in which he lives, and largely under the control of a social environment which he and millions of others like him have constructed and maintained during the evolution of a culture. The direction of the controlling relation is reversed: a person does not act upon the world, the world acts upon him ([19], 211).

Skinner's theory might be extreme, but only because it clearly describes the consequences involved in every scientific explanation of human thought and action. It names the vanishing point ultimately aimed at by every scientific interpretation of the human race. Science knows no subjectivity and no spontaneity. Human behaviour which becomes part of science *eo ipso* also becomes part of Nature. The separation of the human internal ego from its external body, of *res cogitans* from *res extensa*, is only possible up to a certain point. Human subjectivity does not exist – and this is the true kernel of Skinner's theory – beyond Nature, but is part of it and resides within it. By making its physical nature an object, the human being also makes its subjectivity an object – and thus a part of Nature. The strict difference between subjectivity and Nature, which forms the basis for the concept of human dignity, disappears. Briefly: there is no room within a scientific picture of humanity for the idea of human dignity. In its scientific self-interpretation, the human being positions itself "beyond freedom and dignity".

And not just in theory, but also in practice. Bacon and Descartes' programme aimed not only at finding out about Nature, but – and especially – at controlling it. Human beings were to become "maîtres et possesseurs de la nature", and this goal did not only include human nature, but referred primarily to it. Just like the external world, the human body puts barriers in the way of the human subject in the latter's quest to unfold its subjectivity. Its perception and memory capacities are slight, its physical strengths are severely limited and quickly used up, its health is susceptible and its life short. One of the central goals of the Modern Age has therefore, from the start, been to overcome these barriers to subjectivity, one by one. Bacon was very hopeful with regard to future medical achievements and control over the human body; for Descartes, medicine was one of the three main fruit-bearing branches (the others being mechanics and ethics) on the tree of knowledge; Condorcet concluded his presentation of the progress of the human spirit with the perspective of an unlimited human "organic perfectibility" ([6], 289). Thus the human being is no longer just the subject of scientific knowledge and technological control, but – the further the latter progress – also becomes their object.

Nothing could be more short-sighted at this point than the objection that this scientific penetration and technological control only apply to the natural side of the human being, and not to its spiritual side and subjectivity. Hopes of saving the "inner" human being with this kind of dualism have always turned out to be naïve. The human spirit is very much part of this world; it has a natural basis. The subject may not coincide with the body, but neither can it be separated from it. Technological access to the body will therefore not stop there: at some time or another, it will also affect the subject and its spirit. The modern sciences are already party to a number of ways of directly influencing human feelings, thoughts and actions, whether through electrical stimulation of the brain or (bio)chemical means such as endorphine or neurotransmitters. To put it another way: the more human physical nature may be controlled by technology, the more human subjectivity also becomes controllable – and thus the *object* of control.

One could object that the concept of human dignity is introduced into debates on modern biology in order to prevent just that: to prevent the human being from becoming the mere object of technological control, mere "matter" for gene technologists and biological engineers. Yet this presupposes that the point at which technological interventions (which cannot be principally illegitimate, after all) render the human being "mere" matter is definable. A hard kernel within the human being, a normatively binding human "substance" has to be defined. Yet this is not only difficult; (cf. [2]) it would also contradict the philosophical core of the human dignity concept; it would alter its direction by 180° As we have seen, the latter is based upon turning away from all the anthropological theories which postulate a fixed human nature, defining the essence of humanity as self-determination and self-unfolding: i.e. as processes which are not exactly characterised as fixed. It is thus no coincidence that one of the most prominent advocates in Germany of the human dignity argument, as used against gene manipulation, has raised the question of whether the kernel of human essence is really to be viewed as the human spirit elevating it above the rest of Nature, the human capability to take responsibility for independent moral decisions, or whether it is not far more human "imperfection". [4] If this question is answered in the affirmative – and its author leaves us in no doubt that he believes this to be the correct answer – then two central elements of the philosophical concept of human dignity – rationality and autonomy – are obviously thrown overboard, and this in order to be able to reject the prospect of human gene technological self-alteration as illegitimate. The subjectivity of the human being is ranked below the facts of its nature. To put it another way: by giving up on human dignity, an attempt is made to save it.

IV. CONCLUSION

The concept of human dignity is characterised by inner tension and contradiction. I would like to summarise this as three points.

(1) Modern subjectivity, the epitome of which may be viewed as the

concept of human dignity, includes the postulate of a continual expansion of scientific and technological control over Nature. The human being itself is a part of Nature, however, and thus unavoidably also becomes an object of scientific research and technological control. It then uncovers those characteristics constituting its "dignity". The human being interprets itself as a part of Nature – just what it was seeking to avoid with the concept of human dignity. The sharp dividing line drawn by the concept of human dignity between humanity and Nature is systematically blurred by science and technology. Anthropocentrism is put into reverse: the human being turns itself into a part of Nature. The fact that the human being is subject over its own nature necessarily implies that it is also an *object* of its own subjectivity. This leads to a paradox situation: whilst distancing itself from Nature in its role as subject, in its role as object, the human being penetrates Nature with increasing depth.

(2) The concept of human dignity thus undermines its own prerequisites. This effect occurs when human subjectivity begins to become self-referential. We are then faced with a problem which can be formulated as a classical dilemma: *either* we hold onto the concept of human dignity in the sense of an unlimited subjectivity of the genus; then individuals will seek refuge as mere carriers within an unending process of self-unfolding on the part of the the genus. In this case, we have to release human physical nature as "matter" of the self-determination and self-alteration of the genus; and then even human subjectivity will become an object of technological control. The autonomous human being disappears. *Or* we declare our own nature to be sacrosanct and taboo: then we will have to do without certain technological options, just as we will have to do without permanent self-transcendation, returning to that concept of a normatively binding human nature which the philosophical concept of human dignity was supposed to overcome.

(3) In view of this, it would seem a good idea to do away with the concept of human dignity altogether. It would, of course, be easy to give up the *term* "human dignity". But this would not help us in the slightest. Whether this term is used or another one: the concept of human dignity is – as I have attempted to demonstrate – much too deeply embedded in Modern Age thinking for us to dispose of it like a piece of clothing which we no longer like or which no longer fits us. The problems examined within this paper are problems which prevail throughout Modern philosophy and the Modern conscience. It is not only difficult to give up a view of the Self and the World which is anchored

in history; in this case, it even seems to be impossible. For if we were to take the problems surrounding the concept of human dignity which I have described here as reason to overthrow this concept and all the normative orientations connected with it, then this could only be interpreted as an expression of our rationality and autonomy, in other words, as an expression of our human dignity. I can therefore only conclude with a paradox: one cannot hold onto the concept of human dignity without destroying it; one cannot give up the concept of human dignity without confirming it.

Department of Philosophy
University of Münster
Germany

NOTE

* Translated into English by Sarah L. Kirkby

BIBLIOGRAPHY

1. Arendt, Hannah: 1969, *The Human Condition*, The University of Chicago Press, Chicago and London.
2. Bayertz, Kurt: 1994, *GenEthics. Technological Intervention in Human Reproduction as a Philosophical Problem*, Cambridge University Press, Cambridge.
3. Bayertz, Kurt: 1996, 'The nature of morality and the morality of nature', in B. Gustavsson, R. Porter and M. Teich (eds.), *Nature and Society: The Search for Missing Links*, Cambridge University Press, Cambridge.
4. Benda, Ernst: 1985, 'Erprobung der Menschenwürde am Beispiel der Humangenetik', in R. Flöhl (ed.), *Genforschung – Fluch oder Segen?. Interdisziplinäre Stellungnahmen*, Schweitzer, München, pp. 205–231.
5. Berger, Peter: 1983, 'On the obsolescence of the concept of honor', in St. Hauerwas and A. MacIntyre (eds.), *Revisions: Changing Perspectives in Moral Philosophy*, University of Notre Dame Press, Notre Dame and London, pp. 172–81.
6. Condorcet, M. de: 1796, *Outlines of an Historical View of the Progress of the Human Mind*, Philadelphia, reprint on microcard by the American Antiquarian Society, Worchester, Mass. 1964.
7. Diderot, Denis: 1976, 'Droit Naturel', in *Encyclopédie III (Lettres D-L)*, Edition critique et annoté, présentée par J. Lough et J. Proust, Hermann, Paris, pp. 24–29.
8. Forster, Georg: 1966, 'Über die Beziehung der Staatskunst auf das Glück der Menschheit', in *Über die Beziehung der Staatskunst auf das Glück der Menschheit und andere Schriften*, ed. W. Rödel, Insel, Frankfurt/M, pp. 138–169.
9. Hume, David: 1978, *A Treatise of Human Nature*, ed. P. H. Niddich, Clarendon Press, Oxford.

10. Hume, David: 1987, 'Of the dignity or meanness of human nature', in *Essays Moral, Political and Literary*, ed. with a Foreword, Notes and Glossary by E.F. Miller, Revised Edition, Liberty Classics, Indianapolis, pp. 80–86.
11. Kant, Immanuel: 1964, *Groundwork of the Metaphysic of Morals*, translated and analysed by H.J. Paton, Harper Torchbooks/The Academic Library, Harper & Row, New York/Hagerstown/San Francisco/London.
12. Kant, Immanuel: 1968, *Metaphysik der Sitten* (Akademie Textausgabe Bd. VI), de Gruyter, Berlin.
13. Manetti, Giannozzo: 1990, *Über die Würde und Erhabenheit des Menschen*, ed. A. Buck, transl. H. Leppin, Felix Meiner, Hamburg.
14. Meyer, Michael J.: 1992, 'Introduction', in M.J. Meyer and W.A. Parent (eds.), *The Constitution of Rights. Human Dignity and American Values*, Cornell University Press, Ithaca and London, pp. 1–9.
15. Mill, John Stuart: 1977, 'On liberty', in J.M. Robson (ed.), *Collected Works of John Stuart Mill*, Vol XVIII: Essays on Politics and Society, University of Toronto Press and Routledge & Kegan Paul, Toronto and London.
16. Pascal, Blaise: 1900, *Thoughts*, in C.W. Eliot LLD (ed.), *Blaise Pascal*, Harvard Classics, Vol. 38, transl. W.F. Trotter, P.F. Collier & Son Company, New York
17. Pico della Mirandola, Giovanni: 1950, 'Oration on the dignity of man', in E. Cassirer, P.O. Kristeller and J.H. Randall, Jr. (eds.), *The Renaissance Philosophy of Man*, University of Chicago Press, Chicago.
18. Rousseau, Jean-Jacques: 1900, 'Discourse on inequality', in C.W. Eliot LLD (ed.), *French and English Philosophers. Descartes, Rousseau, Voltaire, Hobbes*, Harvard Classics, Vol. 34, P.F. Collier & Son Company, New York.
19. Skinner, B.F.: 1972, *Beyond Freedom and Dignity*, Jonathan Cape, London.
20. Wills, Christopher: 1992, *Exons, Introns and Talking Genes. The Science Behind the Human Genome Project*, Basic Books, New York.

MARTIN HAILER AND DIETRICH RITSCHL

THE GENERAL NOTION OF HUMAN DIGNITY AND THE SPECIFIC ARGUMENTS IN MEDICAL ETHICS

"Menschenwürde", a widely used term in German philosophy, political science and ethics, is a term with no exact English equivalent. It correspondends in general to "Human Dignity", sometimes "sanctity of life", or "security of person", or "personal security" to denote the inalienable autonomy of human beings. The use of "personal security" is frequently found in books on the Philosophy of Law, e.g. by Roscoe Pound ([15]; cf. also [7]), while the term "Human Dignity" is quite generally used in ethical discourse, especially in social and political ethics. Politicians use it as does the International Commission of Jurists and other groups and agencies affiliated to the United Nations. Here it is used as though it were an ethical technical term with a distinct meaning which, however, is not really the case. Still, the use of the term "Human Dignity" in these circles represents more or less the meaning of "Menschenwürde" in the German language. Finally, the term "sanctity of life" is – contrary to a wide-spread opinion – not of theological origin. Helga Kuhse has recently used it in the title of her book [11]. In the following we will use the term "Human Dignity", although this translation denotes only a part of what the German term "Menschenwürde" means in post-war usage.

There is no doubt that the use of Human Dignity is inflationary and that the notion as such is anything but distinct. However, it would display an unduly positivistic attitude toward language if we were to exclude the use of the term from ethical discourse. Admittedly, the term is broad and encompasses a variety of values and rights, all of which have received new relevance and significance after the horrors of World War II and the inhumanities of the Nazi period. It governs implicitly the Universal

K. Bayertz (ed.), Sanctity of Life and Human Dignity, 91–106.
© 1996 *Kluwer Academic Publishers. Printed in the Netherlands.*

Declaration of Human Rights and the many Conventions and Pacts since 1948. Here it is mostly negatively defined by referring to extermination, enslavement, torture, enforced labour, and – going back to the Nuremberg Codes – to human experiments, administration of hypnotic drugs for judicial examinations, secret recording of private conversations etc. These latter points touch upon the area of medical ethics.

Is the reference to Human Dignity at all suitable for medical-ethical discourse? Even though it is not a philosophically clear-cut notion but rather a summary-term referring to a group of values, could it perhaps offer special validities or at least a frame of reference for detailed ethical arguments? The following systematic and historical observations are an attempt to work toward an answer to this question.

I. HUMAN DIGNITY AS A CRITERION IN MEDICAL ETHICS?

It should be clear from the outset that "medical ethics" is not merely "doctor's ethics" but that it is operative in at least three areas: 1. in the doctor-patient relationship, 2. in the wide field of health-policy, including legislation, hospitals, insurances and the industrialized world's responsibility toward developing countries in matters of health and 3. in the area of actual health performance and -behaviour of the public, including attitudes towards one's own body, toward medication and drugs, health education, attitudes toward sickness and death etc. Moreover, a consensus should be presupposed that medical-ethical judgements should not be the decree of an individual person, but ideally speaking, the agreement within of a group that has been in ethical discourse, and secondly, that judgements are not merely based upon sentiments, impressions and personal preferences but on arguments which can be articulated, reproduced and also reviewed by others. Assuming a consensus on these general matters, could one perhaps go a step further and say that despite the enormous progress in medical technology there are not too many medical-ethical problems which are, as it were, absolutely "new"? Genuinely new problems have arisen in the following areas: 1. in the possibilities offered by the development of the modern intensive care unit, 2. in the field of "innovative therapies" both in surgery and in psychiatry, 3. in human genetics and the genome-analysis and the unforseen results of research in this area, 4. in reproductive medicine, 5. in transplantation-medicine, 6. in the problem of macro-allocation within our own countries and also with regard to our relation to underdeveloped countries and cultures.

Thus it is not only in the area of classical or traditional medical-ethical problems but also with reference to these new ethical problems that we need criteria for responsible ethical decisions. My experience in the past twenty years in the U.S.A., in Australia and on the Continent of Europe suggests that in the search for criteria and their argumentative application one should not overestimate the difference between the various philosophical or religious convictions of the participants of the discourse. With the exception of a very few problems (e.g. abortion) the actual ethical discourse is not burdened or disturbed by such differences of background and conviction. Usually the arguments which are legitimate components of a decision are based upon a wider frame of reference and proceed from there toward narrower, more specific applications and in the end lead to a conclusion. Such "frames" can be a general sense of responsibility toward the patient (or the population of a country etc.) or to the patients' rights or, for that matter, to Human Dignity. It is not easy to draw a sharp demarcation line between such "frames" since they all refer to a frame of reference which transcends the detailed criteria or references to parallel experiences which are applicable to the specific problem or case in point. The differences between basic philosophical or religious convictions of the participants lie, as it were, before or underneath such frames of reference. In other words, a reference to a general sense of responsibility toward people, especially patients, or to Human Dignity, or to the autonomy of patients is not the argument itself but it is the platform upon which arguments can be built. Our question here is wether Human Dignity is a specially suited platform upon which medical-ethical arguments can be built or wether it is not because of its broadness and lack of precision. The question is all the more interesting when one considers that Human Dignity is a notion which preceeds Human Rights, i.e. that Human Rights are a judicial concretisation of the more general concept of Human Dignity. Human Dignity as a concept belongs to a pre-political or pre-juridical realm.

If the general character of the Human Rights concept as a "frame-concept" is not seen, the following misconceptions can emerge which are, in the case of medical ethics, especially unfortunate.

1. A generalisation of the concept as though it were not an anthropological principal but an ethical or even legal criterion from which specific conclusions can be deduced. I mention as an example the critical reactions of the media and the general public to the crash-tests with

human corpses, including the corpses of babies in simulated automobile accidents conducted by the department of Forensic Medicine in the University of Heidelberg ("Such tests are a violation of Human Dignity").

2. The squeezing of the concept of Human Dignity into quasi legal terminology as though Human Dignity were not the platform or frame of reference for Human Rights but were a summary of Human Rights in itself. The broad concept of Human Dignity here serves as a vague substitute for legal language, when in fact the constitution or any part of positive law should have its say. It is also taken in lieu of concrete ethical criteria. One may mention as an example the often heard criticisms of "unnecessary prolongation of life" by means of modern medical technology.

3. The general emotional appeal to Human Dignity in form of a "knock-down-argument" in socio-political and especially medical-ethical fields of problems, indicating that a rational analysis and argumentative discourse has not taken place or is avoided in favour of what seems to be a foregone conclusion and all-dominating assertion.[1]

Attention is to be paid, however, to a trans-individual aspect of the use of the term Human Dignity. It has been pointed out by authors on both sides of the Atlantic that the Human Genome Project and also, in a much wider sense, the ecological problems in general, present an aspect of Human Dignity which goes far beyond individual beings who are alive at present. The lives and the fate of future generations is at stake and the traditional focus of Human Dignity on the individual must by necessity be expanded to a much wider scope in the light of what is technically possible today in the area of genetics and, of course, in the devastating and destructive effects of modern technology in general on future generations of humans and, for that matter, also of animals. (The inclusion of animals is at least logically justified by the stringent observation that humans who mistreat animals are thereby violating their own Human Dignity.)

What then is the status of the concept of Human Dignity in medical ethics? So far we have only collected some general observations on a possibly promising and helpful employment of the concept as a basic frame of reference as well as on some typical misuses and erroneous claims as a criterion from which specific criteria for decisions can be deduced. However, a further investigation is necessary. Although historical observations can never take the place of systematic analysis it will be advantageous to have a brief look at the historical roots of the

Human Dignity concept. This will at least illustrate the breadth of the anthropological basis on which a specific ethics can be built, provided one believes that ethics cannot be without such anthropological presuppositions.

II. LESSONS FROM HISTORY

History presents at least four concepts of Human Dignity. They cannot clearly be attributed to certain periods or schools of philosophy, rather they are types which are found in overlapping fashion in various periods and traditions, their distinctive characteristics notwithstanding. In any case the four types were never conceived as legal concepts, at best they were ethical precepts, better still, basic anthropological assertions. The following survey from the time of the Greeks to I. Kant will show what history presents.

1. *The Physis-Concept*

Typical of Stoic philosophy was the thought that the whole cosmos is penetrated by the reason and rationality of God, the *logos*. Nothing in the world is really categorically alien to the rest of the world. Everything is interconnected with all the rest and it is steered and interpenetrated by the *logos*, really everything, nature, human beings, social communities and ultimately even God. This interconnectedness of all things provides the basis for Stoic ethics: One must live in accordance with this concept of *physis*, of nature. The ethical task is to recognize the laws of *physis* in all the realms that constitute the world and to adapt oneself to them since they represent the unchangable will of the godhead. Thus the Stoic ethicist will respect Human Dignity in the same way and for the same reason as he respects laws of nature: Everything in the world, including human beings – all human beings – participate in the *physis* which is penetrated by the *logos*. Stoic philosophy was the first philosophical school in antiquity which favoured a general notion of humanity and humanism, leading even to cosmopolitan ideas and ideals. In this Stoic philosophy was quite different from the basic ideals of Plato and Aristotle whose concepts of society were distinctly non-egalitarian. Later Roman Stoic philosophers concurred with their Greek ancestors in these basic assertions by giving them a distinctive anthropological and ethical edge. One may think of Seneca's statement that *homo est sacra res homini*, a programme that even led to demands for the abolition of slavery, a complete innovation in antiquity with which early Christianity could only in part make its peace.

The Stoic anthropology concerning the inalianable dignity of the human being travelled through the centuries and exercised a significant influence here and there. One can think of Pico della Mirandola in the Renaissance Stoic movement but also of Calvin in Geneva whose first published work was on Seneca. Surely, it was an individualistic understanding of Human Dignity and it had strong idealist components which grew out of the general pantheistic understanding of the interpenetration of the godhead and the cosmos. This thought was only in part translatable into later Christian tradition.

2. *Human Dignity on the Basis of the Biblical Tradition*

Whatever generalisations we may justly articulate about the anthropological assertions implicit in the Bible, one cannot possibly see any consensus with the Stoic idea that God and the world interpenetrate. Rather, God and the world are distinct entities in the Old and the New Testaments. The interesting feature is that the human being mirrors in a certain sense the distinctiveness of God's separation from the world in this that the "image of God"-concept implies that human beings preceive a certain dignity and separate position over against the rest of creation. Obviously – one is reminded of Jean Amery's critique – such distinctive position of human beings over against creation, according to the first three chapters of the book of Genesis, was badly abused and misinterpreted in centuries to come. What resulted from such misreading was a ruthless exploitation of the world of animals, plants and the soil, water and air of the world we live in. It was only in recent decades that the relevant passages in the Hebrew Bible were read in a different light, namely, asserting a stewardship and responsibility of humans over against the non-human parts of creation.

Be this as it may, the biblical books do not conceive of Human Dignity as something inherent within the human, rather it is a dignity imparted to the human being by God, i.e. humans receive such dignity by God who speaks to them and assures them of their destiny and dignity. This basic assertion recurs in modern philosophy e.g. in the works of Spaemann [20] and also in modern philosophy of law and, of course, in theological ethics. The main thought is that Human Dignity is a reality that does not reside automatically, as it were, within the individual, but that society, fellow human beings, ultimately God, are the "speakers" from whom emerge the statement "you are a true human being, you have Human Dignity", regardless wether you are an accepted and succesful member of society or a prisoner, a mentally disturbed or senile person.

3. *Human Dignity by Means of a Treaty: Hobbes and Locke*

The two famous British philosophers have tied their understanding of Human Dignity to the state and to the nature of society. This is what they have in common. They both assert that humans have inalianable rights which are based upon the Human Dignity which is theirs, a status which they ontologically enjoyed before any society or state or contract was around. But while Thomas Hobbes (1588–1679) maintained that the natural status of human existence was cruel, i.e. that there was a *bellum omnium contra omnes* and that a strong, absolute state would have to be created in order to control the aggressiveness of humans, John Locke (1632–1704) had a much more optimistic assessment of the natural status of the human and of human life prior to the existence of states and laws. What is of interest to us here is the point in which the two philosophers agreed: The individual rights and dignities are prior to any social structure, agreement or set of rules. According to Hobbes the state is to watch over the protection of Human Dignity and Human Rights and according to Locke a contract is necessary among humans in order to enable society, ultimately the state, to provide such protection. Locke's concepts have influenced Scottish and English parliamentarianism and also the authors of the American Constitution. Be it the pessimistic version of Hobbes or the more optimistic understanding of Locke, in any case the actualisation of Human Dignity is vulnerable and depends upon willful actions of society and is to be safeguarded by social and political institutions. Here we discover a hidden analogy to the biblical understanding that humans do not possess inherently certain qualities but that they depend upon certain steps taken by society in order to have and maintain qualities such as dignity and Human Rights. In our time it is John Rawls ([16]; cf. [23]) in his "A Theory of Justice" who, not unlike Hobbes and Locke, reconstructs the basis of society by referring back to a hypothetical premordial state of human existence. Humans need principles, rules and structures in order to safeguard equality and a relative stability of justice.

4. *"Never Only as a Means but Always Also as an End"*

Immanuel Kant's (1724–1804) influence on the understanding of Human Dignity and its place in ethics as well as its incorporation into constitutional law can hardly be overestimated. However, Kant criticizes sharply the classical theological foundations of the concept of Human Dignity although in elevating the human race above nature he confirms one of its basic assertions. It is human freedom which indicates and guarantees the prominent and elevated status of the human being over against nature. The finest use of human freedom is to follow the moral law, the second formulation of which is summarized in the well-known phrase to "act so as to treat humanity never only as a means but always also as an end" and that Human Dignity consists precisely in this freedom in which humans are different from everything non-human ([9], A 140). Kant insists not only on the autonomy of ethics (as over against an ethic in which the human being is steered by outside authorities) but also, and precisely because of it, on the autonomy of the human being. This in fact is the reason for and the manifastation of Human Dignity. Thus Kant replaces the classical-theological imago-dei-concept with the idea of an endowment of the human being with reason, a reasoning power that enables him to exercise his freedom and to follow the moral imperative.

It is important to note that Kant does not forbid the use of other people as a means. If that were the case, we could never employ other people to work for us. This we may do but we may never do it exclusively and thereby use others "only as a means". We must under all circumstances and always, even if we employ and "use" other people, see in them an end in themselves and treat them accordingly. The consequences for medical ethics are immediately clear. When conducting experiments there must always be – as we say today – a "therapeutic indication", i.e. the patient with whom we conduct experiments must also have the therapeutic benefit of it. On the basis of this principle the production of human embryos for research would be against Kant's moral imperative. It is at this point, of course, that our modern discussions on contradiction-models versus consent-models with regard to the donation of organs for transplantation come into the picture, also the consent to use one's body for research of any kind after death. Here, the *dead* human body is used exclusively as a means but vicariously for the therapeutic benefit of other, living people as an end. This case was not foreseen by I. Kant. However, provided a consent was declared during the lifetime of those persons, Kant's moral imperative is not violated when we explant organs for the benefit of others or when we perform the above mentioned Heidelberg crash-tests (or traumato-mechanic tests) with human corpses. The problem here arises only with regard to the dead bodies of children who have hardly given consent to such use of their bodies during their lifetimes. Thus the controversy in Heidelberg at this time between the forensic medical colleagues on the one side and lawyers and medical ethicists on the other is only over the legitimacy of appealing to a non-contradiction which is assumed in lieu of an expressly given consent.

III. HUMAN DIGNITY IN UNITED NATIONS-DEFINITIONS AND IN CONSTITUTIONAL LAW

The following observations serve the purpose of showing that the concept of Human Dignity is based upon an anthropological "creed" – not necessarily a religious creed – and that all four historical roots of the concept are operative in modern legal texts or constitutions, all the way from the American Constitution to United Nations-Definitions and post war national constitutions such as the constitution of West Germany of 1949. Elements of the four roots can be traced in each of these texts, namely, a combination of biblical and Natural Law concepts in the U.N.-Definitions and a combination of Stoic and early modern English philosophical concepts in the American Constitution and, finally, a combination of biblical and Kantian concepts in the German Constitution. The strange phenomenon is, however, that the concept of Human Dignity is increasingly treated as a legal term, although it is not that. It is here that ethics and law touch each other and are overlapping in the notion of Human Dignity. While Human Dignity is undoubtedly a basic frame of reference in ethics, based upon a credal assertion concerning the status of the human being, it has found entry into the preambles of constitutional texts without being designated as an originally ethical frame of reference. It operates as a broad axiom from which at least certain prohibitions and negations can be deduced. Such deductions occur often in connection with the double reference to Human Dignity and Human Rights, whereby it is not clear from the outset whether the latter are grounded in the former or whether the two function independently as frames of reference.

1. *The U.N.-Definitions*

The preamble of the Human Rights Declaration runs like this: "Recognition of the inherent dignity and of the equal and inalianable rights of all members of the human family is the foundation of freedom, justice and peace in the world." Detailed studies[2] have shown that the explicit as well as the implicit concept of Human Dignity in this text and in other U.N.-Definitions is based upon biblical as well as Natural Law concepts, i.e. ideas from Stoic philosophy. At the time of their promulgation after World War II it was difficult to foresee that such grounding would, in later decades, provide difficulties. In fact, the rather individualistic concept of Human Dignity (and Human Rights), typical of the Western interpretation of classical Greek as well as biblical traditions was not accepted by the Communist states that had become members of the U.N. They preferred an interpretation of Human Dignity (and Human Rights) in terms of social rights. Moreover, it has become clear that the great number of nations in the Southern hemisphere and in East Asia that have become members of the U.N., cannot see much value or convincing power in the originally Western interpretation of the Greek and biblical traditions. This is also a problem which the WHO has to face when the question of the universalisation of ethical norms in medical ethics is at stake. To be sure, the promulgation of Basic Human Rights in 1948 and in the Conventions and Pacts since that year is to be celebrated as one of the greatest achievements of humankind and it is of the greatest significance for a possible foundation of a universal ethical framework, a "minimal-ethic", as it were. However, Human Dignity plays a strange role in this achievement in this that references to it either appear in isolation in rather vague terms or Human Dignity is couples with Human Rights and then seems to provide a set of norms from which conclusions can be drawn.

2. *The American Tradition*

It is interesting to note that the Constitution in its early form focussed on procedures, i.e. the division of power in Congress, President and Supreme Court and not on Human Rights or Human Dignity. The latter elements were added at the request of Thomas Jefferson in form of the Bill of Rights, referring back, of course, to the basic ideas of the Magna Charta and the English Bill of Rights of 1689 and, of course, to the

Declaration of Human Rights of Virginia of 1776. Here the concept of equality becomes dominant and later, in 1868 with the Fourteenth Amendment the idea of "equal protection". Unfortunately, the Christian churches did not find it easy to support these important decisions, as in general churches were reluctant to commit themselves to the advance of Human Rights.

The influence of John Locke on the Declaration of Independence and on pre-political Human Rights and Human Dignity has often been noted. His – and even Hobbes' – concept of society's obligation to guarantee equal rights and Human Dignity have most certainly influenced these early American Declarations and Amendments. Even though the term Human Dignity does not frequently occur, it is the implicit value in the emphasis on "equality" and "equal protection", a corner stone of democratic practices and also an important idea behind medical-ethical concepts of equal treatment and attention given to different patients.

3. Menschenwürde as the Point of Departure in the German Constitution

The key function of the concept of Human Dignity in the German Constitution is even more explicit than in the U.N.-Declaration. The first article begins with the lapidary statement: "The dignity of the human being is inviolable." The Constitution pursues the clear logical sequence: From the basic fact of Human Dignity follows the recognition of Human Rights and they are positive law in the sense that they can be claimed in court. Here Human Dignity clearly serves as a basic frame reference for Human Rights and for anything that follows in the constitution. The overarching importance attributed to Human Dignity in the German Constitution is undoubtedly a reaction to Nazi ideology which was summarized in the horrid slogan "You are nothing, your nation is all". Thus the inalienable dignity of the individual had to be stressed although interpreters of the Constitution have observed [2] that emphasis on society is not absent in the Constitution, i.e. the Constitution does not advocate a simple individualism.

Human Dignity in the German Constitution has a double function, it constitutes values as well as rights. This is clearly reflected in the Constitution's structure. Moreover, Human Dignity serves as a checking device in the process of actual juridical decision making. The checking device becomes operative when issues of Human Dignity are at stake

[6, 12]. However, Human Dignity is not treated as an absolute notion independent of the situation-bound interpretation. The value of it may be called "absolute", but the interpretation can vary according to the situation: While in 1945 it may not have been a violation of Human Dignity to put up a large family of refugees in a small room, it would be against the principle of Human Dignity to treat a family of asylum seeking people in this way today. For, obviously, Human Dignity is a principle, not an absolute rule, a distinction made by the legal theoretician Robert Alexy [1]. Alexy calls principles of this kind "optimizing offers" which are of universal character, having precedence over rules, but which do not enjoy the definitive unambiguity of rules. They are vulnerable in several respects. They can be in conflict with other principles and they are in need of concretisation by the application of certain rules or laws. This seems to be a helpful analysis of the functions and capacities of the notion of Human Dignity, also applicable to medical-ethical problem cases. For in medical ethics most problems have to be dealt with in the context of the social matrix and also often with reference to the individual story of the patient [17, 18]. Only a relatively small number of medical-ethical problems can be settled in an absolute way, so to speak disregarding the situation, e.g. active euthanasia, deliberate misleading of patients, trading with embryos, definition of death in relation to explantation of organs, etc. However, even some of these and of related problems can not always be sharply defined for the protection of the patients in question. And with reference to the important position of the concept of Human Dignity in the German Constitution it must be observed that the concept did not save parliament from extended and highly controversal debates on the issue of abortion. In other words, the concept was incapable of carrying the burden of proof in complex question of medical ethics and legislation pertaining to it.

IV. HUMAN DIGNITY BY IMPARTATION

The above review of the historical roots of the concept of Human Dignity as well as their utilisation in United Nations Declarations and Conventions and in the American and German constitutional traditions is instructive in several aspects. It has become clear that the concept as such is not a legal one, even though it may be used as such, but that it is a frame-reference in ethics, based upon an anthropological or a religious creed. (Cf. [10]) The Stoic concept of an inherent dignity within the human is the most daring of all explanations of Human Dignity but it remains a mere affirmation that cannot be verified empirically. Empirical and social sciences of our time, advanced as they may be, would be unable to provide convincible evidence that human beings have "by nature" dignity. It would be advantageous, it seems, to resort to concepts of Human Dignity which forego ontological and quasi empirical assertions about an inherent dignity within the human being. Such is the understanding of the dignity of humans and of humanity in the biblical tradition and also the emphasis on the societal responsibility of humans for each other in the systems of Hobbes and Locke. Kant's second formulation of the moral imperative does not contradict these emphasises, rather it can be taken as a formal version of them, truly an expression of formal ethics. Most useful and philosophically tenable seems to be the contention that Human Dignity is not automatically inherent in humans, as it were, but that it is imparted on others by speaking and acting. In other words, there has to be someone who tells me that I have Human Dignity, and by telling me and by acting in accordance with this pronouncement Human Dignity is imparted. Such understanding of Human Dignity by impartation seems to be the only justifiable and philosophically as well as ethically legitimate way of making use of this concept. It provides, as it were, the platform or basis upon which can be built any ethical system or group of criteria with which to deal when it comes to concrete decisions.

It should be clear, however, that such impartation is not left to the arbitrary decision of the person who is speaking and acting toward another human being. Nor is such impartation tied to any moral assessment of the value of another person, thus resembling the most inhumane distinction between "life worth living" and "worthless life" which led to the mass euthanasia and extermination practices during the Nazi time. Such distortions were only possible on the basis of a natural law – and in the widest sense of the word – a Stoic understanding of the inherent value of a human being, making allowances for degrees of worth or of degrees of consciousness or fitness to perform such and such function. This is not to say that the Stoic concept entailed such distortions but it cannot be denied that they can be derived from it. Nor is a biological hierarchy of values or dignities a life option, e.g. by distinguishing between the dignity of lower and higher animals and ultimately of humans. Such distinction by grades leads to a hierarchy of value and worth with which biological anthropologies have always toyed. It suffices to be reminded of Jan Smuts, the South African general, politician and hobby biologist, the founder of "wholism". No, the mere belonging of a human being to the species suffices to speak of a legitimate and necessary axiom that whoever deals with this being is to impart Human Dignity upon it. One may call such axiom a creed and, indeed, this is what adherence of the Biblical tradition would hold, although, it is evident, that one need not be a Jew or a Christian in order to concur with this understanding of Human Dignity by impartation. The validity of this understanding is limitless, i.e. any human being, however distorted, mentally retarded, or even fanaticised or convicted of crime, must needs be a recipient of such impartation. The impartation is unconditional, although this does not mean that the recipiant cannot be punished in case he has comitted a crime. Nor does it mean that the dignity which is to be imparted upon another person automatically contributes toward the solution of problems presented by this person and his situation, e.g. medical-ethical problems. Here we will have to refer back to the limited applicabilty of the concept to the solution of concrete ethical problems. Moreover, on the basis of this concept it seems possible to conclude that a human being void of any human contacts in the known past and the foreseeable future, i.e. in the total absense of anyone who could impart dignity by speech or act to this person, literally does not "have" Human Dignity. This is a border-line thought which may have some relevance in medical ethics as one contrasts two dying patients in an intensive care unit, one imbedded in the circle of family and friends, the other totally and absolutely without anyone to care for him (except the nurses). A case could be made that the two could be treated differently. One must bare in mind however that it is always a great risk to draw ethical conclusions from the analysis of extreme border-line situations.[3]

The social dimension of the concept of Human Dignity as it is here redefined can also be called a dialogical concept. That is to say the dialogical and agreement-character of society's obligation is in the foreground and with this we come into the vicinity of J. Rawls' and J. Habermas' concepts. In the light of this rootage within society's obligation it is useful to imagine what an ethics, especially medical ethics, would look like if it were to operate entirely without the frame-reference of Human Dignity as a principle behind its criteria. It would either be an arbitrary situation ethics or a badly legalized system of ethics, actually an applied form of legalism whereby the medical personnel as well as the patients and their families would be constantly on the lookout

for ever finer legal prescriptions and rules. This would be a highly impersonal, mechanized and ultimately intolerable form of ethics. If, on the
other side, ethics is to have an anthropological basis, consisting of broader
frame-references, Human Dignity as one of them, the ethical criteria
for more specific arguments are well grounded and also controlled. What
is true for ethics in general also holds true for medical ethics.

Ecumenical Institute
University of Heidelberg
Germany

NOTES

[1] [3; 4]; for an assessment of the problem of a direct application of Human Dignity to
genetic technology see [21].
[2] Cf., e.g. [21, 19, 14]; for a general discussion see [8].
[3] One is reminded here of Peter Singer's distinction between vitality and personality,
an issue not to be discussed here, cf. however the new analysis by a member of our
team in Heidelberg [13].

BIBLIOGRAPHY

1. Alexy, R.: 1986, *Theorie der Grundrechte*, Suhrkamp, Frankfurt.
2. Benda, E.: 1984, 'Die Menschenwürde', in E. Benda *et al.* (eds.), *Handbuch des
   Verfassungsrechts der BRD*, Teil 1, de Gruyter, Berlin/New York, pp. 1978–2128.
3. Birnbacher, D.: 1996, 'Ambiguities in the concept of Menschenwürde', in the present
   volume, pp. 107–121.
4. Engelhardt, H.T.: 1996, 'Sanctity of life and Menschenwürde: can these concepts help
   direct the use of resources in critical care?', this volume, pp. 201–219.
5. Fuller, Lon L.: 1964, *The Morality of Law*, 2nd ed. 1969, Yale University Press, Yale.
6. Geddert-Steinacher, T.: 1990, *Menschenwürde als Verfassungsbegriff, Aspekte der
   Rechtsprechung des Bundesverfassungsgerichts zu Art. 1 Abs. 1 Grundgesetz*, Duncker
   & Humblot, Berlin.
7. Ginsberg, M.: 1965, *On Justice in Society*, Penguin Books, Baltimore.
8. Huber, W. and Tödt, H.-E.: 1988, *Menschenrechte – Perspektiven einer besseren Welt*,
   Chr. Kaiser, München.
9. Kant, I.: 1983, *Metaphysik der Sitten*, in *Werke*, W. Weischedel (ed.), vol. IV,
   Wissenschaftliche Buchgesellschaft, Darmstadt, pp. 303–634.
10. Koch, T.: 1991, 'Menschenwürde als Menschenrechte', *Zeitschrift für evangelische
    Ethik* **35**, 96–112.
11. Kuhse, H.: 1987, *The Sanctity of Life Doctrine in Medicine. A Critique*, Clarendon
    Press, Oxford.
12. Münch, I. von (ed.): 1985, *Grundgesetz-Kommentar*, vol. 1, C.H. Beck, München.

13. Nogradi-Häcker, Annette: 1994, *Die Personwerdung des Menschen – Zur Ethik Peter Singers*, LIT-Verlag, Münster.
14. Oestreich, G.: 1978, *Geschichte der Menschenwürde und Grundfreiheiten im Umriss*, Duncker & Humblot, Berlin.
15. Pound, R.: 1922, *Introduction to the Philosophy of Law*, 8th edn. 1966, Yale University Press, Yale.
16. Rawls, J.: 1971, *A Theorie of Justice*, Harvard University Press, Cambridge/MA.
17. Ritschl, D.: 1991, 'Das "Storykonzept" in der medizinischen Ethik', in H.-M. Sass and H. Viefhues (eds.), *Güterabwägung in der Medizin*, Springer, Berlin/Heidelberg/New-York, pp.156–167.
18. Ritschl, D.: 1988, 'Menschenwürde als Fluchtpunkt ethischer Entscheidungen in der Reproduktionsmedizin und Gentechnologie', in T. Schroeder-Kurth (ed.), *Das Leben achten*, Gerd Mohn, Gütersloh, pp. 96–117.
19. Schwartz, B.: 1971, *The Bill of Rights: A Documentary History*, vols. I-II, Chelsea House/McGraw-Hill Book Co., New York/Toronto/Sydney.
20. Spaemann, R.: 1987, 'Über den Begriff der Menschenwürde', in E.W. Böckenförde and R. Spaemann (eds.), *Menschenrechte und Menschenwürde*, Klett, Stuttgart, pp. 295–313.
21. Tóth, J.: 1970/71, 'Human rights and social chance', *World Justice* **XII**, 15–30.
22. Vitzthum, W. Graf: 1987, 'Gentechnologie und Menschenwürdeargument', *Zeitschrift für Rechtspolitik*, 33–37.
23. Wolff, R.P. (ed.): 1977, *Understanding Rawls – A Reconstruction and Critique of A Theory of Justice*, Princeton University Press, Princeton/NJ.

DIETER BIRNBACHER

AMBIGUITIES IN THE CONCEPT OF MENSCHENWÜRDE

I. THE INFLATIONARY USE OF THE CONCEPT OF MENSCHENWÜRDE IN RECENT GERMAN BIOETHICS

The use made of the concept of Menschenwürde in the recent German ethical and legal debate on bioethical issues such as germ line gene therapy, surrogate motherhood or embryo research is an irritating one, and not only to anglo-american observers. This irritation is partly due to the fact that the concept has been used in what is clearly an inflationary way. Partly it is due to the unclarities and ambiguitites of the concept itself, inviting the suspicion that Menschenwürde functions mostly as a "Leerformel" with no fixed content of its own, lending itself to merely rhetorical and opportunistic application.

Frequent reference to Menschenwürde as a normative principle is a distinctive feature of German bioethical discussion. The concept plays a major role, for example, in the moral rejection and legal prohibition of various of the new methods of reproduction and of gene manipulation. In both contexts, Menschenwürde is typically invoked, both by ethicists and lawyers, as a kind of ultimate article of faith rather than as a principle open to rational debate. It typically functions as a "conversation stopper" (Keenan) settling an issue and tolerating no further discussion. The more pluralistic the values of a society become, and the more relativistic its thinking about these values, the more it feels a need for taboo concepts defining, in a negative way, its residual identity. In Germany the concept of Menschenwürde functions roughly in this way. Perhaps this is one of the sources of its absoluteness and of the high degree of emotionality associated with it.

K. Bayertz (ed.), Sanctity of Life and Human Dignity, 107–121.
© 1996 *Kluwer Academic Publishers. Printed in the Netherlands.*

It is not difficult to see, however, that in its application to bioethical problems the concept of Menschenwürde has been used in an inflationary way overstepping all bounds of plausibility. A case in point is the attempt made by Ernst Benda, then president of the Bundesverfassungsgericht, to derive a prohibition of the technique of cloning people right from the "essence" of man. According to Benda, it is an elementary right of everyone not to be genetically the exact copy of one of his parents ([1], p. 224). Characteristically, this kind of "natural law" argument is presented without any further explanation. No consideration is given to the fact that the existence of identical twins makes it doubtful whether genetic individuality is really part of the "essence of man" in any not purely normative sense. Another case is the verdict of the Bundesverwaltungsgericht of 1982 against "peep shows" for the alleged reason that these infringe the Menschenwürde of the women concerned by assigning them the role of pure objects ("entwürdigende objekthafte Rolle") (see [5], p. 95f). This reasoning is, again, highly irritating. If the infringement of Menschenwürde lies in the fact that these women are commercially made the object of male sexual interest, all kinds of paid and agreed to sexual services, including prostitution, should be held to be an infringement of Menschenwürde, whether by women or men (something nobody is seriously considering). A plausible case for an infringement of Menschenwürde could only be made if the women concerned were forced into their role or if the employer were guilty of exploiting a situation of need. In these cases, however, the sexual nature of the services rendered would be without importance, whereas for the court it was exactly the sexual nature of the objectifying use made of the women that motivated its invocation of Menschenwürde.

There are several reasons to reject an extensive recourse to the concept of Menschenwürde in bioethics. One reason is that one gets the impression that the inherent emphasis and the inherent pathos of the concept is exploited simply in order to eschew the difficulties of giving rational arguments for moral and legal injunctions against unwelcomed practices. These difficulties cannot be underestimated. The fact that practices like surrogate motherhood and embryo research are rejected, more or less emotionally, by a great majority of the population – and, probably, by a majority of intellectuals –, is by itself not sufficient either to justify the moral judgment that they are inherently immoral or the penal sanctions imposed, e.g., by the German Embryonenschutzgesetz of

1990. By functioning as a "knock-down" argument the Menschenwürde argument offers an easy way out of this dilemma.

Another reason for rejecting the inflationary use of the concept of Menschenwürde is its tendency to blur all conceptual distinctions. There is a tendency, in some quarters, to use the concept of Menschenwürde in a way that makes it coextensive with the principle of sanctity of life as if protection of *life* were the only or the only central concern in the protection of Menschenwürde (see, e.g., [9] for a case in point). Instead of probing into the complex details of the relation between Menschenwürde and right to life, it is rashly assumed that both principles ultimately coincide, ignoring areas of conflict such as suicide or voluntary euthanasia where the principles have contrary implications. Suicide and voluntary euthanasia (at least its active variants) are clearly ruled out by the principle of sanctity of life, but clearly compatible with Menschenwürde. The right to a free choice of the time of one's own death is even directly subsumed under Menschenwürde in some of its legal explications (see [7], p. 291). Extracorporal fertilization is often alleged to be an infringement of Menschenwürde, but, being a technique of producing life, it is doubtful if it can be thought of as an infringement of the sanctity of life.

The third and most important reason is that an extensive use of the argument of Menschenwürde necessarily weakens the authority and moral emphasis of the concept. This is deplorable because the concept has still an important role to play. It would be a pity if by importing subjective and fashionable contents into its meaning the concept loses its normative force and ends up as a mere expressive gesture, a piece of empty rhetoric full of connotation but devoid of denotation. To counteract this development, one is well advised to reduce the descriptive content of this normative concept to a central and undisputed core meaning that leaves less room for subjective interpretation and stands above the political disputes of the day. If Spiegelberg's dictum is to remain true that "human dignity seems to be one of the few common values in our world of philosophical pluralism" ([11], p. 198) and if the concept is to function – in ethics, law and politics – as a kind of quasi-absolute, its content must as far as possible be free from controversial elements.

II. THE CONCEPT OF MENSCHENWÜRDE: WHAT IT MEANS

This presupposes that the concept does indeed have a role to play and is not, as Schopenhauer ([10], p. 412) argued against its predecessor, Kant's second formula of the Categorical Imperative, an empty formula "insufficient, without proper content and problematical". Does have Menschenwürde a "proper content", and if so, what are its elements?

I believe that there *is* a content to Menschenwürde and that this content can be identified both for the ethical and the legal concept with a collection of inalienable and unforfeitable *rights*. To respect Menschenwürde means to respect certain *minimal* rights owned by its bearer irrespective of considerations of achievement, merit and quality, and owned even by those who themselves do *not* respect these minimal rights in others. What are these rights? There are, it seems, four components: 1. provision of the biologically necessary means of existence, 2. freedom from strong and continued pain, 3. minimal liberty, 4. minimal self-respect. These four components can be looked upon as minimal "basic goods" of which the principle of Menschenwürde says nobody should be deprived. Deprivation has to be understood in this context as comprising both action and omission. Though Menschenwürde is often used in a purely negative sense, the principle works both as a negative and as a positive right. It sets a minimal standard of acceptability both to what is done to people and to what people are allowed to suffer. It sets a limit to inhuman treatment (like torture, slavery, capital punishment), but also to inhuman omissions (like letting others starve, or allowing them to be humiliated or persecuted as members of racial, ethnic or religious minorities). Thus, though the rights postulated by Menschenwürde are only minimal, the efforts required by their effective protection may be considerable.

This is especially true, under certain circumstances, of the first component of Menschenwürde, the provision of the necessary means of existence. This component has been widely neglected in the standard explications of Menschenwürde, but being a necessary condition of the others, it is really the most basic component. The fact that the concept of Menschenwürde is rooted in the essentially liberal tradition of Locke and Kant with its stress on negative rather than positive rights has tended to focus the attention on liberty as the central component of Menschenwürde and to blind one to what the effective exercise of liberty requires as its presupposition. As even the "idealistic" poet Schiller was materialist enough to note in his "Würde des Menschen" (1796), there is little worth in freedom, freedom of pain and self-respect unless the minimal means of physical existence are provided: "Nichts mehr davon, ich bitt' euch! Zu essen gebt ihm, zu wohnen;/ Habt ihr die Blöße bedeckt, gibt sich die Würde von selbst." Menschenwürde is compromised not only by salient human rights violations like torture, war atrocities and the persecution and suppression of minorities, but also by the habitual and inconspicuous practice of moral indifference, e.g. towards starvation in the Third World.

One minimal condition of the exercise of liberty, and a central content of Menschenwürde, is the provision of the means of existence, but another is freedom from pain, at least from serious and continued pain. This component is of particular relevance to the German health care system with its scandalous reluctance to provide pain-killing drugs and devices to patients. It is clearly a violation of Menschenwürde to allow a patient to suffer serious and continued pain when the technical means are available or could be made available at reasonable costs.

The two other components, minimal liberty and minimal self-respect, are too widely recognized to need much commentary and are well-established elements of the legal interpretation of the first article of the German Grundgesetz. If there are controversies they are not about the core content of Menschenwürde in these areas but about where to draw the line between what is minimal and what is not. Thus, there is no doubt that privacy is an essential element of Menschenwürde as well as the prohibition of slavery, torture and humiliating treatment and persecution, but there is, of course, always the question of how these concepts ought to be applied to concrete cases. Kant's second formula of the Categorical Imperative according to which none must be used by others *only* as a means, does not offer very much help in this respect, not only because this formula does not by itself imply anything about is concrete application apart from Kant's casuistic examples (which are too idiosyncratic to be taken into account nowadays) but because it is far too narrow. If the members of a hated minority are lynched by the majority they are not "used only as means", but they are treated cruelly for the sake of cruelty. If prisoners of war are tortured or starved to death it is not necessarily in the rational pursuit of ends such as gaining secret information, intimidation or deterrence, but it may also be because of strong non-rational interests in retaliation, revenge, and cruelty itself. It would be absurd to restrict the principle of Menschenwürde in such a way that it protects

people from inhuman treatment only when those inflicting the treatment think of it in terms of means-ends-rationality – what Max Weber
called "Zweckrationalität" – instead of direct wish-fulfillment. It is true
that historically the concept of Menschenwürde was introduced into
the German Grundgesetz mainly in reaction to the characteristic Nazi
schizophrenia of irrational (archaic) ends and rational (technological)
means. That explains that it has been primarily applied to what the state
(rather than individuals or groups) does to the individual and especially
to what the state does to the individual for allegedly good reasons.
But it would clearly be beside the point to make the protection provided
by the principle dependent either on the degree to which the *state* (rather
than, e.g. a dominant social group) is involved in the infringement of
minimal rights or on the precise *intentions* of the injurer. Menschenwürde
is a principle against *social* tyranny just as well as against *political*
tyranny. And if intentions, and indeed rationality of intention, were
important, Menschenwürde would paradoxically grant protection only
against the technical rationality but not against the arbitrariness and
wickedness of tyrants. Both restrictions are, furthermore, incompatible
with the absoluteness of the protection provided by the principle of
Menschenwürde, understood in a minimal sense, and with the relia-
bility of this protection for those who are not presently in need of it.

So much for the objection that the concept of Menschenwürde is
without content. Another common objection is that it is *redundant* and
does no other work than to put an additional emphasis on certain moral
and legal principles independently recognized. Even if it has some content
and plausibility, the objection goes, this content and this plausibility
are not specific, but parasitic on the content and plausibility of other principles such as the principles of autonomy and non-maleficence.

The answer to this objection depends on what is understood by "other
principles". Indeed, all the goods or rights protected by the principle
of Menschenwürde, at least as an ethical (as opposed to legal) principle, are *also* protected by other moral principles. In this "material"
sense, then, the principle has *no* specific content of its own. What is
specific to it, however, is the *priority* it gives to certain minimal individual rights and claims over against the rights and claims of others,
especially of the state. The principle of Menschenwürde is rather a principle of *ranking* independent goods than a principle postulating a good
of its own. Its point is to secure to the individual a minimal amount of
goods which themselves are postulated by other principles, a "ground

floor" of basic goods which it excludes from political cost-benefit-calculation. Menschenwürde, in brief, can best be conceived of as a social minimum of rights, as a principle that bars weighing of goods and rights *within* its sphere against goods and rights *outside* its sphere. It does not, however rule out the weighing of goods and bads exlusively outside or exclusively within its sphere. If the minimal claims contained in the concept of Menschenwürde cannot be satisfied at the same time (as under conditions of catastrophe), some compensatory weighing is inevitable. But, of course, one of the implications of the principle is the injunction to minimize the incidence of situations in which the need for such a weighing of basic goods and rights arises.

This answer to the redundancy-objection has to be modified for the *legal* interpretation of the principle in the context of the German Grundgesetz. The meaning of the principle of Menschenwürde as a legal norm does *not* seem to be exhausted by the catalogue of human rights listed in the articles immediately following, though a case might be made for the thesis that the contents missing in the catalogue are provided elsewhere in the Grundgesetz, for example in the definition of the German state as a social state in Article 20. The guarantee of Menschenwürde as a legal concept – which is different in function but identical in content with the ethical concept – is clearly more than a mere summing up of the human rights of the articles 1 to 19. This is evident from the fact that according to article 18, there are conditions under which some of the human rights of the catalogue can be forfeited, whereas it belongs to the essence of Menschenwürde that it is never forfeited, as well as from the differences in content. Menschenwürde has additional content over and above the human rights listed, for example the prohibition of humiliating treatment and of persecution. On the other hand, its content differs from the content of the catalogue of rights by covering only a *minimum* of these rights. This latter restriction is necessary exactly to forestall the necessity of introducing the concept of Menschenwürde into ubiquitous conflicts between competing norms. If any infringement of the right to liberty, or to physical integrity, were by itself an infringement of Menschenwürde, Menschenwürde would no longer be able to play its characteristic role as a moral and legal absolute and as an ultimate limit of acceptability.

One further objection to the concept of Menschenwürde should be mentioned, the objection that it is *speciesistic* – "speciesism" understood in the polemical sense introduced by Ryder and taken over by Peter

Singer, i.e. as a criticism of the privileges the concept assigns to human interests over against the interests of non-human animals. With respect to the contents of the concept explicated so far, this criticism is beside the point. (It has a point, however, in relation to what I shall discuss below as the "extended use" of the concept.) Menschenwürde is an inclusive, not an exlusive concept. To accept the principle of Menschenwürde does not mean to privilege the human species over against other species. On the contrary, so far as its contents are applicable to other biological species as well it does not foreclose but rather invites (though under another name) the extension of the same amount of minimal protection to other species. There is no incoherence in the moral stance that postulates that humans should be given a certain minimal protection simply because they are humans, i.e., irrespective of their merits and qualities, and that sentient animals should be given an analogous protection simply because they are sentient animals, irrespective of any value or disvalue they might have for humans.

III. EXTENSIONS OF THE CONCEPT OF MENSCHENWÜRDE

I have given reasons for defining the concept of Menschenwürde in a restricted and deflationary sense, among them the importance of preserving the status of the concept as a moral absolute (or quasi-absolute) and the resultant necessity to keep it out of the common run of ethical value conflicts and political controversies. Now it is time to state that this is easier said than done and that there are obstacles to such a restriction within the concept itself.

The inflationary use of the concept in recent bioethical debate is not only the result of too extensive an interpretation of the concept by moral philosophers and lawyers but is foreshadowed in an inner ambiguity of the concept as it is used in informal contexts (see [2]). Even in its everyday use the concept of Menschenwürde does not only denote a limited sphere of unforfeitable individual minimal rights to be respected irrespective of considerations of merit and quality, but it is also used in two *extended menanings*: one in which it is applied, among others, to the early and the residual stages of human life (human embryos, fetuses, and corpses), and one in which it is applied not to any individuals but to the human *species*. In Germany, the application of the concept to human embryos was notoriously sanctioned by the Bundesverfassungsgericht in its decision against the liberal abortion law of 1975 when it stated that Menschenwürde "is a property of human life wherever it exists" ([4], p. 41), where human life is meant to include the life of the human zygote from conception on, a position echoed in the legal literature (see, e.g., [13], 252). The other extended application – to the human species as such – is exemplified in the common rejection of the production of man-animal-hybrids where the underlying principle is not the consequentialist one that the potential prospects of the beings produced are too uncertain to justify the experiment but the sentiment that such an experiment would destroy the identity of the human species and would overstep boundaries in the order of nature that should somehow be respected.

There can be no doubt that these applications of the concept of Menschenwürde cannot be subsumed under its core meaning, that of minimal individual rights. If the concept is invoked to reject (and to prohibit) the production of biological hybrids it is not because of the prospective rights of the individuals produced (nor because of any rights of the experimenter) but because of an independent principle of species purity. If the concept is invoked to reject (and to prohibit) abortion it is not because of any prospective rights the individuals not carried to term *would* have if they were carried to term (which would be rather absurd) but because of an independent principle that human life as such is endowed with Menschenwürde.

Even in its everyday use, then, and most noticeably in its legal use, there is no unitary and homogeneous concept of Menschenwürde, but rather a family of meanings, the members of which behave differently not only semantically but also syntactically. While Menschenwürde in its core meaning needs an individual subject as bearer, this is not necessary with the extended concepts. With them, there need not be a real subject to correspond to the grammatical subject. This is evident where Menschenwürde is applied to the species as such, but it is also the case in its application to human zygotes and early embryos, entities which cannot reasonably be assumed to be "real subjects". While Menschenwürde in its core meaning postulates minimal individual rights, Menschenwürde in its extended meaning postulates obligations without corresponding rights since there may be no bearer of rights. With Menschenwürde in its core meaning the object of respect and protection is the concrete human being. With Menschenwürde in its extended meaning it is something more abstract: humanity, human life, or the identity and dignity of the human species defined by its specific potentialities. This abstractness of the reference of the extended concept is already conspicuous in Kant when he says in a variant of his second formula of the Categorical Imperative that it is not man but *humanity* ("die Menschheit", the human essence) that should never be used only as means ([6], p. 429).

My thesis is that the distinction between core concepts and extended concepts is of the utmost importance both in ethics and in law. Its importance lies in the fact that these different concepts carry quite different moral weight so that conflating them must either lead to an unacceptably weak protection of individual and concrete Menschenwürde or to an unacceptably strong protection of generic and abstract Menschenwürde.

Menschenwürde in the concrete functions as a moral quasi-absolute. Sacrificing the principle for other values or principles is generally held to be never acceptable, or acceptable only in very exceptional cases. Menschenwürde in the abstract, on the other hand, is generally thought of as a much weaker principle that does not rule out, as the strong principle does, being given up in favour of other values such as individual autonomy or scientific and medical progress. The respect due to a human embryo or a human corpse is a *weak* form of respect ([8], p. 113), much weaker in any case than that due to a human person with its capacities of consciousness and self-consiousness. The prohibition to sell one's children or one's wife is generally perceived to have a much higher priority than a prohibition to sell living human embryos or live human organs, or a prohibition to sell dead human embryos or parts of human corpses. Whereas the principle of Menschenwürde in its core meaning is generally assigned the absolutely highest priority (even higher than that assigned to the principle to protect life), the normative status of its extensions is both lower and more uncertain.

In fact, the difference in *normative* status is mirrored by a corresponding difference in *epistemological* status. Whereas for Menschenwürde in the concrete a strong justification can be given in terms of the interests and needs of the individual concerned, Menschenwürde in the abstract can only be weakly justified, with reference, ultimately, to the sentiments of observers. To justify the minimal rights postulated by individual Menschenwürde one does not need more than the weak assumptions of theories of minimal morality such as Bernard Gert's theory of moral rules, i.e., theories whose essential assumption is the triviality that people indeed have a strong interest in life, physical integrity, liberty, and self-respect, and a strong interest to live in a society which guarantees at least a minimum of these goods. A justification of the contents of the extended concepts is much more difficult. Whereas the strong principle can be justified via a relatively constant set of basic human needs there is no comparatively elementary justification for the extended concepts. Justifications of these are bound to involve more relative and culture-dependent ideas of value and dignity.

This is evidenced by the fact that roughly the same catalogue of basic human rights is part of nearly all constitutions of the world whereas there is a complete lack of unanimity with reference to, say, the moral status of the human embryo. On this point, opinions diverge widely both between and within national cultures. While the German law on embryo protection strictly prohibits any form of embryo research, embryo research is legal in Britain under the conditions that the research is clinically relevant, that the donor of the tissue consents and that the zygote is cultivated in vitro only up to the stage of development of fourteen days. The relevant law passed both houses of Parliament with an unambiguous majority. In Germany, "right-wingers" in ethics, law and politics generally maintain a strong principle of abstract Menschenwürde, whereas "left-wingers" – and even one of the commentaries to the Grundgesetz (see [7], p. 329) – tend to restrict the application of the principle to living human individuals from birth on, basing the obligation to respect human corpses not on the concept of Menschenwürde but on an explicit duty of piety as well as on the duty to respect the will of the deceased. There is, thus, considerable uncertainty about where the ethical truth lies. Personally, I agree with Lamb ([8], p. 113) that living human embryos deserve no less, but also no more respect than human corpses or human organs, with the consequence that if the principle of Menschenwürde is applied to them at all it can have no more moral weight than other principles of piety. But even those who judge differently on this point would not, I presume, go so far as to give the principle of Menschenwürde the same moral weight in this context as in the contexts to which the "core meaning" of the principle applies. The mere extent and intensity of the disagreement on this point should be a warning to all politicians eager to enforce an extended principle of Menschenwürde by penal sanctions.

IV. WHO IS A BEARER OF MENSCHENWÜRDE?

Making a conceptual and normative distinction between the "core meaning" of Menschenwürde and its extensions does not solve the problem of *who* is a bearer of Menschenwürde, but shifts the problem to the question of who is to qualify as a bearer of Menschenwürde in the strong sense. In fact, much of the attractiveness of a unitary and homogeneous concept of Menschenwürde seems to lie exactly in the fact that it does not seem to require any *criterion* of who qualifies as a bearer of Menschenwürde in the strong sense. Its only criterion is biological so that the necessity of giving differential values to various forms of human life apparently does not arise. But this attractiveness, I believe, is really only apparent. Even those who defend a unitary concept cannot do without some measure of differential valuation. Hardly anybody would be prepared to say that the same respect is due to a human corpse and to a living human being. And only very few hardliners among the pro-lifers would presumably be prepared to say that the early human zygote is a bearer of Menschenwürde in exactly the same sense and in exactly the same degree as the adult human person. Even the Bundesverfassungsgericht which made Menschenwürde to bear on the issue of abortion in its 1975 decision found nothing unacceptable in the 219d of the German penal code which explicitly exempts the early human embryo (up to fourteen days) from the protection provided by the legal prohibition of abortion in 218. This differentiation cannot be explained by a potentiality principle since the early and the late embryo have the same potential. Nor can it be explained by the principle of Menschenwürde if this is consistently applied to everything that is specifically human in a purely biological sense.

My answer to the question of who is a bearer of which kind of Menschenwürde is a split one: Menschenwürde in the extended and morally weak sense is a normative property of everything specifically human, not only human life. Some measure of respect is due also to human corpses, to dead human fetuses, and to human organs. In its weak sense, the principle of Menschenwürde is one, though not a decisive, reason against a trade in human embryos ([12]) or in human organs for transplantation.

Menschenwürde in its individual and strong sense, on the other hand, is a normative property specific to human beings with the capacity of consciousness. It cannot be ascribed to the early human embryo, and I very much doubt whether it can be ascribed to the anencephalous born without those parts of the brain parts on which, as far as we know, consciousness depends. All these forms of human life are only bearers of Menschenwürde in the generic and weak sense. The reason for this is that the principle of Menschenwürde in its core meaning is a collection of subjective *rights*. A principle such as this loses its meaning where there is human life (or humanity) but no subjectivity. A being which is specifically human in the biological sense but which lacks the capacity to experience anything or to have even minimal preferences cannot be made better or worse off by anything coming from outside. There is no point in protecting it from the threat of death, suffering, loss of liberty, or loss of self-respect since there is no inside representation of these threats. The threat is a threat only for *us*. Without subjectivity there is no work the strong principle of Menschenwürde could do.

Against this it is commonly argued, especially on the part of Roman Catholic authors, that even sleeping adults are without consciousness and that the position outlined would imply that even they are no proper bearers of Menschenwürde in the strong sense. If, on the other hand, sleeping persons are obvious candidates for Menschenwürde in the strong sense, why not also human embryos of which we assume that they will "wake up" in the normal course of events, developing full consciousness (and self-consciousness)?

The answer to this is that the objection does not sufficiently distinguish between *dispositions* and *potentialities*. The capacity of consciousness is a disposition, i.e., a property that can be owned without being actually realized, and this property is owned by the sleeping person in the same way as by the waking person. It is true, only the waking person *manifests* this disposition, but the sleeping person *has* it all the same. In contrast to this, the early human embryo has the capacity of being conscious only in the sense of a potentiality, i.e., it will come to own the disposition only at some later stage of its development. The reason for making this distinction is essentially that dispositions are supervenient on non-dispositional properties, and that possession of these properties (like having a functioning central nervous system) is a necessary condition of having the disposition. Though there is a problem of drawing a clear boundary between disposition and potentiality for cases in the transition period (a practical problem for neonatologists dealing with very early new-borns), it is clear that the human zygote in the first trimester of pregnancy has the capacitiy of consciousness only in the sense of a potentiality whereas the normal new-born human infant has this capacitiy in the fully dispositional sense. A being with the potentiality to develop a certain dispositional property has not, however, eo ipso the same moral status as a being with the fully developed dispositional property. If its moral status, as with the strong principle of Menschenwürde, essentially depends, as I have argued, on the disposition to be capable of consciousness, it *cannot* have the same moral status. The class of human beings in the biological sense cannot coincide, therefore, with the class of bearers of Menschenwürde in the strong sense. The class to which the strong principle applies is necessarily a sub-class of the class to which the weak principle applies.

This result may seem more devastating than it really is. It does not imply anything in particular about the intensity of protection to be given to those forms of human life which, in my view, are not candidates for Menschenwürde in the strong sense. Strong protection may well be called for by principles other than Menschenwürde (see [3], pp. 152 ff). My point is only that we cannot expect the principle of Menschenwürde to do the job.

University of Dortmund
Germany

BIBLIOGRAPHY

1. Benda, E.: 1985, 'Erprobung der Menschenwürde am Beispiel der Humangenetik', in R. Flöhl (ed.), *Genforschung – Fluch oder Segen?* Interdisziplinäre Stellungnahmen, Schweitzer, München, pp. 205–231.
2. Birnbacher, D.: 1987, 'Gefährdet die moderne Reproduktionsmedizin die menschliche Würde?', in V. Braun, D. Mieth and K. Steigleder (eds.), *Ethische und rechtliche Fragen der Gentechnologie und der Reproduktionsmedizin*, Schweitzer, München, pp. 77–88. Reprinted in A. Leist (ed.), *Um Leben und Tod*, Moralische Probleme bei Abtreibung, künstlicher Befruchtung, Euthanasie und Selbstmord, Suhrkamp, Frankfurt/M. 1990, pp. 266–281.
3. Birnbacher, D.: 1990, 'Embryonenforschung: Ethische Kriterien der Entscheidungsfindung', in C. Fuchs (ed.), *Möglichkeiten und Grenzen der Forschung an Embryonen*, Gustav Fischer, Stuttgart, pp. 127–138.
4. Bundesverfassungsgericht: 1975, *Entscheidungen*, vol. 39, Karlsruhe.
5. Hoerster, N.: 1983, 'Zur Bedeutung des Prinzips der Menschenwürde', *Juristische Schulung* **82.2**, 93–96.
6. Kant, I.: 1911, *Grundlegung zur Metaphysik der Sitten*, Academy-Edition, vol. 4, Reimer, Berlin, pp. 385–463.

7. Kommentar: 1984, *Kommentar zum Grundgesetz für die Bundesrepublik Deutschland*, Series Alternativkommentare, Luchterhand, Neuwied.
8. Lamb, D.: 1988, *Down the Slippery Slope, Arguing in Applied Ethics*, Croom Helm, London.
9. Poliwoda, S.: 1992, 'Keimbahntherapie und Ethik', *Ethik in der Medizin* **4**, 16–26.
10. Schopenhauer, A.: 1949, *Die Welt als Wille und Vorstellung*, vol. 1, in Sämtliche Werke, ed. A. Hübscher, vol. 2, Brockhaus, Wiesbaden.
11. Spiegelberg, H.: 1986, *Steppingstones toward an Ethics for Fellow Existers*, Essays 1944–1983, Reidel, Dordrecht.
12. Stutz, S.: 1988, *Embryohandel*, Zytglogge, Bern.
13. Vitzthum, W. Graf: 1985, 'Gentechnologie und Menschenwürde', *Medizinrecht* 6, 249–257.

THOMAS PETERMANN

HUMAN DIGNITY AND GENETIC TESTS

I. INTRODUCTION

Experience has frequently shown that a threat to basic values is often needed to prompt a re-examination of the nature and significance of the norms required for social life. The dangers associated with new technologies in particular have resulted in intensive reflection on and discussion of the vulnerability of the normative foundations of society.

In the many debates over the opportunities and risks involved in utilising new technologies in Germany, reference has frequently been made to the issue of human dignity, as guaranteed in Article 1 of the German Basic Law: "Human dignity is inviolate. Its respect and protection are the duty of all the forces of the state." From the discussion over the use of atomic energy through the debates on electronic documentation and registration of the populace within the framework of national censuses, to the latest controversies about biotechnology, reproductive technology and genetic engineering, human dignity has frequently been used as "argument and taboo" [19]. The "inflationary use of the human dignity argument" ([12], p. 356) has not always led to greater clarity – on the contrary, in fact, which is why the burgeoning use of human dignity as an issue has been rightly criticised [20].

The debate about the methodologies of diagnostic techniques using human genetic material also addresses the question of human dignity. Both in the legal literature and in numerous reports by commissions and working parties appointed to advise on policy, the question is raised whether genetic tests may under certain circumstances "lead to significant risks to the right of privacy and violate the principle of respecting

123

K. Bayertz (ed.), Sanctity of Life and Human Dignity, 123–138.
© 1996 Kluwer Academic Publishers. Printed in the Netherlands.

human dignity" ([4], p. 41). Following a short note on the relative inde-
terminacy of the term, the sections below will review its polemic use
within the framework of the political and scientific debate in Germany
on genetic testing.

II. HUMAN DIGNITY: AN "UNINTERPRETED THEORY"

There is no extensive operationalisation of the concept of human dignity
in the literature, for example in the form of a list of criteria for human
dignity. The term continues to have many meanings,[1] and this periodi-
cally emerges as a shortcoming in debate. On the other hand, this avoids
the problem associated with listing the characteristics of a life consonant with human dignity – and thereby also , delineating and debasing
lifestyles "lacking in dignity". Through its status as an "uninterpreted
theory",[2] article 1 of the German Constitution makes it possible, in
assessing challenges and hazards to human dignity, to find an answer
which is appropriate to the relevant situation. This means that the term
"human dignity" can become more concrete in the context of reviewing
threats and violations to human dignity ([7], p. 909). The task of oper-
ationalising the term falls particularly to the legislature, and is a major
element in its "decision-making prerogative" ([8], p. 247).

Even if, as noted, it is out of the question to list the components of
human dignity, the quintessence of human dignity is still regarded as
amenable to definition. The "object theory" developed by Dürig, the commentator on German Basic Law, which goes back to Kantian philosophy,
has established itself in German jurisprudence and court decisions as a
basis for identifying violations of human dignity. According to this,
human dignity is violated if the specific individual[3] is degraded to the
status of a "mere object", as an instrument, a "substituable dimension".
The goal is to protect the personal individual value of a human being,
the autonomous right of self-determination ([1], p. 35). The task of
enabling and protecting this may be regarded as a duty of politics.

However, as the German Federal Constitutional Court has held, this
general formula can at best serve as an indicator of where to seek the
violation of human dignity. This is because individuals are frequently
mere objects not only in terms of circumstances and social evolution –
they must also bow to the law without regard to their own interests.
This alone, according to the court, is not enough to constitute a
violation: instead, an individual must be treated in a way that casts

fundamental doubt on his or her autonomy, or which in a specific instance constitutes an arbitrary disregard of the individual's dignity. For this, "disparaging treatment" must be present[4] which objectively expresses a derogatory view of the value of the individual as a person (Entscheidungen des Bundesverfassungsgerichts [BVerfGE] 30, 1 (26)).

III. VIOLATION OF THE AUTONOMY OF THE INDIVIDUAL?
GENETIC TESTS USED AT THE WORKPLACE (SELECTION)

The German courts have held that one form of treatment of individuals which casts doubt on their autonomy, i.e. reduces them to objects, is if their personal characteristics are completely documented and catalogued or if detailed personality profiles are produced of them. If genetic assay is capable of being used to make a comprehensive study of individuals, and particularly of their genetic traits, then the possibility of a hazard to human dignity could not be dismissed. In view of the conceivable situation where genetic testing could be widely used in job-linked examinations, it is occasionally noted that the creation of genetic profiles may help identify susceptibility among employees, e.g. to pollutants, and could be useful as a basis for rejection or dismissal. Other predispositions, susceptibilities or even characteristics unconnected with illness which are relevant to work could be recorded and used by the employers for their own purposes. This would transform prophylactics in the interests of the employee into a process of selection by the employer. This would be, as the Enquete Commission "Opportunities and Risks of Gene Technology" of the German Bundestag put it, "a grave development in the wrong direction" ([6], p. 166).

Allusions to potential abuses are used in debate (although only in isolated instances) to reject categorically all genetic assay of employees, including even analysis for the legitimate purpose of ensuring safety at work. It is feared that such examinations would (because of competition in the labour market and the structural imbalance of power between the employee and the employer) inevitably lead to discrimination against employees with allegedly greater susceptibility.[5] Individuals seeking work who are currently in good health would, it is argued, be generally subjected to a predictive analysis of their genetic predisposition, and would be rejected if there was doubt of their suitability. The genetic assay would reduce the individual to a mere object of specifically economic interests. It is, in this view, incompatible with human dignity "to be exposed as a bearer of the 'commodity of labour power' to a cost-effectiveness calculation extending to genetic makeup" ([15], p. 171).

This argument follows to a certain extent the rhetoric of fear, prefering the worst case scenario over the best case. The weakness is its foundation, namely the assumption of a comprehensive examination of the individual through genetic assay, creating a "fishbowl employee" who is subject to the danger of selection by an employer.

The main objection to this argument (e.g. [10], p. 116 *et seq*.) and the assumed risk to human dignity as a result of genetic tests is that in our current state of knowledge we are unable through genetic assay to identify the entire genetic makeup. According to this line of argument, a possible use of genetic analysis within the framework of labour law would not have as its aim a general screening of the genetic makeup,[6] but would be restricted to screening for individual predispositions for more or less uncertain future developments. In these terms, a job-related diagnosis would not be the human genetic makeup in its entirety but only specific traits or susceptibilities. Identification of a specific predisposition, e.g. to an allergy or miller's asthma, would not be comparable to a general examination, so that it would be difficult to classify a test of this nature generally as a violation of human dignity.

Even so, this line of argument accepts that examinations are conceivable which encroach on the core of the personality to the extent that a violation of human dignity could be held to exist in an individual case. This could above all be the case with respect to the identification of multifactor characteristics, such as intelligence, disposition to alcoholism and aggressiveness, and the use of this information by third parties. However, consideration would demand a review of the concrete methods of examination used and the intended diagnosis and its use. The conceivable impermissibility of individual techniques of genetic analysis would not, however, be sufficient to constitute a violation by such methods of article 1 (1) of the German Basic Law ([9], p. 34).

Further, it is argued, there is nothing fundamentally new in the evaluation of genetic makeup in labour medicine. Conventional techniques of examination are also used to predict the future health of the employee and thus his or her suitability for the job. The development of more reliable and more exact techniques of examination for the purpose of avoiding hazards to employees does not in itself lead to the reduction of the employee to the status of an object ([9], p. 35).

Finally, the argument is encountered in jurisprudencial discussion that genetic testing of employees may under some circumstances constitute encroachment on the right of the individual to develop their personality freely, guaranteed by article 2 (1) of the German Constitution. Authors who take a very broad view of this right fear that certain medical examinations (and thus also specific genetic tests) may well violate the boundaries of individual privacy and thus human dignity.

The German Federal Constitutional Court has consistently asserted in its rulings the right of the individual citizen to an inviolable area of private lifestyle. Based on this, the general right of privacy in combination with article I (1) of the German Basic Law includes the right of the individual "to decide autonomously in principle when and within what limits personal details are disclosed" (VerfGE 65, 1 (42 *et seq.*)). Knowledge of genetic makeup would probably come within these limits just like information on illness generally. Correspondingly, protecting human dignity requires that individuals must have the right to a protected "intimate sphere" in their personality in line with article I (1) of the German Basic Law. This implies the further right to refuse disclosure to anyone, including the state. This right must also extend to "not having to disclose physical defects, peculiarities or afflictions without compelling reason" (Benda, quoted in [7], p. 912).

This brings us to a decisive point in the discussion: are there "compelling reasons" apparent for the use of genetic tests in labour medicine which would permit encroachment on the intimate sphere of the individual personality without violating human dignity. Such a reason could be present where genetic testing would be able to identify predispositions which could place either the employee or others at risk. However, such tests would require a legal basis. Such legislation would have to identify specifically the permissibility of tests, and particularly to link them to "informed consent", i.e. the principle of voluntary action.

This pragmatic line (cf., e.g. [6], p. 168 *et seq.*), which represents the prevailing opinion in the scientific and political debate, is opposed by a categorical argument which emphasises the fundamental dangers which even a legally regulated use of genetic testing could involve. Justification restricting the freedom of self-determination of individuals in terms of safety at work, combined with a legislative framework to prevent abuse, is countered by stressing the permanent danger of abuse, not least because of the weaker position of employees. Proposals for purely selective use, limited to certain occasions and certain types of information, are met by the argument that this would irreparably breach the dam.

IV. INDIVIDUAL'S PERSONALITY DESERVING PROTECTION –
INDIVIDUALS AS SUBJECT TO STATE AUTHORITY:
DNA FINGERPRINTING IN CRIMINAL CASES (IDENTIFICATION)

While there is little use of genetic testing at the workplace in Germany and none of DNA analyses, considerable progress has been made on the forensic use of genetic analyses. The legal basis for this is section 81 a, c of the Criminal Code, although this does not explicitly provide for "blood tests" of this type. The conditions for the use of genetic fingerprinting in criminal cases have been operationalised and developed through court rulings.

The majority in the debate believe that the procedure is permissible, but this is often combined with a belief in the need for legislation containing new and clear regulations on its use ([10], p. 154). Such regulations are planned in the form of an amendment to the Criminal Code, and draft regulations have been submitted by the Ministry of Justice and by the members of the Social Democratic parliamentary group in the Bundestag.

As far as the debate on the risk to human dignity is concerned, two considerations are primarily advanced (cf. [9], p. 135). First, there is the question whether identification of personality-related characteristics does not constitute an unreasonable invasion of the individual's sphere of personality requiring protection. Second, there is the point that the use of DNA fingerprinting could result in the citizen (as accused) being degraded to an object of state authority.

First, let us look at the question of a possible unreasonable incursion into the sphere of the individual's personality requiring protection. Here again, we encounter what is essentially a categorical argument – namely, that an incursion into this inviolable area occurs simply by virtue of the fact that not only is material (blood, tissue) removed which contains cells but these are examined at the genetic level. Even if this material is only partly examined, the biological basis for individual existence can be uncovered. The sample could also be used for research into the characteristics and features of the subject of the examination. There is "no conceivable effective possibility" to prevent a possible misuse ([18], p. 736).

In dealing with this argument, the proponents of DNA fingerprinting accept that even in the forensic use of DNA analysis it would be possible in principle to identify parts of the individual genetic code, because the material is available. However, this would be an abuse, and not covered by the actual purpose of the analysis. The only purpose in using this kind of probes is to study sections between the coded sections of DNA which do not provide any information on the individual personality. All that would be identified and compared is the differing lengths between certain specific DNA sections. The sample generates a pattern of bands typical of a particular individual (similar to a bar code) which does not permit any linking to other coded sequences. As a result, "excess information" with relevance for the personality would not occur – a form of built-in data protection.

Under these circumstances (it is claimed), no incursion into the inviolable sphere of the individual's personality occurs ([10], p. 161 *et seq.*; [2], p. 47). It is not denied that there may be encroachment on the individual's privacy (Privatsphäre) – (which enjoys a lesser degree of protection under German law) – for example, it is possible to establish the paternity of an illegitimate child. However, identification of this circumstance would be covered by the public interest in effective administration of justice, which is in itself sufficient to cover such an encroachment on privacy. As a result, the argument based on violation of human dignity does not (it is claimed) apply. As the German Parliaments Enquête Commission "Opportunities and risks of gene technology" argued, nobody could claim a protected interest for not getting identified as a culprit of a criminal act or as a father (of an illegitimate child). Of course there would be a protected interest, that findings about hidden illnesses or genetically conditioned personal characteristics would not be gathered ([6], p. 176).

A second dimension of hazard to human dignity is seen in the fact that widespread use of genetic fingerprinting in the pursuit of criminals and in criminal proceedings would make the citizen a mere object of external authority (cf. [9], pp. 136–137). Rulings by the German Federal Constitutional Court classify as an obvious degradation the recording of all the personal information on an individual. The courts take a reduction of the individual to the status of an object as given where individuals are documented and catalogued in all aspects of their personality or where personality profiles are generated of individuals (BVerfGE 27, 1 (6), 65, 1 (53)). The decisive factor is that the individuals are deprived of the autonomy which is inherent to them. This would apply just as much to a complete or extensive analysis just as much as to the reduction to functioning under compulsion, in disregard of the individual's spiritual and mental identity and integrity. Generating even partial personality profiles may overstep the bounds.

These court decisions are the basis for the (rarely advanced) criticism of conceivable legal compulsory submission to DNA testing as a violation of human dignity. It is claimed that the performance and use of such a test in criminal proceedings make defendants both "experimental subjects and self-incriminating evidence". This, it is argued, contradicts the recognised principle that "the accused as the subject of the proceedings must never be reduced to a mere object of crime control" ([18], p. 736). If this investigative technique is admitted, prosecution and jurisprudence would have little chance of escaping the fascination of the technological possibilities, and the door would be opened wide to abuse.

Proponents counter appeals to such dangers by noting that constitutional (i.e. not unbounded) use of DNA fingerprinting prevents a situation where the goal of the criminal investigation is success at all costs. Supporters accordingly call for clear legal regulation to prevent possible abuses and make clear to those affected the conditions for and purpose of the procedure. Together with other evidence, the order to perform a DNA analysis (simply to check the identity of a piece of evidence with the accused's blood sample) would both be constitutional and compatible with human dignity ([10], p. 172).

There are accordingly two different types of argument in the debate. One regards the risks of the forensic use of DNA fingerprinting as real enough to object to such use in favour of preserving the rights and human dignity of the accused (even at the cost of a lower success rate in criminal investigations). The other points to the methodological impossibility of recording personality-linked characteristics and also − if an appropriate legal regulation would be given − to the constitutional propriety of the role of DNA analysis in criminal proceedings. Both, it is claimed, accord due protection to the sphere of the accused's personality, which would rule out any violation of human dignity.

V. VIOLATING THE DIGNITY OF THE FOETUS:
PRENATAL DIAGNOSTICS (PREVENTION AS HINDRANCE OF LIFE)

The techniques currently in routine use in prenatal diagnosis cannot be used before the eighth week of pregnancy. The overwhelming opinion in the constitutional literature and on the German Federal Constitutional Court is that the living entity present at this time in the mother's womb has human dignity.[7] This means that the embryo also enjoys the full protection of the German Basic Law (BVerfGE 39 , 1 (41)).

Prenatal examinations of the genetic predispositions of the embryo would therefore be impermissible (even with the approval of the parent) if they degraded the unborn child to the status of a mere object and accordingly qualified as a violation of the human dignity guaranteed under article I of the German Basic Law. The state would then have to fulfil its obligation of affording active protection to human dignity.

A violation of the dignity of the unborn child in the sense discussed above is widely seen in the "abortion reflex" (Abtreibungsautomatismus) associated with prenatal diagnostics ([2], p. 31). A prenatal diagnosis could produce a result which indicates an embryopathic indication. This could lead to abortion and thus infanticide. Individual decisions of this kind would also have to be seen in the context of scientific and social advances. The earlier, more reliable and more differential genetic analysis on unborn infants became, it is argued, the greater the social pressure resulting from the change in social values and consciousness to perform the test and subsequently to abort diseased or handicapped foetuses or even those with undesired characteristics. There would be a tendency for a failure to abort to be regarded by doctors and peers in the same light as the failure to carry out therapy, namely an irresponsible decision ([4], p. 69). Parents would have to justify their failure to abort, rather than the abortion.

The creation of a "testing society" in which the unborn are routinely screened for a range of predispositions would, it is feared, be the possible result of permitting tests (even a few). Genetic tests therewith would open the gate for prenatal "selection of human life through the parents." ([6], p. 150) However, such individual and social practice would appear as a violation of the rights of the embryo.

Many participants in the debate see the integration of prenatal diagnostics and eugenic indication as provoking abuse (e.g. [20], p. 132; [2], p. 18). However, many do not share this view of a fundamental danger to the dignity of the unborn foetus, noting that the use of prenatal diagnostics to provide certainty for parents (even if this use resulted in the termination of the pregnancy) could only be regarded as violating the dignity of the foetus in terms of an extremely categorical view of this dignity. Generally, it should not be assumed that the embryo is subjected to the unconsidered and selfish interests of the parents, degraded to the status of an object and not respected for its intrinsic value: the parental choice of genetic testing with the possible consequence of subsequent abortion is in most cases motivated much more by concern about the future state of health of the unborn child. If parents (and particularly the future mother) opt to terminate a pregnancy on the basis of a diagnosis, they do not disregard the right to life of the child to the point of degrading it to the status of an object, as is clear from the fact that most women have great difficulty in making the decision to terminate and also suffer severe psychological problems after the abortion. It could not be denied that this concern also has selfish elements, but this would not be enough in itself to constitute a violation of human dignity ([9], p. 191).

In addition, there is broad agreement that prenatal tests – for example, when these are used to identify conditions that are not amenable to therapy, with the option of terminating the pregnancy if a pathological condition is present – are ethically justified. Again, diagnostic procedures which benefit the infant – e.g. diagnosis of treatable conditions – do not conflict with principles of human dignity.

It isn't denied, however, that wherever genetic analysis is used as a basis for classification as "nonviable" (lebensunwert): here, people are no longer regarded as an end in themselves, but merely as a means, and are deprived of their claim to respect as objects of irresponsible parental triage. Out of considerations about these dangers many statements emphasize the parents' freedom of choice. But at the same time it is claimed to assure,[8] "that the DNA-analysis in relation to a potential abortion is not being misused for selecting children in accordance to desired or nondesired genetic attributes." ([6], p. xxvii)

The issue of the "abortion reflex" does on the whole raise questions of possible acts and motives which under certain circumstances – and particularly where the practice is widespread – could constitute a hazard to the dignity of unborn life. However, if it is to be qualified as a violation of human dignity, the intent to destroy life would have to constitute an arbitrary act which degrades the embryo to the status of an object. This would be the case if the unborn child was to be killed or aborted in response to the desire of parents who were only prepared to accept a "perfect" child in terms of eugenic criteria ([2], p. 31). Here, the parents would be ignoring the consideration that the unborn child they are dealing with is a person with the same rights as themselves, and that they have a duty as its custodians to take responsible decisions ([9], p. 192).

There are also two incompatible positions with respect to the use of genetic testing in prenatal diagnostics. First, there is the position based on the autonomy of the woman and her right to control her own life (e.g. [10], p. 88 *et seq.*). In the context of the right to life of the embryo, this provides ethical justification for genetic testing under the conditions discussed above. Second, there is the extreme position which asserts an inviolable right to life of the unborn child even if it will be seriously handicapped, and which makes this the ultimate priority. This position necessarily rejects the right of parents to opt for genetic testing, as it sees this as not only violating the right to life but also the dignity of the embryo.

VI. THE RIGHT OF NONDISCLOSURE

Genetic analysis makes possible explanations about inherited pathological conditions and conclusions about predispositions to handicaps and susceptibilities. Such highly sensitive personal medical data deserves special protection, in that it offers the individual concerned information on himself or herself.[9] On the other hand, other parties have an interest in this information. This is why a right to nondisclosure is often cited in the various areas of use of the results of genetic test. This should be asserted by the individual in order to prevent such information being collected for use by a third party. It is argued in the context of compulsory tests that information about genetic makeup should not be forced on individuals. Individual dignity would be "violated if the individuals were not able to decide this for themselves. Dignity always involves an element of autonomy. Where this autonomy is disregarded, human dignity is infringed" ([7], p. 913).

However, it is often far from clear here whether this applies to relatively insignificant results such as allergy or visual weakness or whether diagnoses with implications for life and health such as serious or untreatable illnesses trigger the right to nondisclosure in order to protect individual privacy. Another unsettled question is how extensive an assay of individual genetic makeup must be to justify classification as a possible or actual risk to human dignity.

In extreme cases, the issue of violation of human dignity undoubtedly has merit for debate. This would apply where the individual is informed against his or her will of the genetic basis for an illness which may be fatal or may emerge at a later stage in life. The problem of an inacceptable burden also arises in cases where there is uncertainty (often present in genetic diagnostics) whether the potential illness will emerge or not. The results of the testing often only justify statements which have statistical and epidemiological validity only at the level of groups of individuals. A specific individual involved then has to live with the uncertainty ([7], p. 910). There is general recognition that this constitutes thrusting information on the individual in a way which cannot be justified by other goals or legal maxims.

In view of these problems, it is argued that the "right to nondisclosure" is the only way of preventing the threatened loss of individual liberty and autonomy in life choices. Where the government makes genetic testing mandatory, there must be the possibility of asserting the right to nondisclosure. In this case, ignorance is a prerequisite for freedom ([13], p. 194). This view has now been accepted to the extent that there is general recognition of the need to obtain the consent of the individual affected for the analysis.

At the same time, it is accepted that this agreement will not suffice in every conceivable case to protect human dignity. This is why the principle of consent becomes difficult where genetic tests are involved which the individual is formally entitled to refuse but such a refusal would involve disadvantages. This problem is addressed for example in the context of a situation where considerations of safety at work oblige employees to undergo genetic testing. Although individuals would be free to refuse consent, they would then be faced by the consequences, probably in the form of dismissal. In this case, there would be de jure freedom of choice but de facto compulsion. The problem is also apparent in the (hypothetical) case of a genetic test required for mandatory insurance ([10], pp. 139–140). The applicant would be free to refuse consent, but would then be refused insurance cover. Although private insurance would provide an alternative, there would be no genuine freedom of choice. Under these conditions, consent to a test would be "imputed" and would be de facto compulsory. For this reason, consideration is needed here of the possibility of a violation of human dignity.

In the debate on whether such "obligations" to undergo genetic testing violate both the right to nondisclosure and also human dignity, two positions can be discerned.

First, there is the argument that infringement of an individual's right to nondisclosure is justifiable e.g. in labour medicine examinations where the aim is to protect not only the individual concerned but also third parties (customers, fellow employees). Such justification is also conceivable in the context of mandatory social security: testing (after the necessary information and consent) could be in the interest of the community covered by social security (because of the savings in costs) and would be reasonable.

Another contrary argument (cf., e.g. [7], p. 913) is that it is not primarily relevant that a directive to undergo testing should be concerned with the interest of the individual affected or should seek to save costs. The individual has, it is claimed, a right to choose nondisclosure irrespective of any possible risks to health. It would also apply if the aim of mandatory genetic testing was to identify genetically-linked risks to health for the purposes of preventive medicine. The basic idea of the right to nondisclosure, it is argued, is to preserve the individual's freedom of decision in life choices. From this point of view, mandatory genetic testing for the purpose of restraining costs in the long term reduces the individual to the status of an object of fiscal considerations.

VII. CONCLUSION

This review of certain lines of argument in the debate in Germany has shown how the risk that genetic testing of humans may result in a violation of human dignity has been explored from many points of view. Besides the possibility that the dignity of the unborn child may be violated, the emphasis is on the risk that individuals may have information on their genetic makeup forced on them against their will, for example through diagnostic testing at work or in connection with insurance. In addition there is the potential issue of genetic assays and comprehensive documentation of genetic characteristics with the resulting possibility that the individual could become a mere object which others can dispose of as they will, whether these are state institutions or private organizations or individuals.

The issue of human dignity is handled with restraint in the debate. In contrast to other bioethical questions, such as in vitro fertilisation, surrogate motherhood, embryonic research and genetic therapy [14], the focus in the debate is not on human dignity as an asset requiring protection. Participants in the debate take to heart the idea not to advance violations of human dignity as an argument before other arguments have been fully aired. An example of this is that hazards to the legal rights of those affected by genetic testing are seen as covered by other legal principles ([12], p. 354; [20], p. 138). Individual privacy for example is protected by the general right to privacy and the right of self-determination with regard to information. Labour medicine testing and testing in connection with insurance schemes are also covered by the Data Protection Act, Civil Code and Criminal Code to the extent that comprehensive testing (a "genetic striptease"; cf. [11], p. 379) would not be permissible. Finally, the embryo's life is protected by the constitutional right to life and freedom from physical injury. These questions are at the centre of the debate. With great restraint human dignity is treated as the argument of "last resort" to admonish of the dangers of slipping into an unthinking "society governed by tests" and of the dangers of selection, identification and prevention (as hindrance for life).

Office of Technology Assessment
Deutscher Bundestag
Bonn
Germany

NOTES

[1] "It doesn't seem possible to exactly define both the type or extent of violations to human dignity. Quite extensive scope is open when judging and evaulating such violations" ([8], p. 247). However this unclear situation could less orginate from the inexact definition of the term "dignity" but rather from a general uncertainty and lacking concensus on pertinent norms. But not only the multiple meanings of the term generate difficulties in the debate. The function of this term as an unambiguous "winning argument", i.e. an argument that always overrides all other feasible counter-arguments is also problematic ([17], p. 139). A third problem is the circumstance that recourse to the "human dignity" function in order to obtain specific guarantees, e.g. data safety, has very little value ([8], p. 249).
[2] This term was first used by the former federal president of Germany Theodor Heuss.
[3] I will not deal here with the question whether not only the human dignity of specific individuals but also "the human image in its entirety" ([1], p. 32) have to be protected (cf. the references in [17], p. 144). This occurs in German discussion particularly in the controversy over manipulation of the germ-line and/or embryo research, and affects the question of the possibility of eugenics [8].
[4] The rulings of the German Federal Constitutional Court speak of "humiliation, denunciation, persecution, ostracism" or of cruel, inhuman and humiliating punishment (BVerfGE 1, 97 (104); BVerfGE 45, 1987 (228)).

[5] The trade unions are arguing in a similar way. In their opinion the risks exceed the opportunities. Therefor they usually argue for a general prohibition of genetic testing at the workplace. There is no referring to the danger for human dignity. Also the former Federal Commissioner for Data Protection stated his support for a legal ban. As a reason he mentioned the endangerment of the "informational selfdetermination" (but not of human dignity) (Bundestags-Drucksache [BT-Drs.] 11/8520).

[6] A general screening (serial examination) is unanimously declined in the discussion (cf. e.g. [6], p. xxxi).

[7] The German Federal Supreme Court holds the view that life begins with the nidetion (BVerfGE 39, 1 (37)). It should, however, be noted that the literature contains many reservations on extending the protection of human dignity to fertilised cells, the embryo or the foetus. Another critical question is whether the "means-end theory does not presuppose the personal existence of the other party. An entity can only be treated as a means to an end if it is also capable of being treated autonomously, as a person, and therefore has to be so treated" ([17], p. 149). If the unborn child is not accorded human dignity on the basis of these doubts, this does not mean that the unborn child is left unprotected. Such protection is then based on other provisions in the German Basic Law.

[8] At this point enters the obligation of the state, "to protect and promote this life," and to "guard it against illegal interferences from others" (BVerfGE 39,1 (42)).

[9] There is general agreement that an adult may not be refused the right to obtain information on his or her genetic makeup. If the individual freely takes the decision to be tested, they have a right to proceed – the "right to know" as a basis for their actions and planning their live. "Seen in this way, the use of genome analysis constitutes an opportunity for attaining human dignity." ([7], pp. 910 *et seq.*) The situation is different where the decision to perform a test is not made by the individual concerned. This would be the case for example where a government body ordered genetic screening or in labour medicine examinations where the results are used for decisions on hiring or retaining employees. Here, individuals are forced to waive part of their inalienable right to shape their own lives, a right protected by statute and constitution ([7], p. 911).

BIBLIOGRAPHY

1. Benda, E.: 1985, 'Die Erprobung der Menschenwürde am Beispiel der Humangenetik', *Aus Politik und Zeitgeschichte* **35**, B 3, 18–36.
2. Bundesminister der Justiz (ed.): 1990, *Abschlußbericht der Bund-Länder-Arbeitsgruppe "Genomanalyse"*, Bundesanzeiger 42, Nummer 161 a, Bonn.
3. Bundesminister für Forschung und Technologie (ed.): 1984, *Ethische und rechtliche Probleme der Anwendung zellbiologischer und gentechnischer Methoden an Menschen*, Schweitzer, München.
4. Bundesminister für Forschung und Technologie (ed.): 1985, *In-vitro-Fertilisation, Genomanalyse und Gentherapie. Bericht der gemeinsamen Arbeitsgruppe des Bundesministers für Forschung und Technologie des Bundesministers der Justiz*, Schweitzer, München.
5. Bundesminister für Forschung und Technologie (ed.): 1991, *Die Erforschung des menschlichen Genoms*, Campus, Frankfurt a.M., New York.
6. Deutscher Bundestag, Referat Öffentlichkeitsarbeit (ed.): 1987, *Chancen und Risiken*

der Gentechnologie: Der Bericht der Enquête-Kommission des 10. Deutschen Bundestags, Bonner Universitätsdruckerei, Bonn.

7. Donner, H. and Simon, J.: 1990, 'Genomanalyse und Verfassung', *Die Öffentliche Verwaltung* **43**, 907–918.

8. Enders, C.: 1986, 'Die Menschenwürde und ihr Schutz vor gentechnologischer Gefährdung', *Europäische Grundrechte-Zeitschrift* **13**, 241–252.

9. Forschungszentrum Biotechnologie und Recht, Prof. Dr. J. Simon, Universität Hannover: 1993, *Rechtliche und rechtspolitische Aspekte der gegenwärtigen und zukünftig erwartbaren Nutzung genanalytischer Methoden am Menschen*; Gutachten im Auftrag des Büros für Technikfolgen-Abschätzung beim Deutschen Bundestag, Bonn.

10. Hennen, L., Petermann, Th. and Schmitt, J.J.: 1993, *TA-Projekt "Genomanalyse". Chancen und Risiken gentechnischer Diagnostik*, Arbeitsbericht Nr. 18 des Büros für Technikfolgen-Abschätzung beim Deutschen Bundestag, Bonn.

11. Hirsch, G. and Eberbach, W.: 1987, *Auf dem Weg zum künstlichen Leben*, Birkhäuser, Basel usw.

12. Hoffmann, H.: 1993, 'Die versprochene Menschenwürde', *Archiv des öffentlichen Rechts* **118**, 353–377.

13. Jonas, H.: 1987, *Technik, Medizin, Ethik: Praxis des Prinzips Verantwortung*, Suhrkamp, Frankfurt/M.

14. Kaufmann, A.: 1987, 'Rechtspolitische Reflexionen über Biotechnologie und Bioethik an der Schwelle zum dritten Jahrtausend', *Juristenzeitung* **42**, 837–847.

15. Klees, B.: 1990, *Der gläserne Mensch im Betrieb. Genetische Analysen bei ArbeitnehmerInnen und ihre Folgen*, Rotpunktverlag, Zürich.

16. Klug, U. and M. Kriele (eds.): 1988, *Menschen und Bürgerrechte: Vorträge aus der Tagung der Deutschen Sektion für Rechts- und Sozialphilosophie (IVR) in der Bundesrepublik Deutschland, vom 09.–12. Oktober 1986 in Köln* (= Archiv für Rechts- und Sozialphilosophie: Beiheft 33), Steiner, Stuttgart.

17. Neumann, U.: 1988, 'Die "Würde des Menschen" in der Diskussion um Gentechnologie und Befruchtungstechnologien', in Klug and Kriele (eds.), *Menschen und Bürgerrechte*, pp. 139–152.

18. Rademacher, Chr.: 1991, 'Zulässigkeit der Gen-Analyse?', *Neue Juristische Wochenschrift* **44**, 735–737.

19. Struck, G.: 1988, 'Die "Würde des Menschen" als Argument und Tabu in der Debatte zur Fertilisations- und Gentechnologie', in Klug and Kriele (eds.), *Menschen und Bürgerrechte*, pp. 110–118.

20. Vitzthum, W. Graf von: 1988, 'Gentechnologie und Menschenwürdeargument', in Klug and Kriele (eds.), *Menschen und Bürgerrechte*, pp. 119–138.

LUDGER HONNEFELDER

THE CONCEPT OF A PERSON IN MORAL PHILOSOPHY

What moral status befits human zygotes, embryos and foetuses? How should we treat the mentally severely retarded or patients in an irreversible coma, who are only surviving in a vegetative state? Is everyone who is a human being also a person? For a number of years, these or similar questions have formed the debate on medical ethics, at first only in the Anglo-Saxon countries, of late, however, also on the continent.[1] That such questions are asked is a result of the rapid developments which occurred in the medical sciences over the past decades. We have learned how human life develops at a molecular level, and we know how to handle the procedures and their underlying regularities in such a way that human embryos can be grown in large numbers *in vitro*. For the first time we also have the technical means to keep accident victims or severely ill people alive for an indeterminate amount of time albeit in a vegetative state. How should we use these new possibilities? Evidently, it is the quest for moral orientation, not liberal cynicism, that produces the above questions, even if the way in which they are put, and still more the answers that they are given, are often unconvincing. A survey of the literature shows that the way in which these questions are put is not in itself indicative of a particular moral position.

The concept of a person plays a central part in the debate. The reason is obvious: if the right to life and the right to have one's life protected are linked to the concept of a person, then any answer to the question "Who is a person?" determines what moral behaviour is obligatory. The consequences are not trivial: "If the foetus does not have the same claim to life as a person, it appears that the newborn baby does not either, and the life of a newborn baby is of less value than the life of a pig, a

K. Bayertz (ed.), Sanctity of Life and Human Dignity, 139–160.
© 1996 *Kluwer Academic Publishers. Printed in the Netherlands.*

dog, or a chimpanzee" ([42], pp. 122–123). And, as D. Parfit points out, if zygotes and foetuses count as persons, then every abortion is equivalent to the homicide of a defenceless person ([30], p. 322).

But, in this context, what does it mean to be a person? What are we inquiring about when, in moral contexts, we ask: who is a person? Is the choice of moral philosophy dependent on one's anthropology, or is it the other way around: does the status of a person depend on one's moral point of view? What is the relationship between ethics and metaphysics in this context? And how do we preserve the unity which we need in order to choose the right action when we are confronted with these fundamental moral problems, considering that there is a multitude of metaphysical and moral positions?

To gain some clarity we will proceed as follows: (i and ii) we will look back at how the question is dealt with in modern philosophy; (iii) we will consider the answers proffered in the recent debate and (iv) look at their metaphysical implications; (v) we will suggest a solution to the problem stated above. Our interest is not so much in the particular moral problems, but rather in what role the concept of a person plays in the context of the above problems, and what its use as a concept of moral philosophy suggests for the relationship between ethics and metaphysics.

I.

The word 'person' is derived from the Latin 'persona' which meant mask or role. Thus the origin of the word already hints at its predominant use in contexts concerned with the ascription of actions. In theology, 'person' signifies the way in which God appears to his people as acting. This is seen as a "*sub*-sistent relation", i.e. a relation which carries with it the notion of a *per*-sistent identity. Evidently, the question "Who is the acting subject?" implies the question "What is this Who?", i.e. "What is the acting subject's *nature*?" At least it is this connection, which Boethius establishes when he defines a person as an "individual substance of a rational nature" (naturae rationabilis individua substantia) ([3], c.III); this definition has influenced the entire debate about the concept. The persisting strength of the connection between who and what, between person and nature, in the ensuing debate is documented by Aquinas's observation that it is the "dominium sui actus",[2] the power over one's actions, which implies the special dignity of the individual spiritual nature and calls for the special label of a person.

In modern philosophy, the concept of a person gains its importance in the context of practical philosophy through J. Locke and I. Kant. Though Locke knows and mentions the connection between person and substantial nature, he believes that the concept of a person is burdened with too many theoretical problems for it to be of any use in answering unavoidable practical and forensic questions about the ascription of actions and responsibility ([24], III 11, 15, pp. 516–517). For him, a person "is a thinking intelligent Being, that has reason and reflection, and can consider it self as it self, the same thinking thing in different times and places; which it does only by that consciousness, which is inseparable from thinking, and as it seems to me essential to it . . . For since consciousness always accompanies thinking, and 'tis that, that makes every one to be, what he calls *self*; and thereby distinguishes himself from all other thinking things, in this alone consists *personal Identity*, i.e. the sameness of a rational Being" ([24], II 27,9, p. 335). The rational substance of Boethius's definition remains present, but it carries no weight for the solution of practical problems. "For it being the same consciousness that makes a Man be himself to himself, *personal Identity* depends on that only" ([24], II 10, p. 336). The identity of the acting subject, important for the problem of ascription, is guaranteed by the *identity of the mind*, which, through memory, re-identifies earlier actions as its own and thus can become the bearer of responsibility. Identity is established by a self-identification of the mind through time and thus consists in the continuity which the mind establishes. "And thus, by this consciousness, he finds himself to be the *same self*" ([24], II 25, p. 345). "*Person*, as I take it, is the name for this *self*" ([24], II 26, p. 346). Substance is transformed into consciousness of the self as subject; the place of the *unifying* function of the Aristotelian substantial form is taken by the *synthesising* function of the mind. The unity of the mind is no longer just a *trait* of the person but its *essence*.

According to Locke, this account explains the possibility of morality as a relation from actions onto laws, but does not establish an obligation for human beings to define such a relation at all. It is because of their self-interest – another very important notion – that they should do so; this self-interest is directed towards happiness, a happiness which lies in the future, as Locke can say in view of his belief in a final judgement ([24], II 22, p. 343 and 26, pp. 346–347). To sum up: according to Locke, consciousness, *memory and the interest in one's future happiness* justify the possibility and necessity of behaving morally.

Trying Locke's own criteria proves the limits of his account: it forces Locke to introduce such problematic notions as that of a "possible memory" ([24], II 27, 10–11, p. 336); and it is not metaphysically neutral, but instead replaces the rejected old metaphysics by a new, but no less problematic, dualistic one, which regards mind and body as two substances that, in a human being, form a peculiar union.[3] Moreover, Locke's concept of a person is not up to the practical task which he assigns to it: where he is concerned with the establishment of rights, such as the right to life as it applies to children or the mentally retarded, Locke does not refer to the concept of a person, but to that of a species, which, he claims, carries normative weight because of the associated notions of creation and of God's possession of his creatures (cf. [25]; cf. also [40]). Thus severely deformed people are members of an *intermediate species*, which is to be protected, because it is God's creation ([24], IV 4, 12ff, pp. 569ff). However, if only the relationship between a person and its nature, justified in theological terms, guarantees protection, e.g. in the case of children, what happens if the theological justification is no longer universally accepted? "If an interpretation suggests that human beings have their rights only when they are conscious, then life has become more dangerous" ([43], p. 112). The "theoretical emergency operation" [40] which Locke is forced to carry out for practical reasons casts doubt on the aim for which he undertook it.

Kant agrees with Locke that the concept of a person needs to be divorced from the problematic concept of substance, yet be retained in practical philosophy. But contrary to Locke he thinks that the "identity of a person does not follow from the identity of the I, conscious of all times, in which I recognise myself" ([17], A 365). However, the "concept of personality" as well as that of substance can (and must) be "retained" (ibid.) in practical philosophy. For without it, we cannot understand a human being as a being whose actions are attributable, because the being can relate them to laws. "He to whom actions can be imputed is called person; moral personality, man's independent individuality, is nothing else than the freedom of agent-intelligents, who rank under moral laws. Whence it is evident that a person is subjected to no law except such as he, either alone, or sometimes in conjunction with others, imposes on himself" ([20], p. 223; transl. 172).

Consequently, it is "personality", understood as the "ability" to follow the "special, pure practical laws, i.e. the laws given by one's own reason" ([18], A 155), that is: freedom, understood as the ability to bind oneself to the law, i.e. as autonomy, which constitutes a person. This, however, proves personality only as a practical idea, albeit as one which we must hold onto. For only if we can relate different and changing actions to the "personality" as a subject, which pursues a constant intention following moral standards, those actions can be accessible to judgement (cf. [40]). According to Kant, however, this assumption may not be made without certain metaphysical implications. Only if the empirical human being, the *homo phaenomenon*, participates in the "personality" as a moral idea, i.e. if it is a *homo noumenon*, or, as Kant puts it, if it is a member of the "world of reason", the intelligible world, only then is it possible to think that the "personality" in the human being can become the objective determining factor of its actions (cf. esp. [19], pp. 451ff and p. 436; also [36], pp. 240–241).

Thus, personality is not, as Locke thinks, the product of the specific human activity, but its prerequisite. ". . . , and the dignity of humanity consists just in its capacity of giving universal laws" ([19], p. 440). For, if the human being as a *homo noumenon*, i.e. as the free being capable of taking responsibility, is the "subject of all ends" and therefore an "end in itself" ([19], p. 431), then "this will ideally possible for us is the proper object of respect" ([19], p. 440), i.e. the proper object of respect is not just the fully moral person, but the human being which is capable of being moral, the being which stands *under* the moral law.

Thus Kant relates the personality of the human being, which is a necessary idea, to the nature of that being. "Beings whose existence does not depend on our will but on nature, if they are not rational beings, have only a relative worth as means and are therefore called 'things'; on the other hand, rational beings are designated 'persons', because their nature indicates that they are ends in themselves, i.e., things which may not be used merely as means" ([19], p. 428). However, just how precarious this reference to nature is, comes out when we ask about the respect for other persons and the extension of this respect. Because the respect for a person is related to the "ability to act morally" ([20], p. 441) Kant does not question that respect is due to mankind *as a kind*, and that, consequently, the human being is "obliged to acknowledge the dignity of mankind in each of the other human beings" ([20],

p. 462). However, according to Kant's theory of knowledge, the other human being is only ever given as a *homo phaenomenon*; besides, I cannot know whether I have to respect the other's ends, because I only know whether they are worthy of respect when I can take the standpoint of the first person. For Kant it is obvious that only the principle of *legality* can help in this situation: this principle says, that I have to respect the other's ends as long as they are legal; for legality is based on the principle of the end-in-itself (cf. [36], pp. 248–249). But what about the respect for those who have not yet reached the age of reason? Kant talks without hesitation of "children as persons", and he considers it "a good and even necessary idea for practical purposes to consider the act of begetting as one by which we bring a person into this world without that person's consent" ([20], pp. 280–281). The wording already suggests that Kant is operating with a concept, which, contrary to his own approach (cf. [40]), relates the person to the nature of a rational kind.

II.

It is noteworthy, that the *modern ideas of human dignity and human rights* still contain the two aspects, which Kant had introduced without being able to join them convincingly under his concept of a person: the inviolability of the human dignity and the extension of this dignity to each individual member of the kind.[4] The exposition of the concept of dignity, e.g. in the human rights of the freedom of thought and the freedom of conscience, shows that the basis of this concept is not a "speciesism",[5] i.e. not just membership in the biological species *homo sapiens*, but being a moral subject. On the other hand, being *human* is considered to be the criterion for being a moral subject and for having the requisite dignity. It is not without good reason then, that common language prefers the term 'human being' (which Kant also uses in this sense) over the term 'person'. Neither race, sex or other physical attributes, nor religion, moral acumen or intellectual power are the base of dignity, but solely being human. Consequently, it is part of the concept of human rights that they apply *before* all positive law; this includes the ban on limiting the ascription of dignity. It is obviously a proper characteristic of the concept of human rights that being a person and belonging to the human species are considered as inseparable. To put it in more detail: the irrefutably binding character of the human rights, which is based on the thought that the subject is an end-in-itself, and the universality of these rights, which is expressed in the ban on limiting their application, are inseparable.

Besides the idea that these two aspects of the concept of human rights are inseparable, it is also noteworthy, that the concept, including both aspects, has found world-wide recognition. Despite their contingent source: Western philosophy and the Christian-Jewish tradition, the ideas of human dignity and human rights have found recognition beyond all religious, cultural or moral boundaries, so that even those who do not comply with them are forced to justify or cover up their non-compliance. This phenomenon indicates that we need to distinguish three things with respect to human dignity and human rights: (i) their universally captured *cognitive core*, (ii) the many and varied *'deeper' motivating reasons* underlying this core, and (iii) the equally many and varied forms of the good life into which the core is embedded. That the notion of human rights de facto only appears in conjunction with such deeper reasons and embedded into rich forms of the good life, and that it may not be sustainable without them, is no argument for the claim that the core of that notion has no validity independently of them. Admittedly, the core of the notion needs to be applied narrowly, and the protection of the dignity is to be restricted to that minimum which is to be considered as a necessary condition for the possibility of acting as a moral subject at all. Would the core itself be regarded as a formula for the good life, it would become subjected to the controversy about the concrete forms of the good life. The objection, that the content of a valid notion decreases in proportion to the increase of its universality, is of no importance, because the function of the notion of dignity is not that of a source for the design of a concrete form of the good life, but that of a regulative principle which insures the condition of the possibility of many such designs. Consequently, an appeal to the notion of human dignity is especially convincing, where it uncovers basic infringements upon human rights, i.e. where it is intended to mark the border line.

The notion appealed to can be shown to be valid, because it states nothing more than the possibility to attribute actions and the responsibility for them, a possibility without which acting morally would become impossible. This, I think, is what gives the notion of dignity such world-wide plausibility. For any denial of the claim for dignity, if made with practical intent, presupposes the subject of all ends as an end-in-itself, while at the same time denying such a subject, to use Kantian terminology. To put it in Aristotelian terms: the denial amounts to the destruction of the practical reflexive relation which characterises human acting as acting. Recent moral philosophy would equate the denial with giving up the moral point of view, outside of which moral demands cannot be phrased, nor addressed, nor refused.

Even H.T. Engelhardt's approach, which, in view of a plurality of moral points of view as well as forms of the good life, remains restricted to a notion of "respect for persons" ([5], 43), which alone guarantees that an individual can follow his or her convictions; or C.F. Gethmann's [7] account, which argues for the reciprocal recognition of human dignity from the desire for peaceful conflict solution, are to be considered as proving, or at least supposing, if not the unconditional then at least the irrevocable validity of the core notion.

If we understand *person*, in the way set out above, as an expression of the practical ascription, which is necessarily contained in the reflexive practical relationship between a human being and itself, then *person* is a practical concept, i.e. a concept which is not culled from metaphysics and then applied in moral philosophy, but which has its own genuine practical meaning. In this sense it may be said that moral philosophy is independent of metaphysics, or that it precedes it. This leaves yet to be determined, however, whether the practical concept of a person can be retained as one which is indifferent to metaphysics, or whether it does not at least exclude certain metaphysical theories. To be rather brief: for Kant as well as for Aristotle the ascription of the concept of a person, or moral subject, cannot be retained, unless it is possible to distinguish between putative and true interests, and unless it is possible to suppose an interest in true self-being, i.e. in being a *homo noumenon*, or a rational and free being.[6] But is not this a "higher truth",[7] which cannot be retained under the conditions of plurality which prevail today? If this question has to be answered in the affirmative, we have to ask: whether and how, under these circumstances, the concept of a person and the moral claim attached to it, as they are expressed in the notion of human rights, can be retained at all.

III.

This sets the stage for the most recent phase of the debate about the role of the concept of a person in moral philosophy. How does the concept retain the twofold claim contained in the notion of human rights? And how does it relate to the concept of nature?

If one aims at avoiding all "higher truths", it seems to suggest itself to refer to the concept of interest, understood as reasonable interest, i.e. a concept that accords equal weight to equal interests and thus submits to a simple form of generalisation. However, in its classical *utilitarian version* this concept does not guarantee the dignity of the person. For, as J. Bentham states ([2], ch. XVII, p. 330), the reasons that count against the killing of a person are the same reasons that count against the killing of any other sentient being, i.e. they are the same reasons which generally count against any action which produces more suffering than enjoyment, more cost than benefit. But if the total net benefit, as understood by classical utilitarianism, is the decisive criterion, then two eminent moral claims contained in the notion of human rights cannot be adequately retained: the claim for distributive justice and the claim for self-determination. For, as J. Rawls has shown, the claim for justice presupposes that the parties involved in the claim are persons with a sense of justice, not just an idea of what is beneficial to them, and, as in the case of children, it is sufficient that they have this sense as a capacity ([35], pp. 27–33). The claim for self-determination on the other hand can only be retained, if cases such as "paternalist killings" ([9], p. 83), i.e. killings which are carried out in the long-term interest but against the will of the victim, can be excluded through the introduction of some additional condition. Glover suggests to add a principle of autonomy to the utilitarian theory. However, this would strip the theory of its point. It would be nothing more than a procedure, secondary to the principle of autonomy, by which we could evaluate the consequences of our actions. Such a procedure, by the way, is a useful and necessary part of any Aristotelian theory, too.

To retain the utilitarian approach, as P. Singer and others do, yet avoid the counter-intuitive consequences of its classical version, it is not sufficient to modify the approach; we need to introduce the concept of a person to evaluate the prohibition of paternalistic killings.

This means: we have to compare the preferences, not just the actual interests which the parties involved pursue; for preferences include future interests ([42], 78–81). Thus it becomes possible to justify the prohibition of killing without reference to mere membership of the biological species *homo sapiens*, i.e. without arguing from the standpoint of speciesism, or without recourse to "higher truths". For if it is possible to suppose that persons differ from other sentient beings, because they have preferences about their own future, or prefer to continue living, or have the capacity to desire to go on living, or are capable of choosing to live or die, or have a coherent conception of death ([42], p. 81, p. 97 and p. 131), then the killing is morally wrong and prohibited, because it goes against these preferences and there are no further preferences to balance the action.

The dignity, as expressed in the prohibition of killing, is justified by way of according the status of a person, a status which is equated with or at least made dependent upon the possession of certain characteristics. From this, however, Singer draws the inference that there are human beings, such as embryos, imbeciles and irreversibly comatose patients, who are not persons, because they do not have the necessary characteristics; on the other hand, there are members of certain species of primates, perhaps even dolphins, whales or pigs, that have to be regarded as persons, though they are not human beings, because they possess those characteristics mentioned above, at least in some sense ([42], pp. 93–96). But if, because of the lack of such characteristics, embryos have no "intrinsic value" ([42], p. 118), then their interests are no more or no less to be respected than those of other similarly sentient beings.

However, as M. Tooley saw (cf. [46], pp. 121ff), when he tried to justify the prohibition of killing by reference to a being's long-term wishes and interests, this avoids counter-intuitive consequences such as exceptions from the prohibition of killing for people that are asleep, temporarily emotionally disturbed, or indoctrinated, only if the characteristics mentioned above are combined with the notion of diachronic identity. Only subjects which are conscious of themselves through time can have interests, which are not tied to a particular moment in time, and which thus, according to the utilitarian approach, justify the prohibition of killing.

Diachronic identity can be understood in two ways: as the *persistence of an event* or as the *persistence of a thing*. In the first case, D. Lewis speaks of "perdurance"; in the second of "endurance" (cf. [23], pp. 202ff; cf. also [37]). If one sees diachronic identity of persons in a Lockean manner as the unity of events, i.e. as the unity of person phases, as Parfit does more consistently than anyone else, then the identity of a person remains nothing but a relationship between physical, and even more between psychological, events or phases, which are brought into a relationship of continuity by strong chains of "psychological connectedness", especially in the form of memory (cf. [30], pp. 199–209). A person, according to Parfit, is this psychological continuity, and we do not need some "deep further fact" ([30], p. 323) such as the assumption of a persisting entity.

IV.

Important in this context is Parfit's convincing claim that this *reductionist view of the person*, as a psychological unity of person phases, alone makes the perspective of preferential utilitarianism consistent (cf. [30], pp. 321–347). For if the unity of a life is less strong, i.e. only that of a stronger or weaker continuity of phases, then distributive justice can be related more strictly to the phases and their connections rather than the unity of the life in question. As distributive justice loses its weight, so does the charge, often levelled at utilitarianism, that it reduces the individual to mere number; utilitarianism becomes "impersonal" ([30], p. 336) as it were. At the same time, the reductionist position allows us to homogenise in such a way that the different states of a human being can, without inconsistency, be considered as grades of the connectedness which constitutes the person, and consequently be subjected to the type of evaluation suggested by preferential utilitarianism. "When we cease to believe that persons are separately existing entities, the Utilitarian view becomes more plausible" ([30], p. 342). Foetuses, mentally severely handicapped, and irreversibly comatose patients are then justifiably not regarded as persons. And a human being's responsibility goes as far as the identity provided by his or her memory.

If Parfit's claim about utilitarian ethics and its connection with the reductionist concept of a person is correct, and if the criticism of the concept of potentiality, i.e. the notion that human beings which potentially have the traits that make a person worthy of protection therefore themselves have the right to be protected, is also correct, which is yet to be shown, then the concept of a person in preferential utilitarianism is anything but indifferent to metaphysical theory. On the contrary, it presupposes a metaphysics which has little theoretical or practical plausibility.

What the reductionist concept of a person presupposes is W.V.O. Quine's monist ontology, which is based on R. Carnap's reception of H. Minkowski's four-dimensional space-time-system (cf. [33, 4, 29];

cf. also [37, 38]). According to this ontology, things are reducible to events which persist in time; events themselves can be seen as the tracks of world-points through time, and it is up to our practical interests how we arrange the time units. As the debate has shown [37, 38], such an ontology meets with considerable theoretical and practical difficulties. The most important objection is that which M. Lockwood levels at the Warnock Report ([26], [27], p. 200): that any description of the world given by this ontology is incompatible with the one we presuppose in our natural language ([cf. [33, 4, 29, 37, 38]). For if P.F. Strawson and S.A. Kripke [44, 21] are right in suggesting that names of persons and natural kinds have rigid reference, then, in the case of persons and kinds, we have to presuppose something like entities which persist, even if a large number of their traits change over time. According to this linguistic practice, Aristotle would remain Aristotle, even had he not been born in Stageira and not been Plato's disciple. For any event-based ontology, however, this second Aristotle would be a different one, a "counter-part" (cf. [23], pp. 202ff) of the first, so to speak. But especially in the case of persons, we do suppose that they remain the same, even if they change; that they could live longer or shorter than they in fact do, without ceasing to be the same person. We ask, when the life of a person began and when it ceased; we do not ask, when the person began or when s/he ceased. For an event-based ontology it is the person which persists, and an Aristotle who only lasted to live a further two hours, would have been a different person from the one we know.

There are also some *practical* phenomena which render an event-based ontology problematic. Fear of future suffering is not fear of a *phase* of suffering, but fear of the fact that I *myself* will suffer. And the possibility that there might be persons which come in two-hour-chunks, or which are combinations of formerly alien pieces, violates our common view, which sees persons as beings which call certain feelings *theirs*, which go through developments and have a certain biography, which can be the subjects of blame, and which can be held responsible. One case in particular becomes implausible, a case which counts as the paradigm of human dignity and the freedom of conscience which flows from it: the conscientious decision, e.g. to accept death rather than give up one's personal identity (cf. [13], p. 31). If we add, that the notion of diachronic identity, which replaces *things* with another type of entity, i.e. *phases*, avoids a vicious circle when explaining, how phases can constitute a unity, only by making the constitution of this unity a matter of convention, then we have to agree with P. Simons and E. Runggaldier [41, 37, 38] who place the onus of proof squarely on this re-interpretation of our natural language and our everyday experience.

V.

It is precisely the concern to survive, to which Singer and Parfit refer, which not only stands against the concept of a person presupposed by them, but which demands a different concept altogether. Obviously, my wish to survive cannot be interpreted as the wish that there might be another phase to which I am psychologically related; it can only be interpreted as the wish that *I* survive [49, 50]. Not *whether* I survive, is of interest to me, but *as what* I survive, i.e. as the one who I am. It is *my individual biography*, out of which *my* concern to survive evolves, which cannot be sufficiently explained by Parfit's "pure mental connectedness". The memory which establishes the connection between different mental states is a necessary, but obviously not a sufficient, condition of personal identity. D. Wiggins therefore plausibly claims that the concern to survive *necessarily* entails the concern to keep one's identity (cf. [49], pp. 303–305). Only if I am concerned about myself and my own identity, will I be concerned to survive.

What is true of the practical dimension is also true of the theoretical: conceptual thinking and judging demand the assumption of a subject, to which the thoughts can be attributed, and which can comment on its thoughts and judgements [8], i.e. – to use Kantian terms – they demand the "I think" which accompanies all thought as the point of reference, which gives unity, and without which our theoretical activities cannot be explained. But how can this "I", as which we identify ourselves, as which we are identified by others, and for whose identity we are concerned, be thought of, if it is insufficient to interpret its identity as a "bundle" [8] of phases?

Again we can take our lead from natural language: persons are not just entities to which we refer using words such as proper names or sortals, e.g. 'human being', which refer rigidly and thus presuppose a "continuant"; they also have the feature that we can apply P(erson)-predicates (such as the ascription of actions, intentions, thoughts, emotions and memories) as well as M(atter)-predicates (such as the ascription of bodily features and physical situations) to one and the same individual, as Strawson has shown (cf. [44], pp. 104–112). But if

P-predicates can only be applied to the very same subjects as certain physical predicates, i.e. if P-predicates are necessarily matter-involving (cf. [50], p. 66), then it is clear that a mentalist as well as a physicalist account of persons entails the dissolution of a basic unity, which expresses itself in the fact that the two types of predicates are related to each other and can be applied without distinction. Consequently, Strawson calls *person* a basic concept, which cannot be defined by a combination of more basic concepts such as *consciousness* and *body* (cf. [44], p. 112).

If we stick to our natural language, then 'person' appears to be an expression for the designation of a living being to which we can apply an open list of P-predicates, as well as implied M-predicates; to use Wiggins's words: we consider a person to be a living being "that we have no option but to account as a subject of consciousness (or potentially such) and as an object of reciprocity and interpretation" (cf. [50], p. 69). This means, however, that the terms 'person' and 'human being' have the same reference, and, what is more, mutually interpret each other. It also means, that we cannot understand a human being as the subject of ends without recognising him or her as a person.

This also provides the basis for making sense of the notion of a *potential person*. If we start from the idea that the application of the term 'person' designates a continuant, then it makes a fundamental difference, whether we talk of the possibility of a being coming into existence which has or might have the traits of a person, or whether we talk about the possibility of an existing being developing the traits of a person. Medieval Aristotelians express this by distinguishing between *potentia obiectiva*, a possibility of thought, and *potentia subiectiva*, a being's inherent possibility to develop further traits or actions. In English we might perhaps assign the same distinction to the terms 'ability' and 'capability' (cf. [51], p. 25). No unsolvable "problem of demarcating" ([22], pp. 83ff) the possibility of the sperm becoming a person from the possibility of the zygote becoming a person arises, if we acknowledge the above distinction, unless, of course, we adhere to an event-based concept of a person. Obviously, beyond its mere existence certain conditions have to be met for a foetus to become what it potentially is. But that does not mean, that the sperm or ova, which under certain conditions can also become persons, have the same potentiality, so that non-begetting had to be considered equally reprehensible as abortion. Not to start a process of life which might lead to a person is only "morally symmetrical" ([46],

p. 186) to the disruption of that process, if we do not assume a "continuant" for that process, i.e. if we adhere to an event-based notion of personal identity. It is therefore inconsistent, on the one hand to deny the symmetry, and on the other hand to see abortion not as violating an interest to survive but as prohibiting the *formation* of such an interest, as N. Hoerster does. Such an argument also presupposes that it is meaningful to distinguish a being, before it develops an interest to survive, from that same being after it has done so, a distinction which works only if we presuppose an event-based notion of diachronic identity. If, on the other hand, we presuppose an ontological sameness of the actual person, and if we consider this sameness as an inseparable unity of mind and body, of P-predicates and M-predicates, then it makes sense, and it is indeed imperative, to treat human beings as potential persons, even if they do not yet have the actual traits of a person in the sense of an acting subject to which moral considerations apply, as long as they have the potential to them. As R. Hare points out [10], it is a mandatory generalisation from my interest not to have been aborted to the interest of *all* later persons. However, to think, as Hare does, that this does not exclude that one potential person might be replaced by any other, is only consistent under the assumption of an event-based ontology. It is right to relate the inception of my life to *myself*, and it is also right to regard myself as well as my life as irreplaceable.

Even Engelhardt's account does not succeed in justifying a person's right to protection and its limitations. Following Kant, he says that the respect for persons applies to those human beings which actually have the traits of a moral subject. Other human beings such as children and mentally severely handicapped people (but not zygotes and foetuses) gain the same protection because of social considerations, i.e. relative to persons in a strict sense (cf. [5], pp. 104–127). De facto, however, those persons in a wider, social sense, listed by Engelhardt, are just those which make up the remainder of the species *homo sapiens*. Put in this way, the criterion for the extension of the protection for persons is either the purely biological taxonomy, or the social status, or the more or less chance decision of the persons in a strict sense, or some other unexplained relation from the concept of a person to the notion of nature. That the assumption of a soul constitutes a metaphysical position which is incapable of consensus, is no reason to deny the diachronic identity of a continuing entity, as it is presupposed in our natural language, and as which we see ourselves as subjects of actions. The inherent problems of Engelhardt's account are also apparent in his evaluation of the argument from potentiality: you can consistently assign the foetus a "potentiality to become" and the newborn a "potentiality of", only if you adhere to an event-based concept of a person, which Engelhardt does not.[8]

If we take the diachronic identity of the person to be the endurance of a continuant, we are facing the question, how far we can backdate our personal identity. Under the circumstances described, how far can and must we extend the ascription of the status of a person? The scientific results suggest two things: the human life of a person can be continuously traced back to a beginning, which can be equated with the fusion of ovum and sperm and the formation of a new and individual genome. Every cell of the resulting zygote can at this early stage which is totally open to any further development, i.e. up to the end of implantation, divide into several cells, each of which will develop into a new individual; in rare cases, however, it can also recombine with another, different cell to become a single individual.

What does this mean for the notion of the beginning of the life of a person? The scientific results do not constitute an answer in themselves; they require philosophical interpretation. This is already clear from the fact that the beginning (just as much as the end of life) is not a *point* in time, but a *process* which gains its unity from a teleological perspective. If we suppose that the term 'person' can only be applied to an ontological individual which is undivided and indivisible, then the term can be applied to a zygote, only if the zygote is an individual human being. The possibility of twin formation must then be interpreted as follows: (i) either as the emergence of another individual from a given first, which we regard as a kind of parent cell; but this interpretation meets with the difficulty of how to tell which of the two carries on the original identity; (ii) or as the beginning of two new human individuals, which follow in the steps of the original old one which perished; and this interpretation faces the problem of giving an adequate explanation of the indisputable genetic continuity from the perished individual to the two new beginning ones. Some authors have concluded from these difficulties that we need to distinguish between genetic uniqueness and ontological individuality of the human being, and that the zygote must be considered as genetically unique but not as an individual human being, until the end of the restriction process, i.e. until the formation of the so-called primitive strip, which makes the developing individual appearance visible; and they conclude from this that the existence of the zygote up to implantation constitutes something like a pre-personal phase of the human life, a potential human individual in the sense of having the potential to become such an individual.[9] Several others have argued against this, that the process which begins with conception knows only different levels of development, but no breaks that interrupt the process which starts with the fusion of the two sets of chromosomes, so that genetic uniqueness and numerical individuality are the same and twin formation is a kind of proliferation (cf. [34]).

Whichever way we interpret the scientific results, it is a fact that we are talking about a process, and that the claim for protection is irrefutable inasmuch as this process marks the beginning of a human individual, which has the active potentiality to form the actual traits of a person. It may be asked, though, whether doubts about the *interpretation of scientific results* can have an effect on the *practical ascription* of the term 'person' and therefore on the claim for protection at all. Do we not have to follow an argument which turns on the *benefit of the doubt* – as even Singer knows it albeit not applied to zygotes but to animals (cf. [42]) – i.e. act according to the rule: *idem est in moralibus facere et exponere se periculo faciendi*, which prohibits the hunter from shooting after noticing a movement of which he is not sure, whether it was caused by a human or an animal. If we assume that this rule is valid, which, I think, we may, because we are talking about the protection of a fundamental, though not the highest, good, which is even protected by a prohibition to apply to it a wantonly limiting definition, then we have to choose the *safe way*, i.e. treat the zygote *as* a person, notwithstanding a final explanation for all the questions relating to the beginning of the life of an individual human being. But would not a *probabilistic solution* also be legitimate, as other authors argue? It claims, that in cases of doubt about the applicability of a rule we may choose the less binding alternative as long as we can support that choice with good reasons. Is this not especially the case, when the acting subject (different to the hunter) finds itself in a situation of distress, and one claim has to be weighted against the other (cf. [28, 45])?

This much should suffice in terms of comments on this concrete part of the debate about *the concept of a person in moral philosophy*. For I am here not concerned with the problem of abortion but with the practical concept of a person. I therefore proceed with a summary, which in view of the continuing debate will needs be of a temporary nature.

Person, in a moral context, is a practical ascription by which I identify myself and others as a moral subject. This identification implies the respect for the dignity of the moral subject, i.e. an acknowledgement of its inviolability. The immediate grounds for this inviolability are not certain biological or metaphysical traits, but its status as a subject, capable of determining its own ends and of taking responsibility for them, i.e. its freedom as the condition of the possibility for binding itself to the good. Because being such a subject is part of every moral demand, moral philosophy does not depend on prior metaphysical insight.

The question, how far the acknowledgement of personhood can be extended, is also primarily a question of one's moral point of view. If we only recognise those as persons that are actually able to reciprocate, we may be able to explain morality as fairness and justify a basic level of generalisation, but we will not be able – as G. Patzig rightly argues against Hare (cf. [31, 32]) – to justify morality as solidarity. Solidarity which comprises even the weakest members of society implies a level of generalisation which shuns all special attributes and ascribes inviolability only to the human being as a human being.

This makes it clear that the dignity of the person needs to be related to the notion of specific nature, if the two aspects of the notion of human rights: inviolability and the prohibition to restrict its application, are to be retained. Every interpretation of the concept of a person which restricts the reference to nature is therefore taking the onus of proof. This applies in particular to accounts such as that of Singer, Parfit and others, which presuppose a metaphysical concept which relies on highly problematic premises and is incapable of re-interpreting the reflexive relationship of the moral point of view.

We are therefore obliged to hold on to the concept of a person handed down to us through our natural language, especially in situations when we ascribe responsibility to ourselves and to others. This is where we encounter the concept of a person in its original form: it designates a human individual to which an open-ended list of mental and material predicates, which mutually imply each other, can be applied. This shows an original relation from the concept of a person to the notion of nature, which makes it justifiable, and indeed mandatory, to respect as a person any individual which has the active potency to develop the traits which justify a human individual's claim for personal protection.

However, as the problems relating to the determination of the beginning of the life of a human individual show, the dispute about the concept of a person in applied ethics becomes at the same time a *challenge to philosophy*, i.e. a challenge to deliver what Aristotle and the medieval philosophers in his wake achieved by taking recourse to the scientific results of their times, and what is left open by the modern concept of a person, which implies dignity: to explain convincingly the relation from the concept of a person to the notion of nature which our practical ascriptions indicate. It seems plain that in the debate about how we avail ourselves of the new range of options to act, we will in the near future determine how we see ourselves.

Department of Philosophy, University of Bonn
Institute for Science and Ethics
Bonn
Germany

NOTES

[1] The German public was first confronted with these issues and the controversy surrounding them, when organisations for the rights and welfare of handicapped persons disrupted lectures to be held by the Australian ethicist Peter Singer.

[2] ([1], I 29,1); cf. also ([1], I-II prol.), where rationality and free will as well as being the initiator of one's own actions are listed as the reasons, why a human being is to be considered as an image of God.

[3] Unfortunately, it would go far beyond the scope of this paper to formulate this criticism in greater detail.

[4] Cf. for a more detailed discussion [14, 15, 16].

[5] For the use of this term cf. [42], also [12], pp. 55–69.

[6] For more detail cf. [16], pp. 36ff.

[7] For this expression, cf. [47], p. 156 and [48], pp. 371ff.

[8] Cf. [5], pp. 110 ff; also the discussion of Engelhardt's Foundations by K. Hartmann [11].

[9] Thus not only the Warnock Report, but also several Catholic moral theologians such as [6, 39, 28].

BIBLIOGRAPHY

1. Aquinas, Th.: 1948, *Summa Theologica*, Christian Classics, Westminster, Maryland.
2. Bentham, J.: 1961, *An Introduction to the Principles of Morals and Legislation* (1789/1823), Doubleday, Garden City, N.Y.

3. Boethius, A.M.S.: 1988, 'Contra Eutychen et Nestorium', in Boethius, *Die Theologischen Traktate*, Meiner, Hamburg.
4. Carnap, R.: 1928, *Der logische Aufbau der Welt*, Meiner, Hamburg.
5. Engelhardt, H.T., Jr.: 1986, *Foundations of Bioethics*, Oxford University Press, New York/Oxford.
6. Ford, N.A.: 1988, *When Did I Begin? Conception of the Human Individual in History, Philosophy and Science*, Cambridge University Press, Cambridge.
7. Gethmann, C.F.: 1993, 'Lebensweltliche Präsuppositionen praktischer Subjektivität. Zu einem Grundproblem der "angewandten Ethik" ', in H.M. Baumgartner and W. Jacobs (eds.), *Philosophie der Subjektivität?*, Frommann-Holzboog, Stuttgart, pp.150–170.
8. Gillett, G.: 1987, 'Reasoning about persons', in A. Peacocke and G. Gillett (eds.), *Persons and Personality. A Contemporary Inquiry*, Basil Blackwell, Oxford , pp. 75–88.
9. Glover, J.: 1977, *Causing Deaths and Saving Lives*, Penguin Books, Harmondsworth.
10. Hare, R.: 1988, 'When does potentiality count? a comment on Lockwood', *Bioethics* **2** (1988), 214–223.
11. Hartmann, K.: 1989,'The foundations of bioethics', *Journal of the British Society for Phenomenology* **20**, 166–169.
12. Hoerster, N.: 1991, *Abtreibung im säkularen Staat. Argumente gegen den §218*, Suhrkamp, Frankfurt/Main.
13. Honnefelder, L.: 1982, 'Praktische Vernunft und Gewissen', in A. Hertz *et al.* (eds.), *Handbuch der christlichen Ethik*, vol. 3, Herder, Freiburg, pp. 19–43.
14. Honnefelder, L.: 1986, 'Die Begründbarkeit des Ethischen und die Einheit der Menschheit', in G.W. Hunold and W. Korff (eds.), *Die Welt für morgen. Ethische Herausforderungen im Anspruch der Zukunft*, Kösel, München, pp. 315–327.
15. Honnefelder, L.: 1987, 'Menschenwürde und Menschenrechte. Christlicher Glaube und die sittliche Substanz des Staates', in K.W. Hempfer and A. Schwan (eds.), *Grundlagen der politischen Kultur des Westens*, de Gruyter, Berlin, pp. 239–264.
16. Honnefelder, L.: 1992, 'Person und Menschenwürde. Zum Verhältnis von Metaphysik und Ethik bei der Begründung sittlicher Werte', in G. Mertens and W. Kluxen (eds.), *Markierungen der Humanität. Sozialethische Herausforderungen auf dem Weg in ein neues Jahrtausend*, Schöningh, Paderborn *etc.*, pp. 29–46.
17. Kant, I.: 1903–11, *Critique of Pure Reason*, Academy-Edition, vol. 4, Reimer, Berlin.
18. Kant, I.: 1903–11, *Critique of Practical Reason*, Academy-Edition, vol. 5, Reimer, Berlin.
19. Kant, I.: 1911, *Foundation of the Metaphysic of Morals*, Academy-Edition, vol. 4, Reimer, Berlin, pp. 385–463.
20. Kant, I.: 1903–11, *Metaphysic of Morals*, Academy-Edition, vol. 6, Reimer, Berlin (trl. as Metaphysic of Ethics by J.W. Semple, Edinburgh 1871).
21. Kripke, S.A.: 1980, *Naming and Necessity*, Basil Blackwell, Oxford.
22. Leist, A.: 1990, *Eine Frage des Lebens. Ethik der Abtreibung und der künstlichen Befruchtung*, Campus, Frankfurt/M.
23. Lewis, D.: 1986, *On the Plurality of Worlds*, Blackwell, Oxford.
24. Locke, J.: 1975, *An Essay Concerning Human Understanding*, ed. by P.H. Nidditch, Clarendon Press, Oxford.

25. Locke, J.: 1963, *Two Treatises of Government*, ed. P. Laslett, Cambridge University Press, Cambridge
26. Lockwood, M.: 1985, 'The Warnock report: a philosophical appraisal', in Lockwood (ed.), *Moral Dilemmas in Modern Medicine*, Oxford University Press, Oxford, pp. 155–186.
27. Lockwood, M.: 1988, 'Warnock versus Powell (and Harradine): when does potentiality count?', *Bioethics* **2**, 187–213.
28. McCormick, R.A.: 1991, 'Who or what is the pre-embryo?', *Kennedy Institute of Ethics Journal* **1**, 1–14.
29. Minkowski, H.: 1911, 'Raum und Zeit', in *Gesammelte Abhandlungen*, vol. 2, Teubner, Leipzig and Berlin.
30. Parfit, D.: 1984, *Reasons and Persons*, Oxford University Press, Oxford.
31. Patzig, G.: 1983, *Ökologische Ethik – innerhalb der Grenzen blosser Vernunft*, Vandenhoeck & Ruprecht, Göttingen.
32. Patzig, G.: 1978, *Der Unterschied zwischen subjektiven und objektiven Interessen und seine Bedeutung für die Ethik*, Vandenhoeck & Ruprecht, Göttingen.
33. Quine, W.V.O.: 1960, *Word and Object*, MIT Press, Cambridge, Mass.
34. Rager, G.: 1992, 'Individualität und Personalität des Ungeborenen', in D. Berg (ed.), *Würde, Recht und Anspruch des Ungeborenen*, München, pp. 82–101.
35. Rawls, J.: 1971, *A Theory of Justice*, Harvard University Press, Cambridge, Mass.
36. Ricken, F.: 1989, 'Homo noumenon und homo phaenomenon. Ableitung, Begründung und Anwendbarkeit der Formel von der Menschheit als Zweck an sich selbst', in O. Höffe (ed.), *Grundlegung zur Metaphysik der Sitten. Ein kooperativer Kommentar*, Klostermann, Frankfurt/Main, pp. 234–252.
37. Runggaldier, E.: 1992, 'Personen und diachronische Identität', *Philosophisches Jahrbuch* **99**, 262–286.
38. Runggaldier, E.: 1988, 'Zur empiristischen Deutung der Identität von Personen als Kontinuität', *Theologie und Philosophie* **63**, 242–251.
39. Shannon, Th.A. and A.B. Wolter: 1990, 'Reflections on the moral status of the pre-embryo', *Theological Studies* **51**, 603–626.
40. Siep, L.: 1992, 'Personbegriff und praktische Philosophie bei Locke, Kant und Hegel', in L. Siep, *Praktische Philosophie im deutschen Idealismus*, Suhrkamp, Frankfurt/M., pp. 81–115.
41. Simons, P.: 1987, *Parts. A Study in Ontology*, Clarendon Press, Oxford.
42. Singer, P.: 1979, *Practical Ethics*, Cambridge University Press, Cambridge.
43. Specht, R.: 1989, *John Locke*, Beck, München.
44. Strawson, P.F.: 1961, *Individuals*, Methuen & Co., London.
45. Tauer, C.A.: 1984, 'The tradition of probabilism and the moral status of the early embryo', *Theological Studies* **45**, 3–33.
46. Tooley, M.: 1985, *Abortion and Infanticide*, 2nd ed., Clarendon Press, Oxford.
47. Tugendhat, E.: 1984, *Probleme der Ethik*, Reclam, Stuttgart.
48. Tugendhat, E.: 1992, 'Die Hilflosigkeit der Philosophen angesichts der moralischen Schwierigkeiten von heute', in *Philosophische Aufsätze*, Suhrkamp, Frankfurt/Main, 371–382.
49. Wiggins, D.: 1987, 'The concern to survive', in D. Wiggins (ed.), *Needs, Values, Truth. Essays in the Philosophy of Value*, B. Blackwell, Oxford.

50. Wiggins, D.: 1987, 'The person as object of science, as subject of experience, and as locus of value', in A. Peacocke and G. Gillett (eds.), *Persons and Personality. A Contemporary Inquiry*, B. Blackwell, Oxford, pp. 56–74.
51. Wolbert, W.: 1989, 'Wann ist der Mensch ein Mensch? Zur Frage nach dem Beginn und Ende personalen Lebens', in V. Eid *et al.* (eds.), *Moraltheologisches Jahrbuch 1 – Bioethische Probleme*, Grünewald, Mainz, pp. 15–33.

MARY C. RAWLINSON

ALTERITY AND JUDGMENT: SOME MORAL IMPLICATIONS OF HEGEL'S CONCEPT OF LIFE

Moral philosophers have not looked often to the concept of life as a foundation for moral judgment and moral respect. Greek ethics, focusing on education and the formation of character, treat life as a more or less promising raw material, to be properly reshaped through the application of appropriate pedagogical disciplines. As Christianity develops the conceptual opposition between a sensible and an intelligible world, life appears, not only raw, but also evil. In addition to disciplines of formation or education, disciplines of redemption and salvation proliferate.[1] And, despite some attention to the role of actual sentiment or feeling in supplying both force and direction for moral judgment, notably by Montaigne and Hume, the tradition culminates in Kant's celebrated alienation of reason and feeling: invoking "mutual love" as one of the two indispensible "great moral forces", Kant is careful to point out that this love is "not to be taken as a feeling (aesthetic love), i.e., a pleasure in the perfection of other men; it does not mean emotional love . . . It must rather be taken as a maxim of benevolence" ([5], sec. 25). Thus, respect is not a "feeling" dependent upon actual comparisons among living beings, but a maxim derived analytically from the idea of freedom. The domain of actual life, where moral judgment and moral respect will be deployed, may contribute nothing to their determination. All that can be thought in the concept of life is explicitly put out of play.

In an effort to determine the grounds for the attribution of moral responsibility, as well as the scope of the duty of respect, moral philosophy, rather, turns toward metaphysical definitions of the person. As Hegel argues, this concept arises only in relation to the concept of law: the law invokes an "equality in which all count the same, i.e., as *persons*"

K. Bayertz (ed.), Sanctity of Life and Human Dignity, 161–175.
© 1996 *Kluwer Academic Publishers. Printed in the Netherlands.*

([2], sec. 477). Hegel proves unusual in the tradition in that he explicitly rejects its alienation of life and reason. While valorizing the concept of contract and locating in the State and its bureaucracy the rationally regulated space of mutual recognition, Hegel, nevertheless, presents the domain of life – understood as a domain of action or social practice – as the origin and context of the idea of duty. Moreover, in an analysis of the figure of Antigone, Hegel identifies the apodicticity of duty with the facticity of our blood-relationships, and develops an account of law as delimited by the mutual recognition achieved in an act of forgiveness. This essay outlines Hegel's demonstration of the inadequacy of any moral philosophy that understands itself as deriving from an abstract respect for persons, and it attempts to develop the positive implications of his focus on the domain of life and the concrete recognition of the other in the act of forgiveness. Explicating Terry Pinkard's concept of "sustainability", the essay demonstrates how for Hegel social practices or forms of life prove capable of generating, not only their own norms, but the critique of those norms as well [6]. This "sustainability" depends, as Pinkard argues, on the reciprocity or mutual recognition among the participants in a given practice. Finally, then, the essay questions the mutuality of the recognition described by Hegel, insofar as the subject of recognition is in Hegel's analysis always marked masculine.

I. THE PERSON AND CONSCIENCE

For the most part moral philosophies in the Western tradition base respect for persons, not on the facticity of life, but on the capacity to reason. This theoretical or conceptual gesture itself – the attempt to determine the principles of moral judgment or to establish the scope of moral value and the respect it commands or to attribute responsibility or obligation on the basis of a metaphysical definition of the person as a specific set of capacities, qualities, or properties – is not without moral content. This act itself must be evaluated from the moral point of view: when metaphysical definitions are used in this way, their effects are invariably exclusionary. The definition makes it possible to exclude in virtue of some lack of capacity or quality some set of entities from the sphere of persons entitled to moral respect.

An entity deemed lacking in the power of reason cannot make moral judgments, nor does it enjoy any entitlement to moral respect. It may be an object of other forms of respect, but it is not recognized as an other by the agent of moral judgment, i.e., the moral agent does not recognize in the other his own capacity to rationally evaluate alternatives or to give himself his own law. The unrational entity appears instead as an object to be manipulated "for its own good".[2]

Consider, for example, the results of these arguments in the diseases of drapetomania and female invalidism from the not-too-distant nosological history of medicine. The slave, essentially different from the white man in his lack of an adequate self-regulating rational capacity, must be returned to the master "for his own good"; like a child he must be taught obedience and reliance on those of superior development, lest he develop a mind diseased. And, authorizing the prescription of clitoridectomy were judgments about a woman's rational and intellectual faculties and the relation of those to her reproductive function. Different from man, she was, therefore, to be governed by different rules. Supposedly descriptive concepts of who or what an entity is, thus, generate prescriptive concepts authorising a multiplicity of more or less violent practices upon the living bodies of actual individuals. Thus, in basing moral agency and moral respect on an essentialist definition of the person, moral philosophers, at the same time, invoke concepts of the others who are not like me, the alien others whom I – and the others like me in whom I recognize myself – am obliged to manage and manipulate. Basing moral judgment and moral respect on this essentialism necessarily opens moral agency to moral harm.

Moreover, the concept of the person, as elaborated in the history of philosophy, reflects, not the in-itself of some entity, rather its content is articulated and developed in a set of well-specified problems: the problem of personal identity or identity through time, the mind-body problem, and the problem of the knowledge of other minds. Each of these problems emerges from the abyss of an irresolvable *aporia*. With respect to each problem a duplicity appears to mark the person. With respect to the problem of personal identity: On the one hand, the unity of a single experience must be assumed in order to account for the givenness of objective unities; on the other, the self has no intuition of itself as a unity, and empirical subjectivity, always differing from itself, seems doomed to a selfhood that is no more than a useful metaphysical fiction or the mere result of the mental habit of associating contiguous impressions. With respect to the mind-body problem: On the one hand, experience clearly depends on certain neurophysiological conditions,; on the other, the integrity of the formal structures and determinations of experience is transgressed by any reductive gesture. With respect to the problem of the knowledge of other minds: On the one hand, the particular self-consciousness seems, as Hegel argues, to depend on a relation to the other for its own self-constitution; on the other hand, the impossibility of any immmediate intuition of the other relegates the other to the status of an hypothesis, an existence established only by analogy. Any attempt to define personhood in terms of a set of qualities, capacities, or attributes simply misunderstands the function of the concept, which is not to provide a representation of some in-itself, but to articulate the unity of a field of differences and distinctions , which, worked-through and lived, comprise the being of the person. Certainly, metaphysical definitions of the person command no consensus, and, thus, yeild no adequate foundation for the binding force of the moral ought. This is not because of any contingent and ameliorable incoherence of definitions, but because the form of the metaphysical definition, the attempt to render the person as a set of essential properties, violates the reality of the person as an existential situation, an experience of a complex of undecidables.

From the essentialist definition of the person as *ens rationis* there follows an determination of moral agency as conscience or rational self-judgment according to a law recognized as one's own or under the hypothesis of freedom. The man of conscience believes in himself, in the clarity of his intentions and the veracity of his self-evaluation. He has a "beautiful soul". As an intrapsychic experience, governed by the apodiciticity and immediacy of the 'I am' or self-consciousness, conscience is unmediated by the alterity of any actual other.[3] The self-certainty of conscience, as Hegel argues, falls short as a paradigm of moral experience precisely insofar as its immediacy is not supplemented by a self-externalisation in relation to the other that would displace the inner conviction or law of pure duty as a principle of action. As the consciousness of conscience, the person once again appears duplicitous: the inner law is both absolutely mine, and something in-itself, a universal. Conscience is at once good and evil, judging and judged, universal and particular, absolute and relative to the other, purely formal and specific in its content.

While conscience expresses, not personal desire or interest, but a principle of action apodictically given, it is, nonetheless, an essentially contemplative or intrapyschic experience, for action transforms conscience from an inner conviction, an immediate self-certainty, into a concrete being for another. For the consciousness of conscience the essence of moral life is the self-certainty of conviction, a formal determination opposed to any specific content: "it is in this conviction alone that action is a duty" ([2], sec. 653). The validity of this duty, however, is constituted "solely through the conviction being *declared* [*ausgesprochen*]*.*" Conviction becomes duty only in being publicized, i.e., in becoming something for others. Thus, there arises a complex of differences by which the validity of conscience is itself ultimately displaced. As an actual consciousness, conscience is at once the universal form of conviction, duty as such, and some specific, empirical content; yet, the latter can contribute nothing to the certainty of conscience and must be reduced to the moral maxim which covers it. (Thus it is that judging conscience comes to play the moral valet to the empirical self. [2], sec. 665)

In declaring itself, however, in entering into language and asserting itself as a principle of judgment and action, conscience becomes something before others. And, it is not in the intrapsychic experience – the self-certainty – of consciousness that Hegel locates the paradigm of moral life, but in its confession of itself to the other as essentially evil and guilty in that the specificities of actual duties and actions must always conflict, both among themselves and with the pure form of duty.

His confession is not an abasement, a humiliation, a throwing-away of himself in relation to the other; for this utterance is not a one-sided affair, which would establish his disparity with the other; on the contrary, he gives himself utterance solely on account of his having seen his identity with the other; he, on his side, gives expression to their sameness in his confession, and gives utterance to it for the reason that language is the existence of Spirit as an immediate self. He therefore expects that the other will contribute his part to this existence ([2], sec. 666).

To the extent that conscience clings to itself it can only maintain a "hard heart" [*das harte Herz*] in the face of the other. It fails to realize the reciprocity that is offered in the other's confession, and, thereby, persists "in the contradiction between its pure self and the necessity of that self to externalize itself and change itself into an actual existence" ([2], sec. 668). Conscience puts itself in the wrong by refusing to return the recognition of itself that is embodied in the other's confession, i.e., by failing to see its own conviction from the point of view of the other, as something properly to be displaced.

When the hard heart breaks and the inner law of conscience gives way to the experience of reciprocity, the element of moral experience is finally constituted.

The reconciling *Yes*, in which the two 'I's let go their antithetical existence, is the existence of the 'I' which has expanded into a duality, and therein remains identical with itself,
and, in its complete externalization and opposite, possesses the certainty of itself . . .
([2], sec. 671).

Conscience, immediate self-certainty, "is in general the caprice of the
individual, and the contingency of his unconscious natural being" ([2],
sec. 643). The consciousness of forgiveness [*die Verzeihung*] and
reconciliation [*die Versoehnung*], however, is constituted not as an *immediate* self-certainty, but in and through a complex of displacements. It
experiences both the displacement of itself, by the authoritative conviction of the other, and the inevitable emergence of evil from the conflict
of specific duties. Not self-determination, but what Nietzsche calls the
ability to "shift perspectives" appears as the founding gesture of moral
discourse. Such a discourse arises, not in virtue of an attribution of and
correlative respect for an abstract personhood, but in virtue of a self-
displacement, a 'shifting perspective', wherein I *act* so as to acknowledge
the other's fate or condition as human, and, thereby, as an element in
my own horizon of possibility. Thus, consciousness in its moral life
becomes something other than a judge, an agent of the law, and dispenser
of praise and blame; and, its validity is constituted not by conceptual-
izing the person as an end-in-itself rather than a means, nor by subjecting
inner conviction to a principle of universalizability, *but by acting*, and by
acting so as to embody its displacement by the other.

A moral discourse which derives from the displacement of the subject,
rather than the phenomenon of conscience will not find expression in
the language of imperatives. As a philosophical language, it would
operate always under the aegis of a 'hesitation', mindful of the violence
of action and content to clarify the formal features of our situation.
Empirically, such a discourse would find its body, not in a code of imperatives, but in the actual decisions of mutually reconciled subjects: not
"one ought", but "we have decided, we will do, we have done." The
subject of such a moral discourse would necessarily be plural.

A profound and dangerous error is made in thinking that from the
philosophical analysis of the essential forms and structures of our moral
situation there will follow immediately a series of moral imperatives
for empirical life. Probably, philosophers, however well they might be
able to articulate the conditions of action and judgment, are no better
than anyone else in arriving at determinate moral conclusions; certainly,
they are no better than anyone else in their actions. A moral philos-

ophy oriented by the concept of life and developed from the displacement of the subject is better than one founded on conscience and a
conceptual analysis of duty, precisely because it no longer perpetrates the
illusion that objective moral principles generated abstractly can validly
resolve actual moral conflict, nor does it revert to some calculus of goods.
It recognizes as its orienting value the "yes" which establishes a concrete
relation to the other in the experiences of forgiveness and reconciliation, and it raises the question of who "we" are.

II. SUSTAINABILITY: ACTING BEFORE OTHERS

Hegel's analysis of ethics and morality precludes any attempt to found
moral rules or norms either on some sensible or empirical given or by
appeal to some transcendent standard or in an account of transcendental
conditions. Any attempt to found the normative on the nonnormative,
e.g., on descriptions of how a community in fact operates or on intellectual intuition of abstract objects or on some calculus of goods, is
doomed to failure, for any determination of how one *ought* to behave can
derive only from "acknowledging the *absoluteness* of the right" ([2], sec.
437).[4] Moreover, there is no need for an appeal beyond the domain of
action, for the social practices in which norms are embodied in fact supply
within themselves the necessary criterion of their own evaluation.
By focusing on concrete agents and the forms of life in which they
participate, Hegel reformulates the task of moral philosophy, substituting
for the search for grounds and foundations, what Terry Pinkard calls
the question of "sustainability", viz., the way in which the norms regulating an agent in a given practice either sustain or fail to sustain that
very agency itself [6].

Essential to this account of agency is the concept of mutual recognition. Action, as Hegel argues, is essentially public, both my own and
for another; agency implies an action *with reason* and the expectation
that the other participants in the social practice will recognize those
reasons as justifying the action. Only by participating in public and
communal forms of life is the subject constituted as an agent; and, the
principles governing that agency arise out of the forms of life themselves.

How is it, then, that social practices provide within themselves the
necessary and adequate criterion for justifying norms and ajudicating conflicts between them? Hegel's account of consciousness is a genetic one,
an account of the self-development of consciousness as it attempts to

say what it means, i.e., as it attempts to articulate its own experience. The history of this development is a history of dispelling errors: consciousness asserts a certain account of itself, e.g., that it is identical to sensation, and in attempting to articulate that identity it finds that it cannot make sense of its own experience, that it is led into contradictions which cannot be resolved within that account or which can only be resolved by asserting a new account of consciousness which comprehends those very contradictions. Consciousness is this testing of itself, of its experience against its accounts of that experience, and, as Hegel remarks, in its analysis we do not need to import any of our own "bright ideas". In fact, we don't need to import anything, for the distinction between what consciousness *is* and what it knows or takes itself to be "falls within consciousness", so that the investigation of consciousness is a "comparison of consciousness with itself" ([2], sec. 84). Thus, "consciousness provides its own criterion within itself" for evaluating its knowledge-claims or accounts of itself. The new, more comprehensive account of itself is borne out of the contradictions of the previous shape, and is validated not by reference to any given, transcendent, or transcendental conditions – by nothing external to the experience of consciousness – but just by its success in making sense of that very experience.

The Hegelian account of the normative reflects this general logic of consciousness. To embrace specific norms of practice is to assert a certain conception of who we are and what we are doing or rather where we are going in those practices. Action offers "a display of what is one's own in the element of universality whereby it becomes, and should become, the affair of everyone" ([2], sec. 417). Identity, then, consists not in a set of properties, but in action, in what one does in the context of others. This Hegelian analysis of social norms focuses on "how social practices *make sense* to their participants and how certain ways in which they make sense to them [are] necessary for those participants to sustain those practices rationally" ([6], p. 4). Normative conflict arises, then, when the norms and rules asserted as authoritative for a given social practice fail to sustain that practice: "one finds that . . . the accounts that one gives of what one is doing cease to make sense in that they come to be seen as being immersed in all kinds of dilemmas, selfcontradictions and failures to make good on their promissory claims" ([6], p. 15). When norms and rules taken as authoritative for a given practice prove incapable of "reconciling" these contradictions, the norm becomes "socially impossible", i.e., "it is the norm for a practice in which agents cannot affirm such a norm for themselves, . . . they cannot find themselves 'at home' in it, within the terms of the practice itself" ([6], p. 36). Thus, for example, what Hegel calls 'natural' freedom – the idea that an agent is free when he can do simply what he immediately wants – is such an unsustainable norm, for the agents embracing this norm would "each be compelled to demand things from the other that the other from his own point of view could not find rationally acceptable" ([6], p. 26). The critique of a norm is, thus, generated from within the practice for which the norm is asserted. Or, as is the case with consciousness generally, a social practice "provides its own criteria within itself."

Moreover, as in the dialectical development of consciousness generally, subsequent forms of life emerge out of their predecessors or are necessitated by the irresolvable contradictions afflicting the preceeding form. So too, Hegel's analysis of normativity "understands the authoritativeness of what is internal to a practice to rest in its having come to be required *for us* by virtue of the failures of past accounts . . . it takes the requirements of the practices to be internal to the way that practice understands those requirements to be a result of . . . the necessary failures of past accounts" ([6], p. 28). To offer a critique of the failure of a norm to sustain its practice is, in fact, to articulate, however implicitly, a new norm. It is through this internal self-critique, not in the appeal to a foreign standard, that normative conflict can be resolved and the norms of a given practice or form of life established.

III. MUTUAL RECOGNITION AND THE UNRECOGNIZABLE OTHER

This discussion of sustainability emphasizes, quite rightly, that social practices themselves are constituted by "the ways in which the participants in the practice mutually recognize one another" ([6], p. 13). To participate in scientific practice, for example, I must understand my actions as governed by norms and rules to which others can appeal also and by which their action is also regulated, and I must take my results, my beliefs if you will, as in need of the verification of the others. Professor Pinkard argues that the

self-undermining nature of early modernity [with its twin assumptions that not everyone was equally endowed to grasp moral truths and that obligations could come only from superiors to inferiors] led Hegel to claim that only a fully *mutual* form of recognition *could* be sustainable for the kinds of people that we had come to be. ([6], p. 36)

Yet, the supposedly mutual recognition at which consciousness arrives
in Hegel's account of the self-development of moral consciousness, in
fact, embodies just these assumptions.

Spirit [*Geist*] first comes on the scene with Antigone. She appears
from nowhere in the midst of a critique of the concept of law. Law will
not be adequate to grasp the truth of human action, either because reason
attempts to legislate directly without the mediation of collective forms
and thus produces an arbitrary tyranny, or because it makes the law
contingent on its own test, thereby exhibiting what Hegel calls the
"insolence of a knowledge which argues itself into freedom" by treating
as mere laws, what are in fact "commandments" ([2], sec. 434). Suddenly,
Antigone appears, to serve as the figure of the "ethical disposition" which
consists not in following a law, but "just in sticking steadfastly to what
is right, and abstaining from all attempts to move or shake it, or derive
it." ([2], sec. 437) When Antigone marshalls her arguments for carrying
through the fatal intention of burying Polyneices, chief among them is
that this duty is an essential element of her identity as Antigone. Her
failure to respect this duty would be "dishonorable". Though she faces
death, she has no cause to grieve; death is nothing compared to the real
loss of herself by failing to do her duty. By doing it, she has passed
the test and proven her identity ([8], 1.93 and 450). Hegel's analysis of
Antigone demonstrates that to be who we are is to participate in certain
normative social practices.

This immediate identity of consciousness and the right depends upon
the fact that the ethical dispostion "has put its merely individual aspect
behind it, a mediation that is finished and complete" ([2], sec. 436).
Antigone identifies herself, not as a single, isolated individual, but with
the continuity of blood and with the network of relations into which it
flows. She exhibits just that self-certainty, the certainty of being a moment
of a whole, that is proposed as the hallmark of Spirit.

The world of Spirit, however, turns out to be radically bifurcated.
The two sexes "appear in their ethical significance, as diverse beings who
share between them the two distinctions belonging to the ethical substance" ([2], sec. 459). Woman is identified with immediacy and feeling,
with the body and its blood, with domesticity and the enclosure of the
home (replicating her representation in the enclosure of the womb), and
above all with death and its redemption. The specific act of Antigone,
who gives her life to perform those rites necessary to rescue her brother's
corpse from its vulnerability and degradation as a mere thing, those

rites necessary to respiritualize his body, becomes the duty of woman in general. The man, on the other hand, achieves the universalisation of his particular consciousness *by leaving the family behind*, substituting for the bonds of blood, the self-determining associations of contract in the domain of politics and theory in science. (See [2], sec. 458–9. He "leaves"; she "remains".) Man finds his "actual substantial life in the state, in science [*Wissenschaft*], and otherwise in work and struggle with the external world and with himself" and through this "division" or exteriorization of himself in the formed thing, scientific practice, or symbolic language he achieves "selfconscious being" ([3], sec. 166). Thus, is he constituted as a speaking subject with the right to recognize and be recognized.

The woman, representative of the blood of the family, *has been left behind* in this dialectical progression. As Irigaray points out, Hegel would have us believe "in the delusion of a bisexuality assured for each in the connection and passage, one into the other, of each sex" ([4], p. 217). This "Hegelian dream" is necessary because we must all be put to sleep long enough to forget that here with respect to sexual difference the system violates its own conditions of systematicity. *A moment of consciousness is cast as the dependent or inessential moment and left behind.* While the man can represent and reflect the private domains of domesticity, feeling, and the blood in the discourses and practices of science, thereby superceding, but preserving them, the woman enjoys no correlative opportunity to reflect the public domains of the city, the laboratory, and the work-place. She enjoys only a quasi-subjectivity precisely because she is deprived of that element in which she might go out to the other and come back to herself, viz., language. Hegel deems her incapable of those "higher" activities – science, art, government – that "require a universal element" ([3], sec. 66). She finds herself placed "beyond the pale of language" ([4], p. 235). She has no language of her own; rather, as "guardian of the blood", she is a material condition for the elaboration of the supposedly general masculine language and, in Irigaray's mild understatment, "this makes it hard for her to achieve a *for-itself*." She cannot partake in the specular economy wherein by seeing herself in the other she would be returned to herself in a reflection. She preserves for man the dimension of blood or life from which he has become divided, and, thus, makes it possible for him to redeem or recoup that dimension in a specular reflection. He reflects her self back to her as an object, but she cannot be represented as a subject and cannot represent herself. Thus, she cannot return his reflection so as to complete the circuit of recognition.

Hegel will tirelessly document the persistent error of natural consciousness in distributing between two consciousnesses the unity of an opposition, as the master and slave err in dividing between themselves the moments of freedom and nature. But, with respect to sex this is exactly what Hegel or "we, philosophical consciousness" has done. The material domains of blood, body, and feeling have been dissociated from the discursive domains of labor, science, and politics, and the former have been left behind. And, because of the non-reciprocity of the relation of man and woman, man repeats the master-slave dialectic, exhibiting a freedom that is opposed to nature or life. He is as arrested in his development as the woman he has left behind.

Moreover, the woman is not only an exile, but a threat, the example that would disrupt the systematicity of the system by marking with an irreducible difference the generality of its voice. In his 1822–23 lectures on the philosophy of right Hegel remarks,

The man's dominion is scientific universal cognition, . . . [the] universal, a universal idea, the imagination inspired by reason, the Idea of a universal. These are the man's provinces. There can be exceptions for individual women, but the exception is not the rule. Women, when they trespass into these provinces, put the provinces themselves in danger. ([3], pp. 525–526)

Thus, in his account of the self-development of moral consciousness Hegel not only violates his own conditions of systematicity, but also presents as the initial embodiment of spirit a brother-sister relation that, in principle, cannot be one of mutual recognition, for the sister is prohibited from participating in the very practices in which the relation might be renegotiated.

In developing the concept of sustainability, Pinkard remarks that "recognition involves various contests and struggles for normative status", that "the forms of recognition and their accompanying norms can be one-sided, can privelege certain groups at the expense of others", and that "the conferral of recognition is always the result of a struggle" ([6], p. 34); however, this onesided relation cannot be understood in those terms. It is not pertinent to point out that this initial ethical relation is in fact in Hegel's analysis demonstrated to be inadequate and superseded by more complex forms of social life more adequate to the embodiment of moral consciousness, for it is only the brother who contests this form and leaves it behind. Not only is the repression of the sister necessary to the brother's further development – for he is free to participate in the discursive realms of science and politics, knowledge-production and governance, only because the sister continues to tend the domain of the blood – but also she is on Hegel's account incapable of participating in the project of rationally accounting for our practices. When woman enters the domain of rational discourse"scientific universal cognition" – she "trespasses" and puts the domain itself "in danger". His analysis embodies just those twin, self-undermining assumptions of early modernity noted by Professor Pinkard: that some individuals are governed by norms which in principle they cannot understand and that obligations come from superior to inferiors. What is at issue here is precisely the "we" of the social practices. In Hegel's analysis the "we" is always marked masculine, and is, therefore, "one-sided" and hardly the fully mutual form of recognition that Hegel himself requires.

As Terry Pinkard's analysis of sustainability demonstrates, Hegel is right in arguing that a practice supplies within itself its own criteria for evaluating its norms and rules. Actual forms of life and social practices generate within themselves the claims of duty, and make unnecessary appeals to abstract concepts of the person or universalizability. This Hegelian analysis, however, falls short as an account of normative conflict and its resolution precisely because it invokes an other who can never be reconciled. In asserting that the ajudication of normative conflict is properly generated from within the domain of practice in which the conflict arises, this position assumes that those at odds are engaged in the same discursive practice, that each one is capable of pitting his reasons against those of the others, his narrative against those of the others, and that he will win who offers the narrative most adequate to the sustenance of the practice itself. If, however, one is in principle exiled from these domains, how would one ever enter the fray?

In turning away from the tradition's reliance on metaphysical definitions of the person and abstract, purely formal duties and principles, Hegel turns toward concrete forms of life which prove capable of generating their own principles of valuation. Yet, Hegel's own account of the moral subject remains abstract insofar as that subject depends upon an alien alterity, an unrecognized other. The question of life raises, then, the ultimate question: who is the "we" of moral judgment? In demonstrating how concrete practices and forms of life sustain themselves only in virtue of the mutual recognition of their participants, Hegel has redefined the question of ethics. How must "we" act, if the 'we' is to be genuine, rather than an abstraction limited by an alien, irreducible alterity?

State University of New York
Stony Brook, New York
USA

NOTES

[1] The Christian, primarily Catholic, concept of the 'sanctity of life' actually operates so as to advance this opposition between a devalued world of life and a valorized world of the spirit. The concept is frequently invoked in discussions of euthanasia and abortion, as if it were a defense of life; however, the term life here functions metaphorically: what is defended here is not so much a specific embodied subject as the hypothetical existence of an immortal soul. The right to life of the fetus is necesssarily abstract, unclaimed and unnourished by any actual social relations. Similarly, the injunction against euthanasia denies to the concrete living individual his right of self-determination or denies to his actual others the right to act on his behalf. This concept contributes little to Western moral philosophy, though it is briefly influential in American bioethics in the 50's and 60's. It will not be discussed further here; rather, this essay develops against this abstraction, a more concrete Hegelian account of life as a socially organized form with moral implications.

[2] See, e.g., the extended analysis of the application of this argument to women in [1]. Ehrenreich and English focus specifically on the regulation of women by medicine and psychiatry.

[3] Conscience is, of course, constituted by the virtual other, the other as inscribed through mechanisms of internalisation and identification or through the work of mourning. Even the voice of conscience embodies a *socius*.

[4] For a more detailed critique of the attempt to reduce normative to non-normative judgments, see my "Health, Justice, and Responsibility" [7].

BIBLIOGRAPHY

1. Ehrenreich, B. and English, D.: 1987, *For Her Own Good*, Vintage, New York.
2. Hegel, G.W.F.: 1979, *Phenomenology of Spirit*, A.V. Miller, trans., Oxford University Press, New York.
3. Hegel, G.W.F.: 1991, *Elements of the Philosophy of Right*, Allen E. Wood (ed.), H.B. Nesbitt, trans., Cambrige University Press, Cambridge.
4. Irigaray, Luce: 1985, *Speculum of the Other Woman*, Gillian C. Gill, trans., Cornell University Press, Ithaca.

5. Kant, Immanuel: 1964, *The Doctrine of Virtue*, Mary J. Gregor, trans., Harper and Row, New York.
6. Pinkard, Terry: 1994, 'Historicism, social practice, and sustainability: some themes in Hegelian ethical theory', *Neue Hefte für Philosophie* **35**, 3–41.
7. Rawlinson, M.C.: 1980, 'Health, justice, and responsibility', in T. Engelhardt and S. Spicker (eds.), *The Uses of Medical Knowledge*, Reidel, Dordrecht, pp. 87–102.
8. Sophocles: 1981, *Antigone*, F. Storr, trans., Harvard University Press, Cambridge.

ANTON LEIST

PERSONS AS "SELF-ORIGINATING SOURCES OF VALUE"

I. OBJECTIVE VALUES?

In this article I want to enquire into the sense of the formulation chosen for the title, and to ask where it leads us.[1] The idea involved is clearly from Kant, but here I will not follow the path of exegetical approach; nor is Rawls' political theory the object of my interest. Rather, I want to understand in general the involvement persons have with values. This makes it necessary to dedicate some thoughts to how values come into the world, to the epistemological and ontological status of values, and to the concept of the person.

To orient ourselves around Rawls's formulation: in one sense (a very rough one) it excludes the ontological independence of values, a position not very attractive nowadays. G.E. Moore seems to have held such a position in his earlier years, and it has been brought to life again recently among environmental philosophers. Moore used, and perhaps introduced, the concept of *"intrinsic value"*, meaning an objective, non-relational property of things, as opposed to size or weight ([18], Ch.6; [19], Ch.7; 1959).[2] According to Moore, things somehow have value in them, producing a duty for us to maximize this value, or to behave in such a way as to enlarge the intrinsic value within the world. The practical relevance for environmental ethics of such a position is obvious. If an endangered species has a value in itself, it has to be protected – whatever industrialists or the builders of motorways think.[3] If, on the other hand, value is a relational matter, referring somehow to man, then a positive evaluation by the public would be a necessary condition in order for the endangered species to be saved. And, with the public being either

177

K. Bayertz (ed.), Sanctity of Life and Human Dignity, 177–199.
© 1996 *Kluwer Academic Publishers. Printed in the Netherlands.*

self-sufficient or negligent, the danger looms that some tree, grass or animal will receive little positive evaluation. Or at least not enough in order to win over a motorway.

The motive now clear, how should we regard intrinsic value? Intrinsic value would explain our evaluation of trees, but would lack any other basis. It would, in a sense, explain 'itself by itself'. But why believe in it? In a similar manner as Moore, the German philosopher Hans Jonas ([12], pp. 154–6) more recently used a test of the universe with or without the relevant thing in question: Would it be better (so the test question runs) if something existed or not, judged in isolation from everything else, i.e. without any effect on anything else? Among environmental philosophers this has become the 'last man argument': would we think it evaluationally unproblematic if the last man on earth chopped down some trees which could live on otherwise? Neither Moore and Jonas' formulation of the test, nor the more recent one by environmentalists seems to put us in a position to decide the issue, at least not clearly. Some will say, yes, the universe would be better if one specific tree existed than if it did not; others will go on to say the opposite; and still others will (alas) waive the question as 'merely philosophical', in this context meaning undecidable.[4]

Things might become a bit easier if we realize that the term "intrinsic" is notoriously ambiguous.[5] Intrinsic value could mean

(i)   "something of value in itself, i.e. as opposed to being valuable for something else (instrumentally)",
(ii) "valuable properties inherent in an object", and
(iii) "value in an independent sense, i.e. independently of the evaluations of valuers".

In the first meaning of the term, intrinsic value hardly seems to be evadable. Intrinsic values in this sense are the alternative to instrumental ones, and not all values can be instrumental. The grounding of values has an end somewhere, even if only for the time being. But the intrinsic quality of values in this first sense is neutral in relation to their subjective or objective status. A subjectivist, according to whom values are somehow the projections of valuers, has to accept intrinsic values just as an objectivist does. Values need not be 'intrinsic' in a sense combining the first and third meaning distinguished here. Or to put it another way: a subjectivist need not free himself of the concept totally.

This is one reason why, among environmental philosophers, the objectivist meanings (ii) and (iii) are the more interesting ones. Some philosophers use objective values in order to give protection of nature the only real chance it has of proper foundation. How should we regard this project? Do intrinsic values exist in this ontological sense? The answer depends on the meaning of the term "value". A plausible, Aristotelian or, in a narrower sense, biological answer takes values to be objectively grounded in the natural states of living beings ([21], pp. 19–22). Trees, animals, and living beings in general are – on the basis of their biological set-up – directed towards some state of flourishing. This given, some things (events, states, actions) are 'good for' a being, others are 'bad for' it. Gardeners know what is good for their plants, and also what is bad. Something, i.e. a river, could perhaps be called good, if it is in a good state as far as the flourishing of all the living beings included in it are concerned. And this state would then be an objective one in the sense that it existed *independently of human valuers* and their views on the matter. It would not be an objective one in the sense that states of value existed *without valuers in any sense*: the living beings themselves, or their biological set-up, define what is good for them.

Whatever this biological case of value existing objectively amounts to: it is not linked by an imperative to its own existence, as it had to be in a Moorean sense of intrinsic value. Biological value is grounded in the given biological set-up, and nothing may be deduced from this (to be depicted in a set of statements) regarding 'musts', 'oughts' etc. The biological case perhaps comes closest to the Moorean idea of intrinsic values existing in themselves. The natural world manifests a host of value production 'existent in itself'. But there is no logical consequence from these facts for human action and motivation. There remains a *logical gap*, only to be closed with some additional premises referring to human valuers.

Intrinsic value enjoyed a renaissance among environmentalists, but the postulation of it is also in evidence within bioethics circles. Some of the most severe critics of abortion argue that, without exception, human life may not be ended because of its being 'human life'. Perhaps these antiabortionists are implicitly referring to an additional premise saying that human life should not be ended but rather prolonged (the premise being so evident that mentioning it seems superfluous). At least sometimes, however, the proponents of this position think that an imperative not to end human life springs immediately from the existence of human life itself. Human life represents a good, a biologically grounded plan to be followed, and therefore it should exist. But on closer inspection this conclusion proves to be wrong. It is right that the biological plan built into any living human being (say, a fetus or a newborn child) to a certain extent determines whether the environmental influences which this being will be subjected to will be good or bad. But out of this (and this *alone*) it does not follow that a duty exists to realize environmental good for this being. In addition, we will hardly accept an unrestricted premise containing a duty not to end human life in general: Sometimes we can only save our own lives by killing a human aggressor, and we surely do not want to foreclose this opportunity to continue living.

Drawing from these cases, maybe we can say that objective values do exist in a sense (perhaps a trivial one). But they are not of immediate practical relevance for human action. Instead, values relevant for action have to come from the actor himself or herself. By giving objective values their proper, albeit somewhat restricted place, we have therefore explained part of the Rawlsean quotation. Values of practical relevance can only be 'subjective' in the sense that human subjects in a way 'produce' these values. But there are a multitude of possible ways of doing this. And, of course, we have to see that it can be done at all.

Given this subjectivist turn to values, how can we make a fresh start? In the following, I will first inspect the Kantian approach to valuing, which in a sense is subjective, but not naturalist (2). Some problems involved in Kant's own account are serious, as we shall see. Then I shall attempt a look at the Humean account of values, again discussing it critically (3). My suggestion at the end of this comparison of two classical approaches – as a kind of half-way stand – will be: evaluations are emotionally grounded ways of perceiving, and values are a form of disposition. But several theoretical and phenomenological problems lie ahead of this position. I will confront some of them and try to provide answers to them (4). The Humean account of values will not, however, be able to do justice to some practices of autonomy and the value we assign to autonomy as something valuable in itself (5–6). Finally, I will be arguing for a Kantian-Rawlsian alternative to the Humean account, but merely with mild consequences for bioethics, at least in an ontological sense (7). Seeing human beings as primarily autonomous does not provide an ethically powerful basis for the solving of biomedical riddles by means of a new evaluation of human beings. Rather, as suggested by Rawls in a wider context, autonomy leads to a discursive procedure wherein diverse conflicts of opinion can be solved in form of practical, even if not unanimous, agreements.

II. THE KANTIAN CONCEPT OF VALUE

The Kantian approach regards the idea of rationality (or, in a different formulation, autonomy) as fundamental, not needing to be constructed out of other, still deeper levels in the person (for example, desires or interests). Kant's classical train of thought is as follows. Rationality is a property of some living beings, above all humans. By rationality, Kant means the practical capacity to 'give oneself ends' and to follow them in acting. This sounds plausible enough, but Kant's specificity inheres in the rather original way of formulating the process of giving oneself ends. We are doing this by distancing ourselves from our particular desires and aversions, that is, from our merely contingent ('empirical') human nature. The positive side to this process comes from our giving ourselves a 'moral law', i.e. principles of acting in accordance with their universal application, the application being thought of counterfactually. Values, duties and rights created by this Kantian way of proceeding are 'constructive' insofar as nothing is presupposed in a naturalistic sense, especially not by way of human nature.[6] Naturalistically, human nature is reduced to its desires, dispositions, thrivings, or, more generally, empirically identifiable psychological states. Kant instead identifies human nature, or 'humanity', as a capacity for rational self-direction.[7] In his sense, human nature could be rewritten as rational nature, understood in a 'noumenal', not 'phaenomenal' way. Kantian actors do not seem to base themselves on hardly anything (and especially not on any natural or cultural facts), but seem nevertheless to be productive of quite an amount of liberal culture.

On a closer view, however, Kant introduces his master value of rationality without giving it proper foundation (see [9]). His argument can perhaps be grasped more fully in the famous passage devoted to the rights of animals. The passage goes like this:

Baumgarten speaks of duties towards beings which are beneath us and beings which are above us. But as far as animals are concerned, we have no direct duties. Animals are not self-conscious, and are there merely as a means to an end. That end is man . . . If a man shoots his dog because the animal is no longer capable of service, he does not fail

in his duty to the dog, for the dog cannot judge, but his act is inhuman and damages in
himself that humanity which it is his duty to show towards mankind. If he is not to
stifle his human feelings, he must practise kindness towards animals, for he who is cruel
to animals becomes hard also in his dealings with men . . . . Tender feelings towards dumb
animals develop human feelings towards mankind ([13], p. 239).

The upshot of this is that we do not have duties towards animals as
such, because of their lack of consciousness. In contrast, rational beings
are

called *persons* because their nature already marks them out as ends in themselves ([14],
p. 96).

Rationality as the power to give oneself ends is the value Kant presupposes in arguments like these. The power to set oneself ends is something
persons share, along with their animal-like nature, and they share it
even if they are not making use of it. 'Humanity' or 'rational nature'
is *something only to be manifested* in persons rather than being identical with them. This manifestation relation allows for a more or less
extensive interpretation. Instead, 'rational nature' is something of absolute
value, or of 'dignity' (Würde), and human persons are the object of
respect only because of their capability to bear rational nature.

From the absolute value of rationality as the capacity to set oneself
ends, Kant derives several important and characteristic moral duties. What
becomes clear from these duties is that, similarly to Moore or Jonas, Kant
sees value as including an imperative to existence: something that has
absolute value (such as rationality) should exist, and those who are in
a position to further its existence have the duty to do so. Given this,
Kant's more concrete practical conclusions are easily recognisable as
variants of this general imperative. Among the more concrete moral
duties, Kant mentions the duty not to kill human beings, not to commit
suicide, to further one's talents, to further talents in others, to give
others the freedom to develop their ends in a rational way, and to respect
others as bearers of a rational nature (see [8], pp. 50–5).

Typical for the Kantian position on these duties is that they correspond
with a right for the relevant person, equally absolute as the value which
duties and rights are thought to secure. The bone of contention becomes
clear with a look at the famous distinction between "dignity" or "worth"
and "price" [14]. Dignity in rational nature is an "incomparable" worth,
"exalted above all price". Something with dignity is not to be compared
to something with a price. Among other things this seems to say that

neither a price, however high, nor a value based on needs or desires (the 'phaenomenal side' of men) can justify sacrificing dignity ([8], p. 48). And this can be rephrased, it seems, by saying that anything having dignity has an equally absolute right to exist, a right not to be restricted by any further considerations. Consequentialist versions of rights, in contrast, may be effective when arguing for the security of interests or claims, but by basing themselves on states of non-absolute value they will never be able to derive absolute rights. More concretely, and of quite some relevance for bioethics: a negative right to live, i.e. a right not to be killed, *will not be absolute* for the consequentialist. Some situations may realize themselves where something of comparably higher value can be saved by killing. With the Kantian understanding of rights, this would not be possible.

My purpose so far has been to sketch out Kant's argument concerning the absolute value of rational beings. We now need to form an opinion on it. In the relevant passages of his *Groundwork*, Kant argues as follows: rational nature is the capacity to give ends to oneself, and therefore it too has to be seen as an end in itself – "end in itself" being taken as "valuable", "having worth" etc. In another language this would mean that everyone, because he sees himself as *creating ends* for himself, should also think of *himself as being* an end in itself. But of course this is not an obvious conclusion. *If* humanity is an end in itself, i.e. is valuable, some of Kant's duties might follow from this; but if one sets out with the capacity to give oneself ends, i.e. a psychological capacity, nothing like a value follows from that (in addition to the individual ends, that is). What seems even more appalling is the fact that it is difficult to recognize a specific *moral* power in the capacity for giving oneself ends. Rather, it is a non-moral presupposition for the development of moral capacities. Animals are unable to act self-consciously and to make free choices; and this is the reason why they are unable do develop a morality.[8]

What general conclusion may be drawn from this examination of Kant's arguments? Perhaps the only thing that may be drawn from it is the kind of postulation suggested by Rawls [25]. Persons, says Rawls, are moved by two powers, the power to revise their ideas of the good life critically and a capacity for justice. Rawls brings both together in the remark that persons "recognize that the weight of their claims is not given by the strength or intensity of their wants and desires, even when these are rational" ([25], p. 545). A concept of equal worth "is founded on the equally sufficient capacity . . . to understand and act from the public conception of social cooperation" ([25], p. 546). In my opinion, the move to invest persons with moral powers, if understood empirically, is a sensible one. If, however, this has to be combined with a non-naturalistic conception in which claims are, though "self-originating" with individual persons, not to be thought of as springing from wants and desires, then explanation is seriously lacking. It is the Kantian mark of value to be absolute, i.e. distinct from wants and desires. But Rawls does not make plausible why the postulation of self-critically applied capacities makes them absolute in this sense. I will engage myself to improve in this point on Rawls, trying to stay on the Kantian side of the divide. But first we have to look into an alternative account altogether, the Humean.

III. THE HUMEAN ACCOUNT OF DESIRES

The Humean account of value is naturalist in the sense that it explains values as a kind of psychological state. In the classical version, this state is taken to be a 'desire'. So we first have to clarify desires.

Hume takes desire to be a form of passion, and passions are feelings. If we have an aversion towards or desire something, "we feel a consequent emotion of aversion or propensity" ([10], p. 414). Above all, Hume seems to have in mind pain and pleasure, and additionally socalled "natural impulses" like hunger, thirst, or the desire of punishment to our enemies ([10], pp. 438–9). In all of these cases, *immediate perception* is part of the affect. The context for this concept of desires is to be seen in Hume's reductive analysis of psychic states. Though rich in empirical detail, Hume's theory of psychic states is reductive in allowing only two dimensions: beliefs and passions.

A passion is an original existence, or, if you will, modification of existence, and contains not any representative quality, which renders it a copy of any other existence or modification. . . . In short, a passion must be accompany'd with some false judgment, in order to its being unreasonable; and even then 'tis not the passion, properly speaking, which is unreasonable, but the judgement ([10], pp. 415–6).

Desires in the Humean model of 'belief cum desire' also provide part of an explanation for action: desires being the active element, pushing the actor into action, beliefs being the passive element in form of cognitive orientation. Richard is sending a letter – so he must have the desire to do this (he may have other desires besides this one), and he must have one or more beliefs concerning his doing. In any case, it is not very clear what the beliefs are about, or how they should be formulated. One suggestion for the present case would be, however, that Richard has to believe that putting his letter into a metal box designed especially for this occasion is a way of sending the letter. As one can see immediately, the belief is a complex one, being connected with others. But I will not go into this more explicitly here.

Now, there is quite a list of well-known problems surrounding this Humean conception of desire (see [28]). Most prominent are the following:

1) Desires differ from bodily sensations because of their propositional content. If one desires something, one desires that propositional content. Some desires seem to be directed towards objects, e.g. the desire for a cup of coffee. But on closer inspection this desire is aimed at the having or receiving of a cup of coffee, clearly something to be denoted by a proposition. On the other hand, if you are in pain (bodily sensation), then pain is hardly a propositional content. Propositions are part of a desire, but cannot be felt as such. So desires should not be reduced to feelings.

2) It is not necessary for desires actually to be known, as made clear – above all – by psychoanalysis. Not rarely, our actual deeds bring into the open what it is we desire – and not only to the attention of others, but also to our own. We often make mistakes in the identification of our desires, and sometimes systematically so, denoted by psychoanalysis as 'repressed desires'. Bodily sensations, again, cannot be repressed or hidden because they must necessarily be felt.

3) Not all desires, even if conscious, have to be felt intensely, or have to be felt at all. One can have the desire, for example, to cross the street, but that does not mean that one longs to cross the street. No actual feeling must go along with our acting and its motives.

4) Not all actions are bound to a pro-attitude, as is inherent in the concept of a desire. You write a letter out of a desire to answer your friend, but you don't have a desire to answer the bill sent by your garage. If you do it anyway, you hardly do it out of a desire.

Most of these problems can be answered in a Humean spirit by taking the motivational basis not to be a single conscious event, but a *disposition* to act (see among others [28], [7], pp. 110–1). Terminologically, it seems better to talk of *wants* instead of desires, as will be done predominantly in the following. In doing so we can find answers to these objections roughly as follows:

1) If we talk about dispositions instead of wants we are able to classify according to specific conditions. Dispositions can be classified according to their functional roles, the functional role being given by conditions C and the output (action) A. Functional roles are coextensive to specific propositional contents. Seen as dispositions, wants can thus incorporate a proposition.

2) Dispositional analysis frees us from a phenomenality condition within wants. But it also enables us to be wrong about dispositions. Dispositions are grasped only in actual manifestations, and they may happen to manifest differently than was previously thought.

3) Hume took notice of the "calm passions" existing alongside the more important "violent passions" ([10], pp. 417). Dispositional analysis enlarges on that category. To have a disposition does not foreclose any specific emotive reception towards one's own dispositions, i.e. one's own wants. Actual action under predefined conditions is necessary for identifying a want; feelings of a particular sort may come in optionally.

4) As we have seen, the belief-desire theory has been refuted by acts not including a pro-attitude towards their object. The terminological change from desires to wants takes care of this objection.

Several additional criticisms have been made of the Humean conception of desire sketched so far. I will not go into them here, however, mainly because most of them are relevant for a theory of action, as the belief-desire theory makes plausible. Here I will probe the Humean conception of desires and wants only as a conceptual basis for values. What will a more explicit look at the Humean theory of value reveal, and what are the problems inherent within it?

IV. THE HUMEAN ACCOUNT OF VALUES

At first sight, one could suggest that to value something *is* to want it, thereby identifying values and wants. This would definitely amount to a totally reductive analysis of values, and may raise suspicion from the side of those who regard values as having a meaning of their own. Whatever this criticism amounts to, we ultimately have to accept it. Explaining values with wants is certainly reductive, and additional space must be granted for talk of values. Otherwise, the title of a Humean account of values would be misleading, purgation of values being the more correct one.

Phenomenologically, wants and values present themselves as distinct and irreducible. For example:

1) He wanted to do x, but hardly valued his doing (wanting) x.
2) She wanted to do x, because she thought highly of x.
3) She desires x beyond all limits, but she hardly values x.
4) * He values x, because he wants it.

Here we can see:

a) "wanting" and "valuing" are used in ways that give them different uses or meanings
b) "wanting" can be justified by recourse to "valuing", but not vice versa
c) wanting something and disvaluing it can exist simultaneously.

Obviously a) is implied by either b) or c), but there may be still other relations explaining a).

What do we mean by justifications given to wants by recourse to values? One example would be the following: He wants to be honest with her because he values honesty. Here, certainly, the meaning is not that he wants to be honest because he wants to be honest. Something different has to be fitted into these examples, some kind of (at most) indirect reduction to wanting.

Suggestions of such an analysis could make use of the following elements, all of them to be included within a more complex theory of wanting:[9]

- *general wants*, for example wanting honesty;
- *potential wants*, either general or specific;
- *desirability descriptions*, which specify an object in such a way that a want directed at this object receives some object-specification;[10]
- *desirability perception*, which perceives an object as a want that need not be shared fully, but which must be realized as such.

People sometimes say "I do not want to do x, but I can see the *desirability* of doing x", which I take to be a possible substitute for "I do not want to do x, but I can see the *value* of doing x". Within a position

identifying values as wants, this locution would not make sense. But
the use of desirability descriptions lends meaning to this kind of sentence.
"I do not want to polish my car for two hours, but I can see the desirability of doing this on a Saturday afternoon for others" would be an
example not too far off. As I am referring to mere common knowledge
in this desirability description, a sharing of the want in any sense should
not be necessary. In the latter case, however, I would prefer to speak
of desirability perception. I do not need to have the want myself, but I
must be able to imagine myself in the position of those sharing the
want. There is a well-known terminological distinction between
"empathy" and "sympathy", taking the first to be motivationally neutral,
whereas the latter is motivationally engaged. Desirability perception in
my sense would follow the empathy dimension. We could, of course,
use the term in a motivationally loaded sense, and perhaps this is its more
natural understanding. But we have to find a halfway house between
actual wants and descriptions neutral in value.

The strategic point at the beginning of this section being to create some
distance between wanting and valuing, we now have some material to
hand, with which to rephrase talk of values. Sentences (1)–(4) above
could be made sense of by inserting general wants instead of values.
In contradictions like (1) and (3), particular wants contradict general
wants. In cases of valuing without motivation, one could take the general
want as logically excluding a particular want, perhaps because of that
want's realization being blocked by another want.[11] In some way, the
general want to be honest has to find manifestation in effective wants
and action.

These remarks may be of help in illustrating that a want theory of
value must not be doomed from the start, but they do little more than
that. Wants can be analysed into structures endowing them with more
than the phenomenological condition which we started out with.
Occurrences of the justificatory or contradictory kind can be made sense
of whilst remaining within the realm of a want theory. But this is hardly
enough to make the theory convincing throughout. What we need in
addition is a *positive account*, both of the valuing person and the creation
of values. It still has to be seen whether persons can find an acceptable form of self-conception within this theory, and how we are to view
values springing into existence once deciphered as kinds of wants,
however sophisticated in form. It is surely with these two tasks that
the want theory of value meets its strongest objections. I will collect

objections of this sort under the heading of the '*naturalist-reductionism-objection*'. The following section tries to make explicit what is meant by this and why the want-theory ultimately fails to meet it. My remarks will concentrate on the relation of values and persons, leaving for another occasion to inquire into the more specific creation of values.

V. WHAT ARE VALUING PERSONS DOING?

Persons could be denoted as such a kind of beings that are self-productive. There would not be an overall social culture if persons could not produce their own culture, despite the fact that individual and social cultures may be closely interdependent. If individual cultures are granted, then there are different possible interpretations of how (and to what extent) persons are 'self-productive'. In order to make a clarifying, albeit rather rough distinction, I would like to introduce the categories '*naturalist want-production*' and '*non-naturalist want-production*'. A naturalist want-production will be a restricted one, restricted to at least one *given* want, i.e. a want *not produced*; a non-naturalist production will be unrestricted in this sense, i.e. wants will be universally produced. These categories give different interpretations of what can happen when there is a dynamic change of wants in persons, suggesting two different extensions which the resolution of conflicting wants in a person can have.

The Humean conception of naturalist want-production allows for some internal criticism within a more or less complex want-belief-system, persons being understood as manifestations of such a system. Want-production is accordingly motivated by the optimal fulfilment of wants. Wants will be differentiated into *basic* and *derived* wants, basic wants being those that may not be criticised as irrational, derived wants being those, that are arranged so as to fit instrumentally into a set of basic wants. Basic wants could be internally inconsistent, but the optimizing fulfilment of wants puts pressure on wants to be rearranged hierarchically. Whether one calls such an internally corrected system of wants and beliefs (corrected, amongst other things, due to misinformation, wishful thinking, etc.) 'rational', as do some neo-Humeans, or whether one, like Hume, reserves rationality for empirical truth only, should be of lesser importance.[12] More importantly in the present light, a consistently restructured set of wants will always remain within the boundaries of one or more *given* wants. Non-naturalist production, on the other hand, may be conditional, as is the naturalist one, but not conditional regarding given wants. In the terminology introduced, desirability descriptions could be used in place of basic wants. These descriptions *suggest* objects, actions, results, etc., as the possible objects of wants, but they do not presuppose any actual wants.[13]

The Humean conception should not be linked with hedonism. Hume was clearly not only a naturalist, but a hedonist of sorts, as can be seen from the following:

It appears evident that the ultimate ends of human actions can never, in any case, be accounted for by *reason*, but recommend themselves entirely to the sentiments and affections of mankind, without any dependance on the intellectual faculties. Ask a man *why he uses exercise*; he will answer, *because he desires to keep his health*. If you then enquire, *why he desires health*, he will readily reply, *because sickness is painful*. If you push your enquiries farther, it is impossible he can ever give any. This is an ultimate end, and is never referred to any other object.

Perhaps to your second question, *why he desires health*, he may also reply, that *it is necessary for the exercise of his calling*. If you ask, *why he is anxious on that head*, he will answer, *because he desires to get money*. If you demand *Why*? *It is the instrument of pleasure*, says he. And beyond this it is an absurdity to ask for reason. . . . Something must be desirable on its own account, and because of its immediate accord or agreement with human sentiment and affection ([11], App.I, p. 244).

One could, albeit unnecessarily, interpret this citation as an argument in favour of general hedonism. Of course, the naturalist would rather not be a hedonist, due to some familiar, and – since Mill – even famous criticisms of hedonism. Not only are there different levels of joy or pleasure (besides the dimension of intensity), there are even some unpleasant experiences which are of value to us. One could perhaps put anti-hedonist observations in a nutshell by drawing attention to the general insignificance of pleasure or pain as foundational feelings. Pleasurable experience may go along with the satisfaction of wants, thoughtfully chosen, but this experience itself *is produced* by the chosen want, and not vice versa. And even in the constellation of desires producing pleasure, this production is often a by-product, not the single aim behind the conscientious following out of a specific want. If we are happy, our wants and aims will be accompanied by pleasure, but the hedonic drive is neither foremost in our plans, nor is it predominantly given in relation to pain or milder inconveniences. The upshot is that something like a 'pure' feeling of pleasure is simply not something from which to construct human activities.

Now, the same anti-foundationalist remarks can be directed against basic wants, thus bringing trouble not only to the hedonist-naturalist à la Hume, but also to a want-naturalist. Phrasing itself in terms of desires, wants or preferences is surely the most up-to-date and widespread form of the neo-Humean conception of motivation or rationality (see [7], pp. 84–105). The problem with want-naturalism, again, is that it does no justice to the extent with which we can work on our ability to distance ourselves from our wants, and to see them as desirability descriptions free of actual cravings. Lest I be misunderstood: wants, if connected with bodily feelings, are certainly not to be simply switched off or on. In some cases they may become malleable with time, if worked upon. But feeling a want should not be confused with 'having' a want. *Having a want* for a human person includes some sort of *identification* with it, some grasping with the mind, itself not springing from another want. The naturalist want-theory does not do justice to the extent persons and their self-critical powers are *free from* their given wants and thus capable of working on them. This is what I think Rawls, in a Kantian spirit, meant by his remarks, referred to earlier, that claims are self-originating only if springing from a power irreducible to wants and desires ([25], 545). And this is also what the 'naturalist-reductionalism-objection' mentioned at the end of the last section amounts to. The naturalist want-production is not plausibly backed by observations concerning the human capacity critically to reflect and revise one's own wants and desires. As said earlier, this capacity is not in any way specifically moral, but it is a necessary precondition for moral action and, as will be seen in a moment, moral judgment.[14]

I criticised Rawls earlier for leaving us with hardly any explanation as to the sources of one of his 'highest moral powers', i.e. the capacity for critically revising one's ideas of the good life. Perhaps such a capacity, even if indirectly linked with wants, might still indeed look mysterious. The explanatory element I have brought forward is given in desirability-descriptions, and also in desirability-perceptions if one wants to go a step further. To put things into a larger frame, one could also say that the use of *language*, something of hardly any function in the naturalist want-theory, opens up the possibility of persons being 'self-originating'. Perhaps not all, but surely many wants – and among them the more individual and personal ones – spring from the creative use of language.

A creative process initiated by a power capable of social mediation seems to be necessary in order to explain many wants being of a social and cultural kind. But it seems clear that, again, these remarks do not

add up to something like a satisfying explanation of what a 'self-originating' production of wants as a non-naturalist alternative could amount
to. Whatever such a lengthy explanation may look like, we should remind
ourselves that it will not be bound – as the naturalist is – to something
given in human nature. Linguistic capacities are surely also part of human
nature, and certainly have a natural foundation. But the difference arises
when we think of the endless alterability and revisability of linguistic
structures. N. Chomsky is famous for searching for the existence of a linguistic theory compatible with our 'unlimited' production of sentences.
Other candidates for such a theory of unlimited human creativity lacking,
want-production could be seen as linked with our development of an
endless series of desirability descriptions, both in an individual and a
social sense. Only language, it seems, does have the potential for unlimited use while being securely, i.e. psychologically, founded in manifest
human characteristics. The creative use of language should be seen then
as the most important 'self-originating' human capacity.

VI. A CONTRACTUALIST VERSION OF AUTONOMY

Whatever the psychological details, whether empirical or conceptual,
of persons being non-naturalistic sources of values, there is also a *social
and morally relevant side* to this, which I shall turn to now. The naturalist theory of wants and values does not cohere with the unrestricted
practical dominance we give to autonomy, whereas the non-naturalist
theory reconstructing persons as powers of self-creation neatly fits in
with such a dominance. I will first make more explicit what is meant
by "practical dominance", then follow with the links to want production, and finally suggest an explanation for our accepting the importance
of autonomy as a value. The basis of argumentation will be, as in Rawls,
our shared 'moral intuitions', but instead of arguing for a rather general
political principle of liberty (as Rawls does in his first principle of
justice), I will restrict myself to an illuminating biomedical example.

   What does the statement that autonomy is an extremely powerful value
within our culture mean? Obviously, that we are restricted in our activities which could endanger the autonomy of others, both morally and
juridically. But there are two possible ways of formulating this dominance of autonomy, linked not only with different practical consequences,
but also with different strategies of justification. The relevant distinction here is between an *instrumental* and a *non-instrumental* interpretation

of autonomy. The instrumental one may be the most obvious, or at least the one easily to be combined with the Humean position. In which sense should autonomy be a value? Because we want it, the Humean should say, in the sense of a general want. More explicitly, this could mean that we want autonomy because we want to be free in our activities, to act as we would like to on many occasions. This want is not only in opposition to the alternative of being simply unfree – others restricting our actions against our will and interest – but also to the alternative of being taken care of, perhaps paternalistically – others doing for us as they think best for us. Why do we not like to be taken care of, preferring to decide for ourselves? The 'instrumental answer' would be: because otherwise we would not be able to satisfy our real wants. Take a decision like where to go on holiday: how could somebody else know where we really want to go? (We sometimes just have to make up our minds as we think about it, how should this be possible for someone else?) The instrumental justification would be one referring to wants and their satisfaction *separate* from the want for autonomy itself.

This instrumental understanding does not encompass all of our autonomous practices, to be sure. Sometimes we value autonomous choice not for its ends, but for itself, intrinsically (the first sense distinguished earlier). Someone we love presents us with a gift he has chosen himself, which we do not like. Alternatively, he could have presented us with a gift we like, but not chosen himself. If we are attached to the person, we obviously prefer the first alternative. The same may happen from a first-person perspective. When young, we often dress a bit extrovertly, not to say chaotically, but if this reveals our tastes of the time, it may be better than clothes forced upon us by others. The same also goes for much more important things, like deeply felt convictions or encompassing visions of the good life. How should they receive an instrumental justification at all? How could encompassing ideas, shaping minor wants and goals, be instrumental to single wants, even if taken in sum?

A case in point, of notorious medical and juridical relevance, is the case of a Jehova's witness who declines being saved from his dangerous physical condition by receiving foreign blood. In such cases, bioethical wisdom among Western professional circles usually pleads that he have his will, given that it is informed and not due to external pressure, and that the patient is not a child. We may be uncertain as to whether this view could be linked with an instrumental understanding of autonomy, the autonomous decision on the side of the Jehova's witness being one for his after-life, or for his soul, or for God, or whatever, instead of being purely and simply for autonomy. Nevertheless, the autonomous decision is not instrumental to the leading of a normal life, or to satisfying the usual wants of human nature. What we respect from the outside, in any case, is the autonomous decision of the patient, irrespective of the decision's end, which we may think of as being definitely wrong.

To generalize from this example, we could point out that autonomy could not ultimately be defended in a purely individualistic manner (whether naturalist or non-naturalist), but has to be justified as a *social* value. There is a social side to autonomy, strange as this may sound for such an individualist concept. If we accept the Jehova's witness' decision, we do so because of his autonomy, and one explanation could be that we respect, i.e. in a sense value, his capacity to decide for himself. This reason could be given mutually, and would therefore be, if practically effective, contractualist. (Contractualist reasons being such that the social agreement defines what proves to be a good reason.) In some interpretations, for example among consensus theoreticians, a logical force to accept the central values of a community of persons exists. And Rawls sometimes seems to suggest something similar by using the idea of the "reasonable" as referring to social presuppositions everybody has to accept (see [6] for such a reconstruction). But, in my opinion, this is no more than a terminological manoeuvre. Seeing persons as capacities for critical self-revision is nothing inherently necessary, but one suggestion amongst several. Perhaps we could illustrate why many details of our morality would fit in with each other in a more harmonious way if we respected each other as – above all – autonomous. But this not only has to be made clear enough (something still to be done); it also has to be practically accepted. A constructivist account should, in my opinion, have a definite activist side in it.

VII. THE VALUE OF HUMAN LIFE

In my earlier sketch of the Kantian argument (see section 2) I pointed to Kant's deriving absolute rights from an absolute value of dignity. Although, in a sense, we have meanwhile argued ourselves back to Kant, we will not be able here, I am afraid, to proceed similarly. The practical results of present-day value theory, as for example in P. Singer's bioethics, to many seem heavily counter-intuitive (not to say horrendous), at least in part. Such results spring rather consistently from valuing human life according to the standard of an interest in life, and, more explicitly, according to a want to have one's life continued. Beings incapable of such a want are said to lack an interest in life, and ending their life is then not to be seen as contradicting their life's value. This is roughly how the argument in utilitarian bioethics proceeds (see [27]). Far-reaching, deep moral consequences thus follow from the naturalist identification of life's value with a want to have a life.[15] Beings (including human beings) incapable of having a want to live are then not bearers of value, and – lacking value – killing them seems morally neutral, at least as far as these beings themselves are concerned, which is the most important thing to take account of normally. Consequences of a counter-intuitive sort, indeed.

The original Kantian argument would change the scene because human beings, even if not capable of reason in a specific moment nevertheless would have absolute value as 'bearers' of reason. Even in a reduced form, an absolute amount of value (or value of an absolute kind) could be seen as empowering even its somehow defective manifestations with enough value to be regarded as practically important. But now, if we were to rewrite the Kantian argument as we did with something like empirical psychological capacities, even if it were unlimited in a sense (as in the use of language), moves like these would not be possible. Rather, having a want to continue living seems to be a capacity developmentally of a lesser sort compared with a capacity for critical self-reflection. *Psychologically* grounded autonomy, if seen as the source of human value, will certainly not lead to intuitively nicer results in bioethical contexts than psychologically grounded wants to live a life, presupposing self-consciousness to at least some extent. To be more concrete: fetuses lacking a want to continue living will not be capable of investing their life with some individual sense, or critically 'revising their ideas of the good life'.

A rather different sort of output would be achieved if the value of autonomy were to be invested in a *procedure of social discussion* with the goal of practical decisions. Something like 'practical discourse' with its own ideal structure is obviously needed, *even if* a moral theory *would be* at hand: in hardly any a world will all the citizens be in possession of this theory. Accepting each other as autonomous is the altogether best way of dissolving controversial practical goals and opinions. As a political procedure in the way Rawls views it, the coming to consensual solutions is something which has to become a part of applied ethics, and the practically relevant part may well shift into this political annex. The primary participants within practical discourse will be those who are actually able to prove themselves capable of presenting their own opinion. Those who are unable to do so, because they are too young, too ill, or not human, have to be represented by those who are interested in them. However this is done in practice, it will surely not lead to unanimous agreement. Thus nothing like a universal consensus can be the goal of discourse, but rather *realms of toleration* among those that respect each other for their autonomy. Delineating the boundaries of sensible toleration, as concerning the problem of abortion, for example, is a theoretical task worth working for as far as the value of autonomy is concerned; the realization of this task in a world void of a metaphysics of autonomy – and also void of human beings invested with absolute value – is a practical goal only to be achieved, if at all, at some time in the future.

Department of Philosophy
University of Zurich
Switzerland

NOTES

[1]   The formulation is taken from Rawls, except for one minor change. Rawls refers to persons as self-originating sources of claims ([25], pp. 543–548).
[2]   Moore's repeated dealings with intrinsic value have altered during his career (see [4] for an overview). He later gave up the early position of *Principia Ethica*, which regards intrinsic good as being independent of experience. In *Principia Ethica* Moore wrote that, in order to decide which things have intrinsic value, "it is necessary to consider what things are such that, if they existed *by themselves*, in absolute isolation, we should judge their existence to be good" ([18], 187). But later he came to the conclusion that nothing but experience can be an intrinsic good. "I think now, as I did not when I wrote *Principia*, that the existence of some *experience* . . . does follow from the hypothesis that there exists a state of affairs which is good" (from [4], p. 319). Moore defined "intrinsically good" thus: "it would be a good thing that x should exist, even if x existed quite alone, without any further acompaniments or effects whatever", or "the value a thing possesses depends solely on the intrinsic nature of the thing in question", or "worth having for its own sake" ([20], pp. 93–5).
[3]   Intrinsic, or in slightly different terminology, inherent value has been invoked by environmental philosophers such as [26, 29, 1] and ([3], Ch.8–9). Criticism on these moves came from [23, 30, 31]. Of course, Mackie ([17], Ch.1), is also remembered for his queerness argument.

⁴ See [16] for a more extensive critical treatment of the "last man argument".

⁵ I am following here [24] and ([21], Ch.2).

⁶ But Kant tried to distance himself not only from naturalist positions like Hume's, but also from intuitionists like Leibniz, neither position being appropriately 'constructive'. For his half-hearted criticism of rational intuitionism in the person of Leibniz, see ([25], p. 559).

⁷ See [8] for more explicit treatment.

⁸ Kant of course provides a more elaborate argument for the practical imperative in the second form ("Act so that you treat humanity, as much in your person as in the person of every other, always at the same time as an end and never as a means only" (429)). In the relevant passage of his *Groundwork* he refers to the book's third section, devoted to proving that a rational person has a free will. But the whole argument is a *non sequitur*. Persons having a free will only provide a condition necessary for their moral powers. If persons are seen as sources of free will, then this only grants them the entrance-ticket to the realm of morality.

⁹ I am indebted for the following to ([22], Ch.3), although I may sometimes treat these concepts a bit differently.

¹⁰ There could be different forms of desirability descriptions, as for example the following: ". . . one paradigm of a desirability characterization is a specification of the object of desire in terms which reveal that object to be suitable for meeting some need or want which the agent recognizes himself to have. Such a specification brings to an end the questioning of why the agent has that desire . . ." ([22], p. 38) I take it that the second point mentioned here (the desire's causes or reasons) does not necessarily follow from the first, the object's suitability. I would prefer to define "desirability characterization" as the specification of an object in relation to a desire.

¹¹ See ([22], pp. 77–80), making use of an analysis by Davidson.

¹² Brandt's [2] procedure of 'cognitive psychotherapy' could be taken as a neo-Humean reading of practical rationality. As for Hume, see ([10], p. 415, p. 458).

¹³ Non-conditional want-productions seem logically possible, the pure imperative for being 'authentic', as in Sartre, being followed by some inclination, however deeply felt; but I take it as being implausible on several counts and thus negligible. The topic touched upon here, however, is a tricky one, as will become clear in the following. If wants are not pre-given, then what can be given in (or for) persons as conditionally selecting their want-production without seriously restricting their capacity for self-creation? It seems obvious, lest one involves oneself in Kant's incompatibilism on freewill, that want-productions have to be conditional in a way. But the question remains unclear of which way would be the most plausible one to compatibility.

¹⁴ A more explicit treatment of the capacity referred to here has been introduced into recent discussions by [5]. I will not engage with his still naturalist conception, in part due to lack of space, but also because it seems to me that an analysis of autonomy can ultimately only be argued for by moral intuitions, something done somewhat sketchily by Rawls. See my next section.

¹⁵ I prefer to present Singer's argument like this rather than in conjunction with his concept of a person, because the latter too easily meets the objection of non-naturalist metaphysics (see Honnefelder's contribution to the present volume). Honnefelder's move against Singer is a strange one: firstly unnecessarily pulling his argument into metaphysical

depths, and then criticizing him for his 'theoretically and practically implausible' metaphysics. According to my reconstruction, no theoretically or normatively loaded concept of a person is needed at all for a position such as Singer's. All of his practical points can be argued for with the common sense concepts of preference, want or interest. And someone like Honnefelder should then argue that these concepts are not metaphysical enough, instead of taking the opposite line.

BIBLIOGRAPHY

1. Attfield, R.: 1987, *A Theory of Value and Obligation*, Croom Helm, London, New York, Sydney.
2. Brandt, R.: 1979, *A Theory of the Good and the Right*, Oxford University Press, Oxford.
3. Callicott, J.B.: 1989, *In Defence of the Land Ethic. Essays in Environmental Philosophy*, State University of New York Press, Albany/NJ.
4. Duncan-Jones, A.: 1958, 'Intrinsic value: some comments on the work of G.E. Moore', *Philosophy* 33.
5. Frankfurt, H.: 1971, 'Freedom of the will and the concept of a person', *Journal of Philosophy* 68, 5–20.
6. Freeman, S.: 1991, 'Contractualism, moral motivation, and practical reason', *Journal of Philosophy* 88, 281–303.
7. Gaus, G.F.: 1990, *Value and Justification. The Foundations of Liberal Theory*, Cambridge University Press, Cambridge.
8. Hill, T.E.: 1992, 'Humanity as an end in itself', in T.E. Hill (ed.), *Dignity and Practical Reason in Kant's Moral Theory*, Cornell University Press, Ithaca-London, pp. 38–57.
9. Hoff, C.: 1983, 'Kant's invidious humanism', *Environmental Ethics* 5, 63–70.
10. Hume, D.: 1973, *A Treatise of Human Nature*, L.A. Selby-Bigge (ed.), Oxford University Press, Oxford.
11. Hume, D.: 1975, *Enquiries Concerning the Principles of Morals*, L.A. Selby-Bigge (ed.), Oxford University Press, Oxford (3rd. ed.).
12. Jonas, H.: 1989, *Das Prinzip Verantwortung*, Insel Verlag, Frankfurt.
13. Kant, I.: 1963, *Lectures on Ethics*, Hackett Publ. Comp., New York.
14. Kant, I.: 1964, *The Groundwork of the Metaphysics of Morals*, H.J. Paton (ed.), Harper & Row, New York.
15. Katz, E.: 1987, 'Searching for intrinsic value: pragmatism and despair', *Environmental Ethics* 9, 231–241.
16. Lee, K.: 1993, 'Instrumentalism and the last person argument', *Environmental Ethics* 15, 333–344.
17. Mackie, J.L.: 1977, *Ethics. Inventing Right and Wrong*, Penguin Books, Harmondsworth.
18. Moore, G.E.: 1903, *Principia Ethica*, Cambridge University Press, Cambridge.
19. Moore, G.E.: 1912, *Ethics*, H. Holt, New York.
20. Moore, G.E.: 1959, 'Is goodness a quality?', in *Philosophical Papers*, London, pp. 89–101.

21. O'Neill, J.: 1993, *Ecology, Policy and Politics. Human Well-Being and the Natural World*, Routledge & Kegan Paul, London.
22. Platts, M.: 1991, *Moral Realities. An Essay in Philosophical Psychology*, Routledge, London.
23. Pluhar, E.B.: 1983, 'The justification of an environmental ethic', *Environmental Ethics* **5**, 47–61.
24. Plumwood, V.: 1991, 'Ethics and instrumentalism: a response to Janna Thompson', *Environmental Ethics* **13**, 139–149.
25. Rawls, J.: 1980, 'Kantian constructivism in moral theory: the Dewey Lectures 1980', *Journal of Philosophy* **77.9**, 512–72.
26. Regan, T.: 1981, 'The nature and possibility of an environmental ethic', *Environmental Ethics* **3**, 19–34.
27. Singer, P.: 1993, *Practical Ethics*, rev.ed., Oxford University Press, Oxford.
28. Smith, M.: 1987, 'The humean theory of motivation', *Mind* **96**, 37–61.
29. Taylor, P.W.: 1986, *Respect for Nature. A Theory of Environmental Ethics*, Princeton University Press, Princeton.
30. Thompson, J.: 1990, 'A refutation of environmental ethics', *Environmental Ethics* **12**, 147–160.
31. Weston, A.: 1985, 'Beyond intrinsic value: pragmatism', *Environmental Ethics* **7**, 321–339.

H. TRISTRAM ENGELHARDT, JR.

SANCTITY OF LIFE AND MENSCHENWÜRDE: CAN THESE CONCEPTS HELP DIRECT THE USE OF RESOURCES IN CRITICAL CARE?

I. INTRODUCTION

Terms often play a guiding role in the fashioning of public policy precisely because they are strategically ambiguous. They unite disparate images, hopes, feelings, and aspirations that are at root heterogeneous. They then direct their force towards particular societal concerns. Such terms play a role in influencing outcomes, not a role in leading to resolutions of moral controversies by sound rational argument. In public policy debates it is frequently useful not to scrutinize such terms too closely because these coalitions of images which can move sentiments are then brought into question. Exploring them too carefully creates a tension among ideas and images that have worked in happy but unexamined collaboration. However, as with most coalitions, there come times in which public policy discussions bring the coalition itself into question.

The concepts sanctity of life and Menschenwürde (human dignity) are invoked in a wide range of bioethical debates including abortion, genetic engineering, and euthanasia. This essay will examine some of the meanings of these two crucial terms by focusing on the question of how to limit access to intensive care. Though both concepts will be addressed, the principal focus will be on the notion of the sanctity of life. Though analogies to these concepts can be found in many religious and cultural traditions, the focus will be on these notions within the context of Western Christianity and Western moral theory, for these have provided the prime setting for current debates.

This essay will explore whether concepts such as sanctity of life and

201

K. Bayertz (ed.), Sanctity of Life and Human Dignity, 201–219.
© 1996 *Kluwer Academic Publishers. Printed in the Netherlands.*

Menschenwürde can help guide policy for the use of resources in critical care. All industrialized societies and even developing countries are confronted with an inexorable rise in health care costs tied to new, expensive, and promising ways of postponing death and diminishing disability [15, 32, 40]. In the United States, approximately 1% of the Gross Domestic Product is deployed for critical care alone [21]. Choices regarding the use of critical care can now be made on the basis of various measures of the likelihood of survival, given maximal treatment [4, 10, 20, 22, 23, 25, 26, 27, 28, 38, 39, 41]. It will be increasingly easy to identify classes of individuals who, for example, have less than a 5% chance of survival, but for whom the cost of treatment will usually be in excess of $100,000. Under such circumstances, one will be saving lives at the cost of $2 million per life saved. For those who are elderly and have perhaps only 4 or 5 years of life expectancy, one would under such circumstances be securing additional years of life at the cost of $400,000 – $500,000. Similar questions arise at the beginning of life. If one identifies classes of newborns for whom treatment will generally cost over $250,000 and for whom the likelihood of success is less than 5%, one will be saving lives at the cost of $5 million per life saved. In addition, one can establish prognoses regarding the likely quality of life of such infants, were they to be saved. The public policy question is at what level of cost to general insurance programs ought critical care beds for the treatment of such patients not be provided.

The difficulty is that high-technology care in general, and critical care in particular, can often offer only a small chance of survival, but at high cost. As the populations of developing countries age, a larger proportion is constituted of physiologically brittle individuals who have a lower likelihood or survival, but for whom critical care can still provide a quantifiable benefit.[1] If an attempt is made to provide all health care that may benefit all individuals under all circumstances, health care costs will escalate even further. The proposal to limit high-cost treatment, based on a low likelihood of survival or the low quality of life, is at times criticized as failing to recognize the sanctity of life or the dignity of persons. Others criticize such proposals when they are tied to a system that allows the rich to purchase high-technology care not generally available to the poor.

The question addressed here is whether appeals to concepts such as sanctity of life and Menschenwürde can at all direct the identification of morally appropriate policy in these areas. It may be that these concepts do indeed bring with them sufficient moral content so as to provide the basis for sound rational arguments that can conclude to proper policy. Or, it may be the case that these concepts are useless in secular moral debates because they cannot function coherently outside of particular religious contexts. Or perhaps, insofar as they can give direction, it is more toward the possibility of grounding obligations of forbearance rather than obligations of beneficence.

II. "SANCTITY OF LIFE" AND MENSCHENWÜRDE: CONTENT-FULL BUT PAROCHIAL CONCEPTS

The Oxford English Dictionary offers three meanings of sanctity. The first concerns saintliness or holiness of life. The second, "the quality of being sacred or hallowed", includes the notion of inviolability. The third use, now rare, is equivalent to "holiness" as in the address of a patriarch (i.e., such as "His Sanctity, the Patriarch of Armenia"). The second appears to be the meaning core to the term sanctity of life. The claim contained cryptically in the term sanctity of life would appear to be that life has a certain inviolability or sacredness. This meaning is close to the meaning of *sanctitas* in the Latin, for which Lewis and Short give two central meanings, the first "inviolability, sacredness, sanctity", and the second "moral purity, holiness, sanctity, virtue, piety, honor, purity, chastity" ([24], p. 1626). This Latin root of the English term "sanctity" strengthens the supposition that claims regarding the sanctity of life trade on often unarticulated views that life has an inviolability or a sacredness.

Within a number of religious perspectives, life can have a special sacredness and one may have a duty in particular ways not to harm living entities, in particular, living humans. This latter concern is a fundamental and traditional one in Christianity.[2] Thus, the Didache, which may stem from the first century, enjoins: "Thou shalt not procure abortion nor commit infanticide" ([9], I). The document also condemns murderers of children ([9], V). The concern does not appear to be with some intrinsic quality possessed by humans, but rather with doing God's will and submitting to His laws. This religious concern with life is not an abstract obligation focused on some property of life (i.e., sanctity). Instead, treating life properly is integral to a saintly life, one in conformity with that which is holy. It is only later as concerns with sanctity of life are disarticulated from this living religious experience, and attempts are then made to justify these Christian proscriptions in general secular terms, that claims regarding sanctity of life take on an independent role. But at this point connection with the moral vision in terms of which and within which they had their particular meaning and substance weakens.

Discussions of sanctity of life appear to have entered the English bioethics literature some time in the early 1950's just as the field was emerging.[3] Initially, the reflections appear to concern issues such as euthanasia and abortion. An article, for example, by John Sutherland Bonnell in 1951 develops a criticism of a proposal for the legalization of voluntary euthanasia by noting that Christianity has emphasized "the sanctity of human life and the value of the individual, even the humblest and lowliest, including the afflicted in mind and body" [1]. Bonnell's article is followed by an essay by Joseph Fletcher, which underscores the centrality of persons over the value of mere life, suggesting that one would be better advised to speak of the role that persons have in deciding regarding themselves, rather than to appeal to principles like the sanctity of life [13]. One sees already here the tension between a morality framed by reference to the special value of being human and morality framed by reference only to that which can bind persons as such.

In 1957 the sanctity of life found special articulation in jurisprudential discussions through a book by Glanville Williams [42]. A further important examination was provided by Norman St. John-Stevas in a 1964 volume with the title, *The Right to Life*. There he argued that the Christian attitude toward euthanasia is based on "the principle of the sanctity of life" ([31], p. 43). There is much to suggest that many of the concerns regarding the sanctity of life reflected interests also expressed in the term "right to life" [35]. In any event, in 1967 the term still had a robustly religious significance [30]. Harmon Smith, in commenting on and quoting from Paul Ramsey, speaks to the religious framework within which the notion of sanctity of life was understood by at least a number of scholars [36].

> . . . the question of *when* sanctity attaches to human life is not religiously problematic at all [for Ramsey]: 'One grasps the religious outlook upon the sanctity of human life only if one sees that this life is asserted to be *surrounded* by sanctity that need not be in a man; that the most dignity a man ever possesses is a dignity that is alien to him. . . . A man's dignity arises from God's dealings with him, and not primarily in anticipation of anything he will ever have it in him to be.' ([36], p. 42)

Sanctity of life for Ramsey and those who take similar positions is derived from the fact that God values us.

Attempts were made in bioethics to give the principle of the sanctity of life a less religious significance, as illustrated by Daniel Callahan's 1970 exploration of abortion, where he first narrows the scope of sanctity of life to focus on human life and then attempts to give it a secular meaning. "An affirmation of the sanctity of life which required that one accept a religious view of man's origin would provide a weak base upon which to build a consensus. One then would seem to be saying that there is nothing whatever upon which to ground the sanctity of life save that of religious belief. . . ." ([3], p. 315). Callahan's analysis leads him to understand the affirmation of the sanctity of life as a moral affirmation of "the protection and preservation of human life, both actual and potential" ([3], p. 343). He surveys a diverse collection of rules and rule systems that cluster around the notion of sanctity of life, including "(a) the survival and integrity of the human species, (b) the integrity of family lineages, (c) the integrity of bodily life, (d) the integrity of personal choice and self-determination, mental and emotional individuality and (e) the integrity of personal bodily individuality" ([3], p. 327). Callahan recognizes the ambiguity of the term, but still attempts to derive from it some useful direction and moral sense.

The ambiguities involved in appeals to the sanctity of life led both to criticisms as well as attempts to provide greater analytic clarity. On the one hand, K. Danner Clouser in 1972 concluded that "I find the sanctity of life concept to be impossibly vague and to be a concept that is inaccurate and misleading, whose positive points can be better handled by other well-established concepts" ([5], p. 119). On the other hand, in 1975 William Frankena recognized that "Expressions like 'respect for life,' 'reverence for life,' and 'the sanctity of life' have a currency today they never had before" ([14], p. 24) and attempted to provide a guide to the various meanings that cluster around the term sanctity of life. These he summarized under six points.

1. The sanctity of human life (bodily) is not relevant to the discussion of all bio- and ecoethical questions, but only to those involving the preventing or shortening of human life. Others involve the sanctity of individuality or personality, or quality rather than quantity of life. The sanctity of bodily human life should be distinguished from that of individuality or personality, even if there is a connection.
2. Mere life, whether that of a vegetable, animal, or human organism, has no moral sanctity as such, though it may have aesthetic and

other kinds of nonmoral value, and may be a necessary condition of consciousness, rationality, or morality.

3. Life has moral sanctity, but only where it is a condition of something more, as it is in humans, fetuses, and some animals.

4. This something must be something inherent – consciousness, feeling, reason – in such living beings, but not just being immortal, not something wholly extrinsic.

5. Even then the moral sanctity of human life (bodily) is not absolute, it is considerable, at least from the moral point of view, but it is only prima facie or presumptive.

6. The only tenable view, then, is a derivative, qualified, and noncomprehensive ethics of respect for life (p. 58).

Despite the unclarities and difficulties, philosophers [2], lawyers [33], and social scientists [6] have continued to invoke the concept of sanctity of life. The first American Supreme Court examination of the right to refuse life-saving treatment makes reference to the principle in its holding in Nancy Beth Cruzan vs. Director of Missouri Department of Health [8].

The diversity of appeals and the heterogeneity of the considerations adduced in favor of considerations such as the sanctity of life suggest that there is a fundamental difficulty. If concepts such as sanctity of life and Menschenwürde do indeed have their roots in a Judeo-Christian understanding of a God-grounded concern to reverence life and to recognize the moral standing of humans because they are made in the image and likeness of God,[4] then when appeal to God is no longer made to justify such considerations, the grounding for these concepts may disappear. They may lose both their traditional justification and the reference point in terms of which their content can be specified. If, however, they are to be understood independently of such religious origins and instead in terms of a philosophical argument, at the very least such terms may benefit from a more secular recasting. One might consider a less religious term in lieu of "sanctity of life", such as the "inviolability of life". Even such terms, though, are likely to overstate the ability of any general secular philosophical argument. A strong sense of the inviolability of life would preclude killing insects and using deadly force against would-be murderers. At the most, one could envision secular philosophical arguments establishing claims such as (1) if life is generally valued, one can generally advance grounds to show that more benefits

than harms will come from destroying life or failing to sustain it, and
(2) one ought not to kill persons or that which belongs to persons without
the permission of those persons.[5]

Similar problems beset the concept of Menschenwürde.[6] Consider
Kant's exploration of the concept in *die Grundlegung zur Metaphysik der
Sitten*. There Kant distinguishes between things that have a price and
persons that have a dignity. He argues, for example, that "that which is
related to general human inclinations and needs has a *market price* (Was
sich auf die allgemeinen menschlichen Neigungen und Bedürfnisse
bezieht, hat einen Marktpreis)" ([17], p. 53; [19], IV 434). In contrast,
"that which constitutes the condition under which alone something can
be an end in itself does not have mere relative worth, i.e., a price, but
an intrinsic worth, i.e., *dignity* (was die Bedingung ausmacht, unter der
allein etwas Zweck an sich selbst sein kann, hat nicht bloß einen relativen Werth, d.i. einen Preis, sondern einen innern Werth, d.i. Würde)"
(p. 53, IV 435). Here Kant articulates a concept of Menschenwürde.
"Thus morality and humanity, so far as it is capable of morality, alone
have dignity (Also ist Sittlichkeit und die Menschheit, so fern sie
derselben fähig ist, dasjenige, was allein Würde hat)" (ibid.).

Kant's tie between being human and having dignity can also be seen
in the second version of the categorical imperative. "Act so that you treat
humanity, whether in your own person or in that of another, always as
an end and never as a means only (Handle so, daß du die Menschheit
sowohl in deiner Person, als in der Person eines jeden andern jederzeit
zugleich als Zweck, niemals bloß als Mittel brauchst)" ([17], p. 47;
[19], IV 429). However, the connection between being human and having
dignity is not examined. Kant uncritically identifies humans or humanity
as worthy of absolute worth or as possessing dignity. Yet his morality
is concerned not with humans, but with persons.

Beings whose existence does not depend on our will but on nature, if they are not rational
beings, have only a relative worth as means and are therefore called "things"; on the
other hand, rational beings are designated "persons" because their nature indicates that
they are ends in themselves, i.e., things which may not be used merely as means. Such
a being is thus an object of respect and, so far, restricts all [arbitrary] choice. Such
beings are not merely subjective ends whose existence as a result of our action has a worth
for us, but are objective ends, i.e., beings whose existence in itself is an end. Such an
end is one for which no other end can be substituted, to which these beings should serve
merely as means. For, without them, nothing of absolute worth could be found, and if
all worth is conditional and thus contingent, no supreme practical principle for reason
could be found anywhere. (Die Wesen, deren Dasein zwar nicht auf unserm Willen,

sondern der Natur beruht, haben dennoch, wenn sie vernunftlose Wesen sind, nur einen
relativen Werth, als Mittel, and heißen daher Sachen, dagegen vernünftige Wesen Personen
genannt werden, weil ihre Natur sie schon als Zwecke an sich selbst, d.i. als etwas, das
nicht bloß als Mittel gebraucht werden darf, auszeichnet, mithin so fern alle Willkür
einschränkt [und ein Gegenstand der Achtung ist]. Dies sind also nicht bloß subjective
Zwecke, deren Existenz als Wirkung unserer Handlung für uns einen Werth hat; sondern
objective Zwecke, d.i. Dinge, deren Dasein an sich selbst Zweck ist und zwar ein solcher,
an dessen Statt kein anderer Zweck gesetzt werden kann, dem sie bloß als Mittel zu
Diensten stehen sollten, weil ohne dieses überall gar nichts von absolutem Werthe würde
angetroffen werden; wenn aber aller Werth bedingt, mithin zufällig wäre, so könnte für
die Vernunft überall kein oberstes praktisches Princip angetroffen werden) ([17], pp. 46–7;
[19], IV 429).

Kant provides a morality for persons, not a morality for humans.

Because Kant's morality focuses on what humans ought to do qua
persons, qua rational beings, not on what humans ought to do because
of their nature qua humans, qua biological beings of a certain sort, Kant's
arguments can be seen to have disturbing implications in areas where traditional Christian concerns regarding sanctity of life and the dignity of
humans proscribe interventions such as abortion and infanticide. After
all, the question arises whether fetuses, indeed infants, should be regarded
as persons, as rational beings. Without a convincing argument regarding
the moral significance of the potentiality of humans to become persons,
or establishing that human fetuses and children are in some sense already
persons, moral agents, fetuses and infants will at best, for Kant, have
the moral standing of animals.

Kant's attempt to provide a general secular rational defense of human
dignity thus reveals a difficulty. In order to give content and substance
to the concept of dignity, Kant must import a particular understanding
of what should count as proper human status and conduct. He must establish a particular view of human dignity, such that, among other things,
he can argue that masturbation is evil on a par with self-murder.[7] Either
Kant must find in rationality a particular content, which would appear
to involve the choice of one view of moral rationality over others. Or,
Kant must see rationality as having no general canonical moral content
and settle instead with understanding moral rationality as assessing
reasons for and against different choices in terms of the particular moral
views of that particular moral agent while allowing others to do likewise.
As a consequence, absent endorsing a particular content that is not
integral to rationality as such, as if it were so integral, it does not appear
possible for Kant to provide an account of what it means to respect

persons beyond gaining their consent. Which is to say, Kant requires a particular view of the value of rationality in order to gain the moral content he seeks. Yet he cannot secure the content he wants for his view of morality by an appeal to moral rationality as such. These difficulties put at jeopardy those humans who are not persons.[8] They also undermine traditional proscriptions against acts condemned morally despite their involving mutual consent, such as consensual intercourse between competent adult siblings. The concept of Menschenwürde, if it is to be more than a reminder not to use persons without their consent, must depend on a particular vision of proper human conduct.

In anticipation, this indicates that, when one turns to the issue of developing policy for the allocation of scarce resources for health care, one cannot derive content-full guidance in secular debates from appeals to the sanctity of life or Menschenwürde. At most, such appeals may remind those involved that they should consider the values associated with human life and should gain the consent of persons prior to using them. But no information will be given by virtue of such appeals about how one should rank various values, including various values regarding human life. Nor will one receive guidance about how a concern to respect humans should direct persons in framing concrete policies concerning the use of resources for critical care. Indeed, the centrality in secular moral debate of persons as moral agents over human nature as normative would suggest that policy for the allocation of resources for critical care can at best be created through agreement in particular associations and cannot be discovered by appeal to general moral principles.

III. THE ATTEMPT TO GIVE SECULAR SIGNIFICANCE TO RELIGIO-ETHICAL CONCEPTS: WHY THIS FAILS

The modern Western philosophical project has in great measure been to provide a secular foundation for the general lineaments of Western Christian morality and to secure a secular foundation for the authority of the state, without a reliance on the particularities of Christianity and without confessing the Judeo-Christian God. Against the fragmentation following the Protestant Reformation, and given the bloodshed of the Thirty Years' War and the Civil War in Britain, one can understand this hope. Reason without reliance on particular religious understandings was invoked in order to provide a content-full, canonical morality binding men and women in political structures that were to draw their authority from being that to which rational individuals should consent. This project appeared plausible in the wake of the Middle Ages, with its strong faith, not just in Faith, but in reason's capacity to discover the general content of the natural law, the *jus naturale*.[9] The Christian synthesis of the West presumed not just that grace could direct, but that right reason could disclose the content of a morality binding moral strangers, that it could show individuals the rationally authoritative character of those political institutions, which would be shown by philosophical argument to be just.

This project was appealing as well, given what rational argument promised. If sound rational argument could show which moral choices were appropriate and which were inappropriate, then one could (1) dismiss those who disagreed as irrational and even (2) visit them with coercive force, against which they would have no rational grounds for protest, because the force was used to bring them to rational behavior consonant with their nature as rational. They would be brought into conformity with their true selves. Initially in these arguments, a philosopher's God continued to play a role. In the shadow of the Middle Ages, one recognized that it was inviting to have an intellectual focus, which combined a metaphysical and epistemological grounding for morality. The Christian theology and philosophy of the Western Middle Ages were able to unite in their appeal to the Deity the genesis of morality, the justification of morality, and the motivation for being moral. The genesis of morality was drawn from a God whose general purposes could be discovered by reason. God, in being both the supreme instance of rationality and the ground of being, gave full justification for the character of the morality disclosed. Moreover, violation of a morality endorsed by the Deity was not simply irrational, but clearly improvident in being against the very structure of reality and inviting of eternal sanctions. The standpoint of God was the intellectual vindication of the ultimate reasonableness of morality, so much so that Kant concluded that we must act as if God exists.

As modern Western philosophy became increasingly secularized so as to evacuate any of the marks of Deity from the standpoint of the disinterested observer or ideal hypothetical decision-maker, it became more difficult to bring together what had traditionally been expected regarding the concordance among the genesis of morality, the justification of morality, and the motivation for being moral. As Europeans became more "enlightened", their particular traditional morality came more to be regarded as a cultural artifact. Yet they continued to seek sub-stantive moral guidance from philosophy. However, there turned out to be as little unanimity in philosophy regarding the justification of morality or its content as there was in Western European religion after the Reformation. Worse yet, it is difficult to show why it is always prudent for individuals to take the moral point of view and act disinterestedly, especially when such actions are clearly against the self-interest of those individuals. A cleft opens between what it means to act prudently and what it means to act morally. Moreover, there appears to be no gener-ally securely establishable content-full understanding of what morality requires.

One can understand why in principle, not only in fact, this is the case. In order to make particular moral judgments, one must already possess a moral point of view. The difficulty is that there are so many moral points of view. For example, it is not possible to determine which moral understanding or public policy choice maximizes benefits over harms without first knowing how to rank or compare diverse benefits and harms (e.g., liberty consequences, equality consequences, prosperity con-sequences, and security consequences). Nor is this difficulty remedied by appealing to a preference utilitarianism. It is not simply that one cannot easily compare intensities of preferences, one with another. There is the difficulty of knowing how to compare present versus future prefer-ences, impassioned versus rational preferences. Different choices ground different accounts of morality.

Nor are such difficulties remedied by appealing to a hypothetical choice theory because the chooser or choosers must be fitted with a particular moral sense, set of intuitions, or thin theory of the good. If the hypothetical chooser or choosers are unprejudiced with respect to any particular moral vision, it(they) will have no moral sense and the chooser(s) will not be able to choose. If it(they) is(are) able to choose, this is precisely because it(they) possesses one among the many moral senses or thin theories of the good. One purchases universality at the price of content. One purchases content at the price of universality. Particular moral choices depend on the prior endorsement or accep-tance of a particular content-full understanding of moral rationality, moral discourse, proper moral choice, proper moral intuitions, or proper moral sensibilities, however thin the understanding. One must presuppose what one means to prove. That is, to endorse one particular set of moral choices, one must have already assumed a particular moral vision.[10]

The impossibility of justifying in general secular terms a canonical, content-full morality is the epistemological condition at the root of post-modernity. The recognition of this impossibility frames post-modernity. There is no canonical secular content-full morality for applied ethics uncontroversially to apply in principle and in fact. Still, much of applied ethics, and bioethics in particular, proceeds by ignoring these foundational difficulties. It is as if no one wished to admit that the content of the central canonical concepts that frame contemporary bioethics cannot be justified in general secular terms. One finds instead appeals to intuition or appeals to consensus without an account as to why certain intuitions rather than others should be decisive, or as to why a near but not unanimous consensus conveys moral authority. Any solution to these foundational difficulties cannot provide us with a secular morality with the content, force, and scope which our history has made us expect. If one cannot with justification appeal at least to a philosopher's God, one will not get a firm connection between what rationality can justify, with respect to moral content, and what rationality can motivate with respect to moral action. If one cannot with justification appeal to a unique canonical intellectual standpoint that can deliver moral content, one no longer has one governing moral point of view but an unprincipled choice among many [11].

Post-modernity is defined by this richness of possibilities, and the impossibility of rationally justifying in general secular terms one content-full moral vision as canonical. One cannot discover who is a moral authority in general secular terms by an appeal to a canonical, content-full moral vision. In these circumstances, we are forced to draw authority not from a canonical vision of right action, but from the consent of individuals collaborating in limited projects. This is the case not because individuals, individualism, autonomy, or a limited state is highly valued. It is rather that the appeal to individuals as the source of secular moral authority is the only mechanism by which to justify a common moral authority among moral strangers, when all do not hear God in the same way and when secular reason cannot disclose the content-full canonical morality.

This approach to resolving controversies among moral strangers (i.e., attempting to derive moral authority not from a canonical moral vision, but from the consent of those collaborating) does not promise to provide us with a key to understanding claims regarding the sanctity of life or Menschenwürde. Persons in this secular context are contrasted with their bodies, with their nature as humans, and with that which in secular terms might be invoked as a basis for *human* dignity. In a secular context human nature is the result of random mutation, selective pressure, genetic drift, the constraints of biological, chemical, and physical laws, as well as various random events and catastrophes. Human nature can in this context be regarded as more or less well adapted in particular environments to particular goals that persons may have. Like other physical circumstances, human nature will be a matter to be altered, changed, and improved to meet the goals of persons. In this secular context human nature or human life does not appear as sacred. Nor can humans be understood in general secular terms as having any of the deep dignity that can be appreciated within a religious perspective. Instead, one discovers persons who, as the source of moral authority, can be invoked in order to resolve moral disputes by common agreement, and therefore by common authority. Persons are special, but they cannot be seen in general secular terms as having holiness or sanctity. Secular ethical arguments can maintain the moral centrality of persons, but this centrality is not dependent on the sanctity of life or the dignity of humanity. It rests instead on the circumstance that only persons can raise and resolve moral controversies and convey moral authority for joint action through providing permission. This has significant implications for critical care policy.

IV. CAN CONCEPTS OR PRINCIPLES SUCH AS SANCTITY OF LIFE AND MENSCHENWÜRDE HELP DIRECT THE ALLOCATION OF SCARCE MEDICAL RESOURCES FOR CRITICAL CARE?

Within traditional Western religious moral visions, sanctity of life and Menschenwürde announce barriers against direct abortion and direct infanticide. It is unclear, though, how far in a traditional Christian religious context they can provide a basis for precluding the establishment of upper limits for the use of communal funds in saving lives through critical care. Indeed, within the Christian moral vision where the preservation of physical life through medical treatment is not an overriding obligation, the very notion of saving life at all costs may involve the risk of rendering human physical life improperly sacrosanct and distorting the Christian moral vision that has traditionally focused on goals other than merely sustaining physical existence. In a secular context when these concepts are brought to bear on public policy debates concerning limiting the use of critical care, they provide little guidance.

Imagine confronting the choice between two policy options with regard to the use of critical care units. The first option would require providing critical care to all individuals who themselves or through their surrogates request treatment. Critical care would be required even if survival is unprecedented, costs considerable, and the likely quality of life limited (e.g., the patient would likely survive only in a persistent vegetative state). The second option would require providing critical care only if (a) the costs per year of life saved would likely be less than a particular amount (e.g., $100,000), and (b) the quality of the patient's life would include minimal sentience. In each case, let us grant that the policy has been democratically enacted and the policy funded through commonly owned goods.

The difficulty (though a difficulty not unanticipated, given the foregoing reflections) is that both sanctity of life and Menschenwürde as secular concepts are so ambiguous as to support starkly contrasting choices with respect to these two options. The principle of the sanctity of life, for instance, can be interpreted in the following two opposing fashions.

1. The principle of the sanctity of life requires one to save human life at all costs, including the use of critical care when there is little likelihood of success, the costs are considerable, and the likely quality of life minimal.
2. The principle of the sanctity of life requires that one preserve the values associated with human life and these can be put at jeopardy if one tries to save life at all costs.

The first interpretation of the principle of the sanctity of life builds on general concerns to preserve life, often associated with sanctity-of-life arguments, and would favor the first policy choice. The second interpretation of the principle of the sanctity of life builds on values supported by human life or achieved through living a human life and would favor the second choice. The second choice can in fact be seen as more in accord with the traditional Christian approach to the use of scarce resources articulated in Pope Pius XII's famous remark:

[N]ormally one is held to use only ordinary means – according to the circumstances of persons, places, times, and culture – that is to say, means that do not involve any grave burden for oneself or another. A more strict obligation would be too burdensome for most men and would render the attainment of the higher, more important good too difficult [29].

The Roman pontiff's point can be understood in both secular and religious terms. If one attempts to save life at all costs, among the costs are likely to be moral costs.[11] The assumption of these costs can be regarded as offending against the sanctity of life, because the sanctity of life is here interpreted not to be achieved simply through biological life, but through a self-conscious moral life. There is a recognition that if one makes the mere prolongation of life an overriding good, the place of other goods will be disturbed.

A similar point can be said regarding the principle of human dignity (Menschenwürde). It can also receive discordant interpretations:

1. the principle of respecting human dignity requires saving human life at all costs;
2. the principle of respecting human dignity requires respecting humans as free risk-takers who may collaborate and gamble in ways that will make them unequal even with respect to suffering and death.

The second interpretation will support a two-tier health care system (e.g., a basic tier for all subvened by the government and a private luxury tier available through private insurance and direct payments) in which access to critical care will be limited within the basic tier. How can one determine which one ought to endorse? Answers insofar as they are directed by considerations of human dignity will depend on different views of what it is to have human dignity.

The difficulties with appeals to the sanctity of life or human dignity are then multiple. First, the principles of the sanctity of life and of Menschenwürde are more principles as chapter headings bringing together a number of moral concerns having a historical or family relation than they are principles as foundations or grounds determining a set of choices. This appears to be the case in both religious and secular settings. Second, when disarticulated from their religious origins, the principles become even more unmanageable. In a general secular context, it is impossible to establish a canonical content-full notion of sanctity or dignity. At most one finds the centrality of persons as the source of moral authority for cooperative ventures of which health care is an instance. As a consequence, evacuated of its moral content, the second interpretation of the principle of human dignity is all that can be endorsed. It is not endorsed because it allows the good to be achieved, but because it recognizes persons as the source of moral authority when God is not heard and reason fails to disclose the canonical content-full vision of human flourishing. The language of human dignity must give way in general secular morality to the centrality of persons.

Secular appeals to the sanctity of life or to Menschenwürde will not secure grounds to prohibit policy for the use of critical care resources that would limit access to societally supported high-cost treatment for classes of patients unlikely to survive or to survive only with loss of sentience (e.g., a life in a persistently obtunded state). It will be impossible in general secular terms to discover how lines between acceptable and unacceptable levels of success, or acceptable and unacceptable levels of quality of life should be drawn, because such a disclosure depends on discovering the canonical ranking of values, goods, or goals. This, though, as we have seen, is not available through general secular reasoning. At best such lines must be created by common agreement regarding the use of societal resources, allowing by default (i.e., because one cannot establish the rightness of interfering) those who wish to purchase additional treatment to do so if they are so inclined and able. In secular debates principles such as the sanctity of life and human dignity are best recast in terms of achieving the values persons affirm and in terms of recognizing a secular obligation (by default) regarding non-interference in the peaceable choices of persons. On the other hand, within religious contexts these terms will function best if desecularized and embedded in the context of a particular living tradition that can give them specific content.

Center for Medical Ethics and Health Policy
Baylor College of Medicine/Rice University
Houston, Texas
USA

NOTES

[1]   One of the more useful clinimetrics for predicting survival, the APACHE III score, which scores from 0 to 299, awards 24 points toward the likelihood of death for being 85 or older [20].

[2]   The concern not to take life without due cause and even then with an appreciation of the significance of such actions is rooted in the Old Testament. One finds, for example, the following commands in the Noachite code: ". . . you must not eat meat that has lifeblood still in it. . . . Whoever sheds the blood of man, by man shall his blood be

shed; for in the image of God has God made man" (Genesis 9:4,6). It is the articulation of these concerns by Western Christianity within its philosophically, indeed metaphysically formed tradition that most directly touched the West.

3 There are publications prior to 1950 exploring "the sanctity of life". A book, for instance, which appeared in 1921 develops an argument for the security of property through an appeal in part to the sanctity of life [16]. See also [43] for a more religious treatment of sanctity of life. There is evidence that a significant source for the interest in the principle of sanctity of life in the bioethics literature comes from a recasting of concerns regarding reverence for life as articulated by Albert Schweitzer. William Sperry, the Dean of Harvard Divinity School, states in a 1948 New England Journal of Medicine article that "reverence for life" is the ethical basis of both the profession of medicine and the ministry ([37], p. 988). He develops this contention around a quote from Albert Schweitzer, which appropriation drew a critical response from Joseph Fletcher ([13], p. 206).

4 In an *Encyclopedia of Bioethics* article on the value of life, Peter Singer [34] attributes a religious foundation to concerns regarding the sanctity of life.

5 Caveat lector. This is as much as can be secured in general secular terms. The author as an Orthodox Catholic knows that more is in fact contentfully required and does indeed bind (e.g., he recognizes the serious moral evil of abortion and euthanasia).

6 The author recognizes that the concept of Menschenwürde has a special place in post-World War II discussions of ethics. The German Grundgesetz in point one of the first article in Section I on Grundrechte states: "The dignity of man shall be inviolable. To respect and protect it shall be the duty of all state authority. [Die Würde des Menschen ist unantastbar. Sie zu achten und zu schützen ist Verpflichtung aller staatlichen Gewalt.]"

7 "The ground of proof surely lies in the fact that a man gives up his personality (throws it away) when he uses himself merely as a means for the gratification of an animal drive. But this does not make evident the high degree of violation of the humanity in one's own person by the unnaturalness of such a vice, which seems in its very form (disposition) to transcend even the vice of self-murder" ([18], pp. 86–87) ["Der Beweisgrund liegt freilich darin, daß der Mensch seine Persönlichkeit dadurch (wegwerfend) aufgiebt, indem er sich blos zum Mittel der Befriedigung thierischer Triebe braucht. Aber der hohe Grad der Verletzung der Menschheit in seiner eigenen Person durch ein solches Laster in seiner Unnatürlichkeit, da es der Form (der Gesinnung) nach selbst das des Selbstmordes noch zu übergehen scheint, ist dabei nicht erklärt" (*Metaphysik der Sitten*, [19], VI, 424–426).

8 I have explored these difficulties with Kant's account at elsewhere ([12], pp. 68–71).

9 This faith in reason is well captured in the Roman Catholic requirement that one believe on faith that one can show the existence of God without an appeal to faith but by reason alone. "If anyone shall have said that it is not possible to know certainly the one and true God who is our Lord and Creator by the light of natural human reason through those things that have been made, may he be anathema." *Constitutio dogmatica de fide catholica, Canones*, II. *De revelatione*, 1, from the Fourth Session of the Vatican Council, 24 April 1870.

10 I have developed these arguments at greater length elsewhere. See, for example, [11, 12].

[11]   An excellent exploration of the appropriate distinction to be drawn between ordinary and extraordinary, proportionate and disproportionate treatment is to be found in a dissertation by Daniel Cronin [7]. Both his as well as subsequent analyses of these issues within Roman Catholicism do not embed them within the contemporary context of governmentally supported health care systems.

BIBLIOGRAPHY

1. Bonnell, J.S: 1951, 'The sanctity of human life', *Theology Today* **8**, 194–201.
2. Brody, B.A.: 1975, *Abortion and the Sanctity of Human Life: A Philosophical View*, MIT Press, Cambridge, Mass.
3. Callahan, D.: 1970, *Abortion: Law, Choice and Morality*, Macmillan, New York.
4. Chang, R.W.S., Jacobs, S. and Lee, B.: 1986, 'Use of APACHE II severity of disease classification to identify intensive-care-unit patients who would not benefit from total parenteral nutrition', *The Lancet* **1**, 1483–1486.
5. Clouser, K.D.: 1973, ' "The sanctity of life": an analysis of a concept', *Annals of Internal Medicine* **78**, 119–125.
6. Crane, D.: 1975, *The Sanctity of Social Life: Physicians' Treatment of Critically Ill Patients*, Russell Sage Foundation, New York.
7. Cronin, D.A.: 1958, *The Moral Law in Regard to the Ordinary and Extraordinary Means of Conserving Life*, Pontifical Gregorian University, Rome.
8. Nancy Beth Cruzan vs. Director of Missouri Department of Health, 111 L.Ed.2d 224, 110 S.Ct. 284, 58 U.S.L.W. 4916 (U.S. June 26, 1990).
9. 'Didache': 1965, in *The Apostolic Fathers*, trans. K. Lake, Harvard University Press, Cambridge, Mass., vol. 1, pp. 311, 313.
10. Eagle, K.A. *et al.*: 1990, 'Length of stay in the intensive care unit', *Journal of American Medical Association* **264**(8), 992–996.
11. Engelhardt, H.T., Jr.: 1991, *Bioethics and Secular Humanism*, SCM Press, London.
12. Engelhardt, H.T., Jr.: 1996, *The Foundations of Bioethics*, 2nd ed, Oxford University Press, New York.
13. Fletcher, J.: 1951, 'Our right to die', *Theology Today* **8**, 202–212.
14. Frankena, W.K.: 1977, 'The ethics of respect for life', in O. Temkin, W.K. Frankena and S.H. Kadish (eds.), *Respect for Life*, Johns Hopkins Press, Baltimore, pp. 24–62.
15. Health Care Financing Administration: 1987, 'National health expenditures, 1986–2000', *Health Care Financing Review* **8**(4), 1–36.
16. Hillis, N.G.: 1921, *The Better America Lectures*, Better America Lecture Service, New York.
17. Kant, I.: 1976, *Foundations of the Metaphysics of Morals*, trans. L.W. Beck, Bobbs-Merrill, Indianapolis.
18. Kant, I.: 1964, *The Metaphysical Principles of Virtue*, trans. J. Ellington, Indianpolis, Bobbs-Merrill.
19. Kant, I.: 1903/1911, *Kants Gesammelte Schriften*, Preußische Akademie der Wissenschaften, Berlin.
20. Knaus, W.A. *et al.*: 1991, 'The APACHE III prognostic system', *Chest* **100**(6), 1619–1635.

21. Knaus, W.A. *et al.*: 1982, 'A comparison of intensive care in the U.S.A. and France', *The Lancet* **2**, 642–646.
22. Knaus, W.A. *et al.*: 1981, 'APACHE – Acute physiology and chronic health evaluation: a physiologically based classification system', *Critical Care Medicine* **9**(8), 591–597.
23. Lemeshow, S. *et al.*: 1987, 'A comparison of methods to predict mortality of intensive care unit patients', *Critical Care Medicine* **15**(8), 715–722.
24. Lewis, C.T. and Short, C.: 1980, *A Latin Dictionary*, Clarendon Press, Oxford.
25. Luce, J.M.: 1990, 'Ethical principles in critical care', *Journal of American Medical Association* **263**(5), 696–700.
26. Murphy, D.J. and Matchar, D.B.: 1990, 'Life-sustaining therapy', *Journal of American Medical Association* **264**(16), 2103–2108.
27. Murphy, D.J. *et al.*: 1989, 'Outcomes of cardiopulmonary resuscitation in the elderly', *Annals of Internal Medicine* **111**(3), 199–205.
28. Parno, J.R. *et al.*: 1984, 'Two-year outcome of adult intensive care patients', *Medical Care* **22**(2), 167–176.
29. Pius XII, Pope: 1958, 'Address to an international congress of anesthesiologists', *The Pope Speaks* **4**, 395–6.
30. Ramsey, P.: 1967, 'The sanctity of life', *The Dublin Review* **241**, 3–23.
31. St. John-Stevas, N.: 1964, *The Right to Life*, Holt, Rinehart and Winston, New York.
32. Schieber, G.J. and Poullier, J.-P.: 1991, 'International health spending: issues and trends', *Health Affairs* **10**(1), 106–116.
33. Shapiro, M.H. and Spece, R.G., Jr.: 1981, 'When, if ever, does "Allowing or helping a person to die" vindicate the sanctity of life?', *Bioethics and Law*, West Publishing Co., St. Paul, Minn., pp. 677–79.
34. Singer, P.: 1978, 'Value of life', in *Encyclopedia of Bioethics*, Free Press, New York, vol. 2, p. 823.
35. Smith, A.D.: 1955, *The Right to Life*, University of North Carolina Press, Chapel Hill.
36. Smith, H.L.: 1970, *Ethics and the New Medicine*, Abingdon Press, Nashville.
37. Sperry, W.L.: 1948, 'Moral problems in the practice of medicine with analogies drawn from the profession of the ministry', *New England Journal of Medicine* **239**, 985–990.
38. Stein, R.E. *et al.*: 1987, 'Severity of illness: concepts and measurements', *The Lancet* **2**, 1506–1509.
39. Strauss, M.J. *et al.*: 1986, 'Rationing of intensive care unit services', *Journal of American Medical Association* **255**(9), 1143–1146.
40. Strosberg, M.A. *et al.*: 1992, *Rationing America's Medical Care: The Oregon Plan and Beyond*, Brookings Institution, Washington, D.C.
41. Wagner, D.P. *et al.*: 1983, 'Identification of low-risk monitor patients within a medical-surgical intensive care unit', *Medical Care* **21**(4), 425–434.
42. Williams, G.T.: 1957, *The Sanctity of Life and the Criminal Law*, Knopf, New York.
43. Young, D.R.: 1932, *The Sanctity of Life: Secular and Mid-week Services*, Epworth Press, London.

JOHN C. MOSKOP

NOT SANCTITY OR DIGNITY, BUT JUSTICE AND AUTONOMY: KEY MORAL CONCEPTS IN THE ALLOCATION OF CRITICAL CARE

As several of the authors in this volume point out, the concepts of sanctity of life and *Menschenwürde* (human dignity) have played significant roles in the theology, philosophy, and jurisprudence of the Western world. These concepts refer to basic values which have been and continue to be both widely and deeply held. The aim of this volume is to investigate whether and how sanctity of life and *Menschenwürde* can help to address pressing moral issues in contemporary medicine. In the title of his contribution, H. Tristram Engelhardt, Jr. [4], poses a specific question, "Can these concepts help direct the use of resources in critical care?" Engelhardt's answer is that their usefulness for this purpose is very limited. He argues that these concepts lack the conceptual clarity and rational grounding necessary to dictate specific and compelling answers to our moral and policy questions regarding the proper use of critical care.

In this commentary, I will first review Engelhardt's critique of the use of these concepts to address the problems of allocating critical care. I will not attempt to challenge his conclusions, since I am in substantial agreement with them. Instead, I will try to carry Engelhardt's arguments one step further. In the last section of his paper, Engelhardt argues that each of two different interpretations of the concepts of sanctity of life and *Menschenwürde* suggest different positions on the provision of critical care. I will argue that even if we give each of these two interpretations of the concepts a clear meaning and assume that each is fully rationally grounded, neither can accomplish the proposed task of determining what level of resources should be devoted to critical care.

221

K. Bayertz (ed.), Sanctity of Life and Human Dignity, 221–228.
© 1996 *Kluwer Academic Publishers. Printed in the Netherlands.*

As Engelhardt and Keenan [8] point out, 'sanctity' is first and foremost a religious term. In religious contexts, it clearly denotes the state of being holy, sacred, worthy of reverence, and also, therefore, inviolable and sacrosanct. Christianity traditionally views human life as sanctified by God and, thus, proscribes murder and suicide. Theologians like Paul Ramsey describe the sanctity of human life as "alien" because it is conferred by God and is not the result of any intrinsic property of human beings [9]. If Ramsey is correct here, however, those who do not share a belief in the divine conferral of sanctity on human life will be left with no other, non-theological reason for ascribing this special value to human beings. Pluralistic societies will want to promote respect and tolerance for the strongly held religious beliefs of their citizens, but because such societies do not presuppose a single religious belief, their ability to institute public policies based on purely religious values will be strictly limited.

In an attempt to avoid this and other drawbacks of the traditional theological interpretation of sanctity of life, Keenan [8] appeals to a more secular, general, and all-inclusive sense of the term which is based in human experience and which emphasizes human stewardship for all life. Engelhardt notes that several American philosophers, including Callahan [1], Clouser [3], and Frankena [5], have also attempted to recast the concept of sanctity of life into secular terms. The result, however, is a much more qualified and ambiguous concept without a clear theoretical foundation. The ambiguity and uncertain justification of the secularized concept of sanctity of life renders it unable, Engelhardt argues, to give specific conclusions regarding the allocation of resources to critical care.

The concept of *Menschenwürde*, or human dignity, is clearly similar to that of sanctity of life in both meaning and origins. Both concepts ascribe a special moral standing to human beings and a consequent responsibility to protect them from harm. Though *Menschenwürde* also has roots in Christian theology, Engelhardt focuses on its role in the moral philosophy of Kant. Engelhardt argues that, although Kant may not be entirely consistent on this point, his theory requires that *Würde* or dignity be ascribed *only* to rational beings, and hence not to all humans. Dignity, in other words, properly follows rationality, not humanity, and hence cannot, at least on this Kantian interpretation,[1] justify protection of all human life.

After examining the meaning and origins of the concepts of sanctity of life and *Menschenwürde*, Engelhardt poses a deeper theoretical problem common to both. A major project of modern Western philosophy, Engelhardt claims, has been to find a secular, rational foundation for the basic tenets of Christian morality. This project is doomed to failure, according to Engelhardt, because there is no self-evident or rationally demonstrable meta-principle which enables us to choose among the many moral points of view available both within and outside of the Christian tradition.

Various formal constraints or conditions on ethical theories have been proposed, to be sure. Rawls ([10], pp. 130–135), for example, asserts that moral principles should be general in form, universal in application, able to order conflicting claims, and publicly recognized as the final court of appeal in practical reasoning. Even if all of these formal constraints on ethical theories are valid, however, Engelhardt can respond that they do not rule out all theories but one.[2] Thus, various moral points of view are possible – the moral claim that *human* life has a special sanctity or dignity, for example, has been criticized as a kind of prejudice by proponents of animal rights. One cannot make particular moral choices without adopting a moral point of view, but the adoption of a moral point of view is itself an unprincipled choice of one among many. For this reason, Engelhardt concludes, one cannot justify imposing on others one moral point of view based on a particular understanding of sanctity of life, or *Menschenwürde*, or any other substantive moral position. One can only attempt to convince others to accept one's own point of view, or, failing that, seek areas of convergence among several different moral perspectives. Persons *are* special, Engelhardt acknowledges, but not because special sanctity or dignity is somehow inherent in human nature. Rather, persons are special because, as moral decisionmakers, they are the sources of moral authority.

In the final section of his paper, Engelhardt points out that different interpretations of the concepts of sanctity of life and *Menschenwürde* can be used to defend very different policy options regarding the use of critical care. If, Engelhardt suggests, one interprets the sanctity and dignity of life to apply to all human life, then one will likely be inclined to make critical care available to all without limits. If, in contrast, one interprets the sanctity and dignity of life to apply primarily to the values of self-conscious, rational, or spiritual life, then one would be inclined to limit the provision of critical care in order to maximize those values. In the rest of this commentary, I would like to examine more closely

these proposed implications of different senses of sanctity of life and
Menschenwürde.

I believe that Engelhardt's major arguments regarding the ambiguity
and lack of a rationally demonstrable theoretical foundation for the
concepts of sanctity of life and *Menschenwürde* are persuasive. Let us
suppose, however, that Engelhardt is mistaken about at least some of
these conclusions. In other words, let us suppose that the concepts of
sanctity of life and *Menschenwürde* can be given a clear, univocal, and
substantive meaning *and* that the moral force of the concepts can be
established by rational argument. *If* that were the case, could these
concepts dictate answers to our questions about the use of resources
for and within critical care medicine?

A definitive answer to this question would require consideration of
every conceivable interpretation of the concepts of sanctity of life and
Menschenwürde, a task of Herculean proportions. I would, however,
like to begin this task by considering several interpretations of these
concepts which are, I think, both plausible and widely held, namely the
two interpretations of the concepts Engelhardt outlines in the final section
of his paper. I will argue that even if we assume the moral significance
of either of these interpretations, they still offer very little or no specific
guidance regarding the allocation of resources for critical care.

The first of the interpretations of sanctity of life and *Menschenwürde*
suggested by Engelhardt can be stated as follows: "All human life,
without exception, is sacred (has dignity); therefore, one must save human
life at all costs, and never take action to injure or shorten a human
life." This interpretation, Engelhardt notes, ascribes sanctity or dignity
to all biological human life; it is held, more or less strongly, by large
numbers of pro-life and anti-euthanasia advocates in the United States.
The meaning of the concept seems clear. Assuming, for the sake of
argument, that it is also morally compelling, what, if anything, follows
regarding the provision of critical care?

The answer to this question is, I believe, that nothing follows
from these concepts, so interpreted, regarding the provision of critical
care. Interpreted in this way, the concepts of sanctity of life and
Menschenwürde require an absolute commitment – support all human life
at all costs. Such an absolute commitment to support human life can
be fulfilled only if adequate resources are available to do so. It is,
however, painfully obvious that the present if not permanent condition
of our world is such that we cannot fully support all human lives. In such

a world, we are forced to make painful choices, knowing that some will suffer and die no matter what we do. We have to choose, for example, between investing our limited resources in relief for Bosnian war refugees, or starving Somalian children, or homeless street people in New York City, or any number of others in dire need.[3] We may be convinced that every one of these lives is sacred and deserves our full support, but that belief cannot help us choose which cause to support when we can support only a limited few.

The same scarcity of resources relative to needs exists within our health care budgets. Within health care, the potential to save lives, through increased provision of services and the development of new treatments, is virtually limitless. What is limited is the amount of financial and human resources available to provide those services. Thus, in the United States, the world's largest investor in health care, more than $800 billion was spent in 1992 for health care, including a sizeable percentage of that amount for critical care services, yet 15% of the U.S. population, or 35 million people still lack ready access to health care through insurance [2]. In fact, the major payers of America's health care bills, namely, government, employers, and patients themselves, have in recent years all sought ways to limit their responsibility for this increasing financial burden.

Professing the sanctity and dignity of all human life may initially incline one to call for expansion of lifesaving critical care services, as Engelhardt suggests, but if the total amount of resources available for health care is limited, increasing our investment in critical care will require a reduction in the amount available for other health care services. If, as Engelhardt also suggests, critical care is often provided for persons with only a small chance of survival, then the cost per life saved may be very high. Increasing critical care services at the expense of other, more cost-effective preventive or primary health care programs will likely mean that *more* people will suffer and die than before the redistribution of resources. Upon closer inspection, then, it appears that the goal of protecting all life may not be well served by expanding critical care services.

The injunction "Protect all life," even if it is our devoutest wish, cannot offer much practical guidance when resource scarcity makes the protection of all life an impossible goal.[4] At most, it reminds us that where we think continued life is desirable, we have some responsibility to respond, but it does not tell us how to respond. As David Hume points out, where scarcity exists, we must turn to "the cautious, jealous virtue of justice" to guide us in the allocation of our limited resources ([7], p. 15). There is, sadly but perhaps not surprisingly if Engelhardt is right about the limits of moral theory, no consensus about the correct principle or principles of distributive justice. Debates continue to rage among proponents of libertarian, egalitarian, and contractarian principles of justice. In examining and attempting to apply these principles of justice to medicine and critical care, however, we are at least asking the right question, namely, "How should material benefits and burdens be distributed among various individuals within a given society?" This seems a much more fruitful starting point for discussion of the allocation of resources to critical care than the simple but unsatisfiable injunction to protect all life.

Perhaps the second interpretation of the concepts of sanctity of life and *Menschenwürde* suggested by Engelhardt can fare better than the first as a guide to determining the proper amount of critical care services. Engelhardt describes this interpretation in several somewhat different ways, but for the sake of clarity I will state it as follows: "Sanctity of life (human dignity) is achieved in self-conscious autonomous moral lives; one ought, therefore, to foster and not interfere with the choices and actions of autonomous persons." This is obviously a very different interpretation of the concepts of sanctity of life and *Menschenwürde* than the first. It appears to be much further removed from the religiously inspired sense of the sanctity of all human life and much closer to the Kantian sense of the dignity of persons as self-legislating moral agents. On this interpretation, sanctity of life and *Menschenwürde* are defined in terms of another fundamental moral principle, namely, respect for autonomy. Engelhardt, in fact, defends respect for personal autonomy as the sole source of moral authority in the post-modern world. Assuming then, that this interpretation is clear, and that respect for autonomy is rationally defensible as an alternative to coercion, can it dictate specific conclusions regarding the allocation of resources for critical care medicine?

Once again, I believe, the answer to this question is that the second interpretation cannot tell us how much critical care should be provided or what amount of resources should be invested in critical care. This interpretation cannot dictate specific conclusions about the provision of critical care because both nations and individuals ascribe different priorities to critical care. In the early 1980's the United States had five to ten times as many intensive care beds as the United Kingdom, according to a comparative study performed by Schwartz and Aaron [11]. Does the United States invest too much in intensive care, or does the British National Health Service invest too little? Or, perhaps both invest too much – or too little. How could one conclusively resolve this question? Like Engelhardt, I believe that no one level of support for critical care can be rationally demonstrated to be superior to all others; thus, we must choose from a variety of morally acceptable options.

Interpreting sanctity of life and *Menschenwürde* in terms of respect for autonomy may, however, yield several implications regarding *how* decisions about resource allocation should be made. We should, Engelhardt argues, allow individuals the freedom to enter into private contracts to provide and receive critical care services, if they are so inclined and able. We should also, I would add, require that societal choices regarding the *public* provision of critical care services be made openly and democratically. What specific decisions are reached will then depend on the beliefs, values, resources and traditions of the different decisionmakers. Such decisions will not be easy, since they will give priority to some needs over others. I believe, however, that it is better to address these problems directly in an ongoing public discussion than insisting on the impossible goal of protecting all life at all costs or allowing the expressions sanctity of life and *Menschenwürde* to be used as ideological weapons or conversation stoppers.

East Carolina University
School of Medicine
Greenville, N.C.
U.S.A.

NOTES

[1] In another contribution to this volume, Honnefelder [6] defends a broader conception of *Menschenwürde* which includes within its scope both actual and potential persons. Whether *Menschenwürde* can justifiably be extended on this view to all human beings, including those whose potential for mental life is severely limited, is less clear.

[2] Rawls ([10], p. 131) himself acknowledges that his constraints are relatively weak and that the several traditional conceptions of justice satisfy them.

[3] These examples, of course, raise the question whether social responsibilities within and outside of one's national borders are different. One may very well have greater responsibilities to one's fellow citizens or to those living within one's national borders than to

others, but the interpretation of sanctity of life and *Menschenwürde* currently under consideration does not place any such qualifications on the responsibility to save human life.

[4] It is arguably not even a morally justifiable goal in all cases, since at least some individuals may, for good reasons, such as the relief of suffering, prefer that their lives not be unduly prolonged.

BIBLIOGRAPHY

1. Callahan, D.: 1970, *Abortion: Law, Choice and Morality*, Macmillan, New York.
2. Clinton, B.: 1992, 'The Clinton health care plan', *New England Journal of Medicine* **327**, 804–806.
3. Clouser, K. D.: 1973, ' "The sanctity of life": an analysis of a concept', *Annals of Internal Medicine* **78**, 119–125.
4. Engelhardt, H. T., Jr.: 1996, 'Sanctity of life and Menschenwürde: can these concepts help direct the use of resources in critical care?' in this volume, pp. 201–219.
5. Frankena, W. K.: 1977, 'The ethics of respect for life', in O. Temkin, W.K. Frankena and S.H. Kadish (eds.), *Respect for Life*, Johns Hopkins Press, Baltimore, pp. 24–62.
6. Honnefelder, L.: 1993, 'Persons and human dignity: on the relationship of metaphysics and ethics in the justification of moral values', in this volume, pp. 139–160.
7. Hume, D.: 1957, *An Inquiry Concerning the Principles of Morals*, Macmillan, New York.
8. Keenan, J. F.: 1993, 'The concept of sanctity of life and its use in contemporary bioethical discussion', in this volume, pp. 1–18.
9. Ramsey, P.: 1967, 'The sanctity of life', *The Dublin Review* **241**, 3–23.
10. Rawls, J.: 1971, *A Theory of Justice*, Harvard University Press, Cambridge, Mass.
11. Schwartz, W. B. and Aaron, H.: 1984, 'Rationing hospital care: lessons from Britain', *New England Journal of Medicine* **314**, 52–56.

VOLKER VON LOEWENICH

SANCTITY OF LIFE AND
THE NEONATOLOGIST'S DILEMMA

The term "sanctity of life" may be misunderstood in a religious sense, for instance if translated literally into German ("Heiligkeit des Lebens", "holiness of life"). The contemporary discussion about sanctity of life does not focus on an ethics derived from a particular religion. Sanctity of life is actually used in the sense of "inviolability" or "untouchability" of [human] life ("Unantastbarkeit des Lebens"). The problem to be dealt with is: under what circumstances, if at all, may the right to live or the duty to preserve human life be questioned, for instance in the field of intensive care medicine in the case of the newborn infant.

The discussion about sanctity or inviolability of human life was recently triggered by the Australian philosophers, Peter Singer and Helga Kuhse, and by the publication of their books "Practical Ethics" [10], "Should the Baby Live?" [8] and "Sanctity of Life" [7].

Singer and Kuhse question the sanctity of life, arguing that this concept is no more than a consequence or even a relic of our Jewish-Christian tradition. In fact, the 5th Commandment of the Decalogue, "Thou shall not kill", is part of Mosaic Law, adopted by Christian and Islamic morality. But is it so singular?

Singer and Kuhse support their opinion that sanctity of life is merely a part of Jewish-Christian morality by giving examples of some exotic peoples where respect for human life is, if not fully abolished, at least very limited. This argumentation is interesting but it overlooks the point that human life is, at least in principle, protected in nearly all other human communities, sometimes restricted to the clan, strain, people or religious community in question.

Admittedly, it is a matter of belief whether this sanctity or inviola-

229

K. Bayertz (ed.), Sanctity of Life and Human Dignity, 229–239.

bility of life is founded in divinal law, whether something akin to an intrinsic moral law (e.g. Kant's "moralisches Gesetz in mir") exists within all human beings, or whether the protection of human life is merely a social necessity. In my personal opinion, the existence of an intrinsic morality seems rather plausible. From the hypothetic viewpoint of a remote observer, we may suppose that something like the principle of sanctity of life exists in all human beings.

Singer and Kuhse's argumentation contains two essential points:

1. On the one hand, they question the sanctity of life; on the other, they base the prevalent ban on killing upon the "preference to live" felt by animals and human individuals alike (preference-utilitarianism).

2. They do not accept protection of life for all members of the human species, instead limiting this protection to humans fulfilling the criteria necessary to be a person according to a definition, e.g. that of Immanuel Kant. In Singer and Kuhse's opinion, humans not fulfilling these criteria of personhood are unable to have preferences. Therefore, a preference to live cannot be violated by killing them. Their most provocative conclusion is their lack of any objection to killing handicapped infants up to an age of four weeks.

Ad 1:
Singer and Kuhse are absolutely right to deny any deductibility of sanctity of life. Their aim is to offer an ethics free of methaphysics and axioms. But they overlook the fact that their own conception is not free of axioms either: e.g. the concept of personhood. This concept of personality, as well as the principles of utilitarianism and preference-utilitarianism, have been developed in the minds of philosophers, not deduced from facts and never confirmed by empirical investigations. They are hypotheses, useful to work with but not facts. They may be accepted or rejected. Indeed, so-called metaphysics-free ethics are not at all free from axioms.

On the other hand, Singer and Kuhse are not right in deducing from the non-deductibility of sanctity of life that sanctity of life does not exist.

The manner in which Singer and Kuhse use empirical data is not correct either. Citing some very exotic examples of restricted respect for human life does not take into account that the vast majority of mankind accepts something more or less akin to sanctity of life.

Ad 2:

In the argumentation of Singer and Kuhse, the right to live is only to be respected if the individual is able to plan his or her life in a rational way, to anticipate preferences, to consider his or her own existence in a reasonable way, i.e. if the individual, whether human being or animal, fulfills the criteria of personhood as defined by, for example, Immanuel Kant.

It has to be stressed that this definition of personal life as worthy of protection is nothing more than an axiom, a set-up originating from philosophical considerations, based neither on facts nor experience and lacking any empirical confirmation. This objection is understandable from the viewpoint of a biologically and anthropologically thinking physician, but it will possibly not be accepted by many philosophers who – without any doubt – have the right to think as they deem logical, without their theoretical constructions needing to fit into human reality. These considerations by a man at the front may seem primitive, but I shall try to support them:

Some day, a great many of us will lose our ability to anticipate preferences, to plan our lives rationally, to fulfil the philosophers' criteria of personhood. Should we then also lose fundamental rights, like human dignity and the right to live, as guaranteed for example in Articles 1 and 2 of German Basic Law?

Newborn babies are communicative, very young, but human counterparts. They are able to respond if spoken to, not verbally of course, but understandably and reproducibly. Even very immature infants respond to the human voice, speech or tactile contact. One example: One of our intensive care nurses recently told me about an 800g infant called Antonio who was on a respirator, repetitively fighting against the machine and suffering from respiratory distress: "If Antonio was obviously in distress, I opened a port of his incubator and spoke to him, and then he stretched his body out fully". This means that Antonio relaxed when he heard someone communicating verbally with him.

Immature infants also learn to fear painful interventions, anticipating what will happen if they are touched in a manner known to them, e.g. tapping the radial artery during a blood gas analysis.

There can be no doubt that most adult human beings are normally more communicative, rational and conscious in the planning of their lives than young infants, but where is a well-defined limit to be set? If we simply observe what happens biologically, then the only definition which is not arbitrary is that human life and individuality begin with the joining of the nuclei of female and male gametocytes. This definition was introduced into the German law for the protection of human embryos (Article 8, Embryonen-Schutz-Gesetz [ESchG], January 1st, 1991). All other definitions are either based on historically or biologically obsolete conceptions, for example the definition of legal (!) personhood within the German penal and civil codes, or they are adapted to suit pre-existing opinions, for instance in the context of the abortion debate or the – in part – very emotional discussion about the "Erlangen baby" [4, 13]. Human life contains no steps, but is a continuum from beginning to end. This was pointed out very clearly by Ludger Honnefelder during the Bielefeld conference "Sanctity of Life", upon which this volume is based.

One remark should not be omitted: Singer and Kuhse will go down in history for having stimulated the discussion about sanctity of life very effectively, questioning the types of life which deserve to be respected. Violent and degrading reactions in Germany towards these Australian philosophers were possibly due to the fact that Singer and Kuhse held up a mirror to our society and its thinking. In the German newspaper, "Die Zeit" (Oct. 15th, 1993), the German philosopher and journalist, Andreas Kuhlmann, Doctor of Philosophy and disabled by an infantile spastic palsy, wrote a remarkable sentence: Singer and Kuhse formulated the correct questions but gave the wrong answers.

We are now reaching our first conclusion:

There is no reason to withhold fundamental rights, such as human dignity and the right to live, from human beings, irrespective of how old, large, intelligent, competent or rational they may be.

Accepting sanctity of all human life, the neonatologist's dilemma becomes as follows:

Should he always preserve the life of his patients, irrespective of their prognosis, the severity of their illness, their suffering, immaturity, disability, or even their obvious inability to survive?

This problem may be understood as the fundamental question regarding the physician's duty ("aerztlicher Heil-Auftrag"):

This involves, for instance, three tasks:

1. To cure disorders, or at least to further the healing of illnesses.
2. To relieve suffering.
3. To preserve and also to prolong life.

Ad 1: There can be no question that curing disorders whenever possible is an aim to be taken for granted within all medical work. It therefore requires no further comment.

Ad 2: Without any doubt, relief of suffering is also a fundamental medical task. During the last few decades we have learned that even very young infants, newborns and prematures suffer from pain and fear, a fact that had not been realized earlier. We shall therefore treat consistently any suffering, even in extremely immature patients who are unable to survive. Although they are often unable to signalize their suffering, we have to take notice of the fact that they are suffering, and that they need effective treatment against pain and distress.

Ad 3: The consequence of accepting sanctity of life is that we have to preserve life as long as possible.

But the task to preserve life and the task to relieve or prevent suffering may sometimes compete with one another. In other words: May life be or become not worth living?

In Germany this question is a very delicate one. The term "lebensunwertes Leben" can have two different meanings: "unworthy life", the meaning it had in the Nazi ideology, and the above-mentioned sense of "life not worth living". The difference is the point of view: The first term represents the view of a society unaccepting of handicapped or disabled human beings; the second is the result of empathy with the hopelessly ill and careful consideration of the level of harm, pain or handicap with which the therapist will be justified in burdening his patient.

As a neonatologist looking back on more than two decades in my profession, I always used to be confronted with attacks from the public, from colleagues in other medical fields, from (nearly exclusively Protestant) clerics, sometimes from philosophers, all complaining that neonatologists would preserve life that is damaged, life of unacceptable quality, for instance that of increasingly immature infants, without any regard for the burden thus imposed upon the individual and his or her parents. Although these complaints mostly came from people who were not very well informed about the prognosis of these infants, or who only saw the worst results, e.g. paediatric neurologists, these reproaches made it clear that preserving life without taking the infant's outcome into account were obviously unacceptable.

Since 1989, heavy, sometimes fanatical attacks have unexpectedly come from the opposite direction: specialized Educationalists for the

handicapped, journalists and representatives of a rather militant association for the handicapped have accused clinical paediatricians in Germany
of eradicating disabled infants. These attempts to criminalize neonatology
were, of course, incorrect, resulting from a misunderstanding of certain
publications or even from being insufficiently informed (and an unwillingness to inform oneself!). Although it was easy to refute these
complaints and slanders, they undoubtedly had one positive effect:

It was very clearly pointed out that a strong feeling exists that human
life should not be ranked according to sanity, fitness or potential pro-
ductivity, disability or incurable injury, and that damage is an integrated
part of human life. It was also made clear that handicapped people feel
discriminated against by a discussion about whether it would have been
better not to preserve the lives they have to live.

How should the neonatologist, the physician practising neonatal intensive care medicine "at the front", handle this dilemma? How should he
act between these antipodes of opinion, attacked by a distrustful public
and plagued by his own doubts as well? Where may he get his orientation from?

One may rely on Christian ethics, on the Commandment of charity.
But this morality will only be accepted by people firmly settled within
Christian tradition.

There have been many attempts to create an ethics free of metaphysics
and axioms, and therefore acceptable to everyone. But, from the viewpoint of a non-philosopher, the fact that that these forms of ethics are
not at all free from axioms either has always been overlooked:

The so-called ethics of consensus tries to integrate the plurality of
opinions, beliefs and feelings of all the people involved. This is an
absolutely honorable attempt, but, by respecting a wider and wider
plurality of opinions, beliefs, etc., i.e. axioms, the resulting consensus
necessarily becomes more and more diluted in content, approaching
zero.

The Frankfurt School of Philosophy created "Diskurs-Ethik", a
variation of consensus ethics. Robert Spaemann, in his book "Glueck und
Wohlwollen, Versuch ueber Ethik" [11], criticised this concept, in my
opinion with very convincing arguments. If there is a so-called ideal
discourse, i.e. one in which all the participants wish for the same things
and in which they all know the right way to reach the goal, then the
discourse abolishes itself. If this is not the case, then each discussant will
introduce his own opinion, conviction, ideology or religion into the

discourse and will try to carry through his own position. The discourse then acquires a group-dynamic dimension. I personally remember the practice of discourse during the students' revolution of 1968, where it was claimed that arguments should be brought up and that everyone should then accept and realize the best ones. The only problem was that everyone was convinced that his or her own arguments were the best, with some people not hesitating to enforce their opinions violently.

The question remains of where the physician may obtain some orientation. From the point of view of a practising neonatologist, regularly confronted with the need to make decisions regarding the "be or not to be" of his patients, it is easy to understand the Hamburg theologist Traugott Koch [6] when he wrote: That orientation will never come from the situation because the situation itself is doubtful. An obvious need exists for a conviction which may offer orientation in a doubtful situation. This means that we have to realize that our decisions do in fact depend upon axioms, beliefs or convictions which are unavoidable, indispensible and which may or may not be accepted.

Once again from the point of view of clinical practice, one suggestion might be the Hippocratic oath, advocating care of the patient's best benefit, whatever that may be, and that the patient not be harmed. This principle is not far removed from utilitarianism, and yet with one essential difference:

Whereas classical utilitarianism aims towards the greatest possible amount of common and, to a lesser extent, individual happiness, the Hippocratic oath focusses solely upon the individual involved.

However, realization of the Hippocratic oath may become very problematic:

What is the best benefit for my patient? What could represent the harm I have to avoid or relieve? The danger behind these considerations is the danger of subjectivism.

What degree of handicap, incompetence, suffering, etc. is so intolerable that a life burdened by so much harm should not, or even shall not be inflicted upon a fellow-creature? What is the therapist's point of view, what is the patient's point of view?

One example: A mongoloid infant means harm and grief for the parents, irrespective of their otherwise positive attitude towards the child. From the child's point of view, no problem exists at all, the youngster himself or herself feels wonderful, oblivious of any difficulty regarding his or her existence.

Other examples: A newborn infant is suffering from the most severe form of Ichthyosis congenita, desquamation of the horny layer of the skin is impossible, resulting in a cuirass-like body-surface which is breaking apart everywhere and causing terrible pain. The infant cries awfully, has difficulties breathing, is unable to open his mouth or eyes. We remove the plaques using pealing ointments, have (still) to treat repetitive skin infections and to keep the infant totally dressed with bandages. The infant looks combusted, his looks horrify the unsuspecting, he has an evil smell and everybody can see the abnormality of this human being. Was it justified to keep this infant alive? Another newborn baby with the same disorder, crying from pain but with more pronounced breathing difficulties we chose not to save by artificial ventilation and skin-pealing. We only gave analgetics and a deep sedation, so that he died after a short time. Was this policy better or worse than the first one? Was the right to live, and therefore the right to be kept alive, not respected sufficiently in the second case? Was the suffering in both cases too cruel to be inflicted upon the patient?

More than a decade ago, an extremely premature infant had to be ventilated for several weeks. She developed severe chronic lung disease as a sequela of her primary lung disorder, of the high concentrations of oxygen she had needed, and from the high ventilatory pressures applied. After two weeks of spontaneously breathing high concentrations of oxygen, the infant caught a mild respiratory infection sufficient to make her respirator-dependent once again. She was then on different respirators for eleven years, growing up in an intensive care ward. Her mental development was severely retarded, she was and is still brought up by nurses because her parents totally rejected her. Her parents further became aggressive, filling the hospital shelves with accusatory correspondence, writing polemic letters to national journals about local hospital paediatrics and paediatricians, and involving the courts. About a year later we were confronted with the same situation in another baby. This time we decided not to ventilate the infant de novo, and the baby died. Which one of the two decisions was correct? Which one may be better justified?

In 1986, a committee of lawyers, philosophers, theologians and physicians made an attempt to offer some help in the decision-making process when proceding with severely damaged newborn infants, e.g. defining limits for the duty to preserve their lives. The aim of this conference was not to work out guidelines, but to aid the thinking process. The result was the "Einbecker Empfehlung", published in 1987 [5], English translation [9]. Initially hardly noticed, this recommendation was heavily criticised in 1990 by the educationalists cited above. Why? Obviously the critics had not read the "Einbecker Empfehlung" carefully enough, as their arguments clearly showed. On the other hand, the recommendation could, admittedly, easily be misunderstood by medical lay-persons. The weak points were examples intended to explain the meaning of the recommendation. It is remarkable that, at the Einbeck meeting, paediatricians were against citing examples, whereas the lawyers insisted on giving at least some examples. The new edition of the "Einbecker Empfehlung" eliminated all examples. It had been elaborated upon in 1992 by authors who had learned all too well that each kind of casuistic recommendation will necessarily be misleading, and that each single decision has to be worked out individually.

In fact, there is no generally acceptable guideline available by which one might decide whether the life of a patient should be preserved or not for the best benefit of the patient.

Of course, many thinkers have endeavoured to establish guidelines. But the results differ widely and obviously depend upon the authors' personal experiences. The lawyer Adolf Arndt, for example, later called the "crown lawyer of the German Social Democrats", has, since 1946, vehemently refuted any questioning of the preservation of human life, or even discussion of this topic [2]. In the light of the cruelties which happened in Germany during the preceding years, this might be understandable. A well-known paediatrician of the same generation, Klaus Betke, published a comparable opinion from the clinician's point of view [3]. Christoph Wilhelm Hufeland (1762–1836), physician and friend of J.W. Goethe, wrote that there cannot be any exception to the physician's duty to preserve human life, otherwise the physician would become "the most dangerous man in the State". Even taking into account the fact that, in his time, Hufeland had practically no means of preserving critically ill human life and was therefore not able to acknowledge the problem with our current acuity, this attitude is too short-sighted. It ultimately means preservation of human life at every price, inexorably, without compassion. Because today we are able to preserve even severely damaged human life, we have the duty to consider whether our medical work will be done in the interests of, for the "best benefit" of the critically or hopelessly ill patient, or whether it will harm him or her.

However, the answer to this cannot be given objectively. We have to live with the uncertainty of all medical prognoses, and with the – at least partially – subjective decision-making of all the persons involved, for instance the therapist or the medical team who has to realize the decision thereafter. This subjective moment, from which no decision may be deemed free, may indeed be dangerous. One does not like to oblige infants with severe cerebral malformations to live their reduced lives. But what about infants with partial infarction of the brain? They will certainly have to suffer significant handicaps, and yet we are unaware of the ultimate severity of central nervous incompetence within a particular individual at the moment a decision has to be made.

We have to be very aware of the fact that our decisions may be wrong, in one direction or the other.

There is another danger: Not preserving damaged life may be a comfort to our conscience. If a patient has died because his life was not preserved by us for carefully considered reasons, we can feel that our decision was justified and good for the patient. But there is no means to correct this opinion. If, on the other hand, the patient survived thanks to our medical help, and the patient now lives in a lamentable state, than I will be horrified seeing him and his situation, and I will feel guilty. If another patient survived and is now doing well, I will possibly be horrified by the thought that I had considered not keeping him or her alive.

Actually there is another danger that we should be aware of: We are increasingly confronted by law-suits against physicians due to a bad outcome of patient treatment. When patients die, there are very few malpractice-suits. The fear of being sued, sometimes to an extent which would endanger the physician's professional and personal existence, should never, but in reality can, play a role in decision-making. This, in my personal opinion, is a very alarming development.

A very often published argument against questioning the preservation of human life is the paradigm of the "slippery slope", where one could slip into a loss of respect for human life. This paradigm is justified as explanation of why passing the limits of a norm can lead to further and more acute frontier-crossings. But in another sense this paradigm is not good:

We are not walking along the top of a slippery slope, we are walking along a ridge where we can fall down both sides: loss of respect for human life on one side and, on the other, inexorable preservation of human life without compassion.

Department of Neonatology
University Children's Hospital
Frankfurt a.M.
Germany

BIBLIOGRAPHY

1. Akademie für Ethik in der Medizin, together with German Society for Pediatrics and German Society for Law in Medicine: 1992, 'Grenzen der Behandlungspflicht bei schwerst-geschädigten Neugeborenen. Einbecker Empfehlung – Revidierte Fassung 1992', *Ethik in der Medizin* **4**, 103–104.
2. Arndt, A.: 1976, 'Das Verbrechen der Euthanasie', in A. Arndt (ed.), *Gesammelte juristische Schriften*, E.-W. Böckenförde and W. Lewald, Beck, München, pp. 269–284.
3. Betke, K.: 1982, 'Intensivmedizin: Indikationen und Grenzen', *Monatsschrift für Kinderheilkunde* **130**, 353–357. (commented by von Loewenich, Volker, 357–358)
4. Bockenheimer-Lucius, G. and Seidler, E. (eds.): 1993, *Hirntod und Schwangerschaft – Dokumentation einer Diskussionsveranstaltung der Akademie für Ethik in der Medizin zum "Erlanger Fall"*, Enke, Stuttgart.
5. Hiersche, H.-D., Hirsch, G. and Graf-Baumann, T.: 1987, *Grenzen der ärztlichen Behandlungspflicht bei schwerst-geschädigten Neugeborenen*, Springer, Berlin/Heidelberg/New York *etc*. (Contains the "Einbecker Empfehlung").
6. Koch, T.: 1987, ' "Sterbehilfe" oder "Euthanasie" als Thema der Ethik', *Zeitschrift für Theologie und Kirche* **84**, 86–117.
7. Kuhse, H.: 1987, *The Sanctity-of-Life-Doctrine in Medicine – a Critique*, Clarendon Press, Oxford.
8. Kuhse, H. and Singer, P.: 1985, *Should the Baby Live?*, Oxford University Press, Oxford.
9. Seidler, E.: 1987, 'Recent developments in perinatal and neonatal ethics: an European perspective', *Seminars in Perinatology* **11**, 210–215 (Contains the english version of the "Einbecker Empfehlung" from 1986).
10. Singer, P.: 1979, *Practical Ethics*, Cambridge University Press, Cambridge.
11. Spaemann, R.: 1989, *Glück und Wohlwollen. Versuch über Ethik*, Klett-Cotta, Stuttgart.
12. von Loewenich, V.: 1992, 'Ethische Fragen in der Perinatal-Medizin aus neonatologischer Sicht', in Hegselmann, R. and Merkel, R. (eds.), *Zur Debatte über Euthanasie*, Suhrkamp, Frankfurt am Main, 2nd ed., pp. 128–152.
13. von Loewenich, V. and von Loewenich, K.: 1993, 'Die Diskussion um das "Erlanger Baby" ', *Hessisches Ärzteblatt* 3/1993, 104–106 + 115.

KEVIN WM. WILDES, S.J.

THE SANCTITY OF HUMAN LIFE:
SECULAR MORAL AUTHORITY, BIOMEDICINE,
AND THE ROLE OF THE STATE

INTRODUCTION

In his commentary on English law Sir William Blackstone notes that
the law of nature, the moral law, is one of the foundations of the civil
law. The civil law builds upon the moral duties one has to God, neighbor,
and the self and it fulfills a *remedial* function when people violate the
natural laws or fail in their moral duties ([5], pp. 42–55). This view of
the law is ably articulated in our own century by Lord Patrick Devlin
who understood that the state should "compel a man to act for his own
good" ([13], p. 136). This view of the state presumes that society can
know what the good for each man is and has the moral authority to
enforce it. However, moral controversies, such as those in biomedicine,
which turn on disputes over differing understandings of the good, moral
virtue, or moral duty, raise questions about the moral authority of the
state to resolve such controversies.

Controversies regarding different accounts of the moral life confront
us with the question about how state authority is to be justified in the
face of such moral diversity. Jean-Francois Lyotard has described this
condition the postmodern one in which "the grand narrative has lost its
credibility" ([20], p. 37). For ethics the postmodern predicament precludes contentfull moral disputes being settled by rational argument
outside a particular moral tradition. This predicament complicates the
question of the proper role of state authority in biomedicine since Western
political and moral philosophy have long held that one function of government is the enforcement of morality. If a canonical morality cannot
be discovered, or agreed to by all, then one is left only with particular

241

K. Bayertz (ed.), Sanctity of Life and Human Dignity, 241–256.
© 1996 *Kluwer Academic Publishers. Printed in the Netherlands.*

moralities including disparate understandings of sanctity of life and Menschenwürde. In such a circumstances it will be impossible to justify state enforcement of a particular, contentfull morality for at least two reasons.

First, confronted by a pluralism of moralities, rather than a single moral narrative, moral terms will take on different meanings in different moral perspectives. Moral phrases like 'Menschenwürde' and 'sanctity of life' are here very instructive. For some 'sanctity of life' will mean the intrinsic value of biological existence while for others it will mean the value of conscious rational life ([11], 1990).[1] In a secular state, with many different moral visions, moral terms can lose their argumentative power and become foundations of babel.

Even though there is no shared moral narrative to define moral terms like 'sanctity of life' they are, nonetheless, often invoked to justify coercive state interventions. In discussions of choices regarding human tragedies at the end of life, or possible medical interventions to reshape life itself, 'sanctity of life' is invoked to support arguments for a governmental role in protecting that sanctity. The Missouri Supreme Court, for example, in its decision in the *Cruzan* case asserted that the protection of the 'sanctity of life' was a *state interest* such that the state should prevent the withdrawal of feeding and hydration [12]. Here lies a second difficulty, namely if there are different understandings of moral terms such as 'sanctity of life' there is no justification for lending coercive state authority to enforce one over another.

As one examines the concepts of 'sanctity of life' and 'Menschenwürde' in bioethics, one discovers the fragmentation of moral language. This essay will show that this fragmentation of moral language undercuts the secular authority of the state to enforce a particular morality. Secondly, this essay will address the attempts to remedy this fragmentation in bioethics are subject to the same difficulties. Appeals to middle level principles [4] and casuistry [19] fail when confronted by the fragmentation of moral language and the plurality of moral visions. Thirdly this paper addresses a landmark American case in bioethics [11, 12] which brought questions of state authority with regard to a particular view of 'sanctity of life' to a federal court decision. Fourthly, the implications of the postmodernity for ethics and the limited authority of the state will be drawn out for critical care medicine.

The moral fragmentation of postmodernity provokes radical shift in our understanding of the state. The implications of this shift, for political theory and for biomedicine, cannot be explored fully in this paper. However, by probing the concept of 'sanctity of life' in relationship to the practice of critical care medicine, one can begin to reconceptualize the moral authority of the state.

THE FRAGMENTATION OF MORAL LANGUAGE

The dilemma of contemporary moral language is really twofold. Because moral language is tied to particular narratives and communities the standards for meaning and interpretation are also tied to particular visions. Confusion develops because while people may use the same words the moral terms and phrases often have very different meanings. For some 'sanctity of life' may mean the maintenance of biological life at all costs while for others it may mean the protection of a certain quality of life.

The second dilemma in contemporary culture is that moral language, shorn from its foundations, becomes incomprehensible to those who use it [21]. For example, when people use terms such as 'sanctity of life' they may not even understand the language they use because they do not understand the narrative from which it comes. The example of Captain Cook's encounter with the Hawaiians is instructive in two very important respects. First, it presents the epistemological dilemma created by moral pluralism. Cook's party was shocked because the Hawaiians thought it permissible, in some circumstances, to live together without benefit of marriage. However, the Hawaiians were upset by the fact that European men and women ate meals together. This was a practice which, for them, was taboo ([10], pp. 91–95). The clash of moral viewpoints makes clear the epistemological challenge of determining which moral standards *ought* to bind men and women. For a moral argument to be rationally convincing people must share the premises on which the argument is built. Without such basic agreement the arguments will be rationally unconvincing. MacIntyre captures this dilemma in the title of his book *Whose Justice? Which Rationality?*. The moral conclusions one reaches depend on the assumptions one makes about moral rationality and the moral world.

The Cook example is instructive in another respect in that when asked by the Europeans to explain the taboo the Hawaiians could not. They knew that it was taboo for men and women to eat together, but they were unable to explain *why* this was the case. The postmodern condition is not only marked by moral pluralism but by a condition in which moral language has been separated from its foundations. Outside a particular moral framework one cannot say what sanctity of life or Menschenwürde mean or why they are important.

The particularity of moral language is an affront to Western moral philosophy which long believed that moral content could be discovered by reason. The modern philosophical hope has been that reason could discover a contentfull secular morality which all men and women share in virtue of their nature as reasonable agents. The conceptual difficulty lies in justifying the basis for a particular view of moral rationality. Without particular moral commitments moral arguments remain vacuous. For example, the first principle of the natural law (Do good and avoid evil) ([3], I-II, 94, a.2), needs content if it is to guide practical reason. One needs to specify a ranking of goods and harms if one is to know what is the good to be done and the evil to be avoided. However, absent a canonical ranking of values which is shared by, or binding upon, all a contentfull moral argument cannot be developed ([15], Chapter two).

Bioethics has sought to avoid the conceptual difficulties encountered by theoretical models of moral reason by appealing to casuistry and the model of middle level principles as alternative strategies.

BIOETHICS

Bioethics confronts a scylla and charybdis. Without a content and a particular view of moral reason one cannot hope to resolve the moral controversies of biomedicine. However, with a content one looses universality by becoming parochial and the solutions to moral dilemmas will only satisfy those who share the initial premises. Unless one can resolve this dilemma it will be impossible for the state to impose particular moral practices in biomedicine short of using coercion. Bioethics has tried to navigate these waters by adopting different strategies. These strategies have been deployed to give content to moral terms such as sanctity of life and Menschenwürde and to justify the intervention of the state in biomedicine.

One of the best known strategies is the appeal to middle level principles without foundation in any particular theory [4]. The appeal to middle level principles, however, is fraught with problems. First, there is the difficulty in determining what the principles mean. One suspects, for example, that when Beauchamp and Childress claim to have agreed on their four middle level principles one is witnessing a slight of hand. For a preference utilitarian, like Beauchamp, 'autonomy' will be understood as the liberty to achieve certain goods. For a deontologist, like Kant, autonomy is concerned not with the pursuit of heteronomous desires but with acting in accord with the demands of reason imposed by the moral law. Childress, even though he does not endorse a Kantian deontology, relies upon right making and wrong making criteria which are independent of consequences. So while both Beauchamp and Childress speak of 'autonomy', the words function in radically different languages and will, in addition, have quite different meanings. In the same way while both may use the phrase 'sanctity of life' it will have very different meanings. For a preference utilitarian, it will mean, in part, respect for individual choices. The preference utilitarian may argue that out of a respect for the sanctity of life one must honor the preferences of individual agents. In this view, sanctity of life could justify the practice of euthanasia when chosen by competent individuals. This understanding of the sanctity of life would be used to oppose the use of state authority to forbid euthanasia. In stark contrast, for a deontologist, the sanctity of life would be interpreted to mean a reverence for life such that the deontologist would support the use of state authority to stop euthanasia.

A second difficulty with the middle level principles approach is that it is never clear how the principles should relate to one another. Each of the principles is conceived of as prime facie binding. If the principles are all of equal weight, it is not clear how we are to decide, in a conflict situation, which principle to follow. A final difficulty facing this approach is that there is no clear justification as to why this particular set of principles should be canonical. One might wish to include other principles, such as 'sanctity of life', which have been omitted from their list. Indeed, Beauchamp and Childress think that 'sanctity of life' can be subsumed under the principle of nonmaleficence. However, we have no agreed upon starting point which allows us to select the principles we think should be followed or to give the principles content.

To resolve these dilemmas, the middle level principle model would have to be recast in the context of a theoretical account which would define the meaning, relations, and justification of the principles. In Western philosophy principles have traditionally been a part of a comprehensive structure which begins with some first principle(s) and moves to secondary (middle level) principles (see [3], I-II, q. 94, aa. 1–6). Seeking to avoid the difficulties of theoretical accounts, Beauchamp and Childress have attempted to excise the secondary principles from any type of comprehensive structure. They hope in this way to avoid the challenge of providing foundations. However, shorn of theoretical and contextual moorings the principles become ambiguous in meaning and impotent in resolving moral dilemmas.

A second attempt to offset the difficulties of moral theory have been an effort to revive moral casuistry [19]. Jonsen and Toulmin offer an historical account of casuistry and allege that an appeal to casuistry can resolve moral controversies in our pluralist, postmodern secular world. However, they never develop an account of how secular casuistry should function. Traditional casuistry was built upon the analysis of particular cases and their resolutions within a concrete contentfull moral tradition. Within that tradition certain cases and their resolution could be regarded as paradigmatic for moral dilemmas. The difficulty with a secular casuistry is that there is no way to decide which cases are to function as the paradigm cases because again, like the appeal to middle level principles, casuistry, shorn of a moral viewpoint cannot select the paradigm cases to make the machinery run. Even if a secular casuistry could develop a set of paradigm cases outside of any particular theoretical and cultural framework or content, there is still no non-arbitrary way to describe the moral dilemmas in need of resolution and choose which case should be the model to resolve it because one's description of a case will depend on one's moral point of view [26].

Since casuistry, according to Jonsen and Toulmin, builds its moral principles from paradigm cases then any contentfull moral principle, such as 'sanctity of life', will depend on the paradigm cases selected to resolve a controversy. In the controversy over euthanasia paradigm cases will give very different meanings to the 'sanctity of life' principle. One might describe a case of a competent patient, with terminal illness, who is in intractable pain. If one applies a paradigm case against the intentional taking of innocent human life, then the 'sanctity of life' will have a meaning quite different than if one applies a paradigm case which views euthanasia, in such cases, as an act of mercy. 'Sanctity of life' will have very different meanings depending upon the paradigm case which is chosen. The crucial difficulty, for a secular casuistry, is that there is no non-arbitrary way to decide which paradigm to apply.

The difficulty for secular bioethics is that physicians, patients, citizens, and biomedical scientists frequently meet as moral strangers: without a common starting point or shared moral vision. There is no starting point, in theory, principles, or casuistry, which is not particular or arbitrary in some way. Each must presuppose some moral content as well as a particular notion of reason to develop resolutions to moral dilemmas. Yet the content is purchased at the price of universality. The proposed solutions in bioethics to the moral fragmentation of the postmodern age face the same difficulties of moral fragmentation as do more general moral theories.

If one cannot appeal to a particular concept of God or to a particular understanding of moral rationality or to a particular understanding of nature, in order to ground bioethics, there is only one source left: the authority of moral agents. If one cannot discover an authoritative moral vision to ground moral judgments then one must appeal to persons as the source of moral authority. When we meet outside a particular understanding of morality, we have only each other to whom we can appeal in order to resolve moral disputes and in order to frame the fabric of moral interactions. It is for this reason that one finds the salience in the postmodern world of such practices as free and informed consent, the free market, and limited democracy (i.e. governments that recognize robust rights to privacy, areas where contentfull moral views of the majority cannot be imposed on those in the minority). Absent agreement on external moral standards the only moral standards possible are those derived from the agreement of persons as moral agents.[2]

If one is interested in resolving issues peaceably without foundational recourse to force and with moral authority that can be justified in a general secular world, then moral authority can only be derived from the agreement of persons as moral agents. In such circumstances one cannot discover who is *a* moral authority, but only who has been put *in* moral authority. One can not rely on reason to discover moral authority, but must rely on the moral will to create moral authority. The necessary condition of mutual respect (the non-use of others without their consent) is not grounded in a value given to autonomy, liberty, or persons but is integral to the project of controversy resolution when God is not heard by all in the same way and when reason has not succeeded in establishing a general, canonical contentfull moral vision. The condition of mutual respect is a minimal condition for the resolution of moral controversies between moral strangers. A morality for moral strangers requires one not only to refrain from using others without their consent, but also to acknowledge them as agents who can agree to or refuse to collaborate. In the secular context of moral strangers one acknowledges, for example, that others may have very different notions of 'sanctity of life' or, perhaps, no recognizable notion at all.

Appeal to mutual respect allows us to understand in general secular terms (those which do not depend on a moral vision) when force and coercion can be justified in general secular terms. Moral strangers may use each other only when they act with commonly conveyed moral authority. Those who use others without consent loose a commonly justified basis for protest when they are met with punitive or defensive force. Limited democracies draw upon the morality of mutual self respect to provide protection from and punishment for the unconsented-to use of persons (e.g. murder, rape, burglary) as well as to insure the enforcement of the contracts. Bioethics has developed a number of procedural mechanisms (e.g. advanced directives, informed consent, institutional review boards), which are designed to help men and women of diverse moral viewpoints collaborate. These are procedural solutions which enable health care institutions and practitioners to navigate the plurality of moral commitments. While recognizing the rights to privacy they can create, through common consent, endeavors such as a basic health care system (limited solidarity).

Here rights to privacy are not celebrated because of any positive value assigned to such rights. Rather they mark out the limits of plausible moral authority of the state to intervene in the peaceable consensual actions of individuals. It is within the exclave that rights to privacy mark out that differing interpretations of sanctity of life can flourish.

THE STATE

In the West the state has often been advanced as the guardian of society's moral culture (e.g. [13]). From the betrothal of Church and state by Constantine through contemporary legal decisions (e.g. [12]) the state has been understood by many as the protector of public moral culture. However, the inability of reason to discover a contentfull canonical morality raises the question as to which moral culture the state is to protect and enforce. In view of the contemporary fragmentation of moral views, and the foundational epistemological difficulty of discovering a common moral view, the role of the secular state, as the protector of public moral culture is brought into question. The secular state provides a framework in which moral strangers, those with differing and diverse moral points of view, can meet and cooperate peacefully. While the moral authority of the state is limited it is nevertheless crucial to a secular society. The state becomes central to protecting the rights and exchanges of moral agents. It must punish those who unjustly take from others by force or deception. It must enforce agreements that have been freely made. The state becomes an agent for allocating commonly held public resources. However, given the plurality of moral values, and understandings of moral rationality, the state will not have moral authority to enforce a substantive moral vision of sanctity of life or Menschenwürde since it cannot be known what the vision should be.

The recognition and understanding of the postmodern predicament leads to a recasting of the way the moral authority of the state is understood: the secular state should be morally neutral with respect to contentfull understandings of the good life including contentfull understandings of sanctity of life and Menschenwürde. The state can, in general secular terms, have the authority to protect its citizens from unconsented-to encroachments, to enforce agreements, and to distribute commonly held resources. Yet there is a reluctance to embrace the limited moral authority of the state. The reluctance stems, in part, from the remaining bits and shards, to which people cling, of a once dominant and coherent morality. These pieces of the past, common words used by different moral communities, often give rise to the hope of a common morality as a basis for the state's moral authority. Jonsen and Toulmin, Beauchamp and Childress take common moral language and common morality as their point of departure. Indeed Jonsen and Toulmin take the work of the National Commission for the Protection of Human Subjects as evidence of a common morality on which to found their casuistry [23]. However, moral terms, while used in common, often have meanings which can only be understood within a particular moral context. Such common moral terms, like sanctity of life and Menschenwürde, are the remnant of an earlier, coherent moral view which can no longer be canonical for a secular, morally pluralistic society. In such a society there will be no one contentfull view that is taken as canonical by all. Secular moral discourse becomes evacuated of all content and the hope to give the state greater moral authority seems misplaced. However, in discourse with moral friends, moral language will have content and meaning which cannot be articulated in general secular terms alone.

A FAMILY OF ATTITUDES

The phrase "sanctity of life" is not restricted to the realm of Roman Catholic or Christian morality [9]. However it is a phrase which is part of a family of views which are rarely carefully articulated [7]. This family of expressions – sanctity of life, dignity of human life, and solidarity – as commonly used in the West, finds their common lineage in Christian understandings of creation and man as the "imago Dei" (Genesis 1:26). The Fathers of the Church were influence by the view of the Old Testament which emphasized God as both the beginning and the destiny of man. Since *all* life is created, what is it that distinguishes human life as "sacred"? Human life has a "unique" status in that God impresses onto man His image and resemblance and therefore makes the human being above all the other beings which are God's creatures but not really a mirror of the Creator. Man is part of the creation but he is also distinguished from the rest of creation as he is to rule creation. All life, since it comes from God, has a sacredness about it and demands respect for it belongs to an-Other. The special dignity and sanctity of human life comes from bearing the image of God and the responsibility to rule like God.

In the human the divine is expressed in the world. Irenaeus best captures this patristic sense when he wrote: "Glorian enim Dei vivens homo, vita autem hominis visio Dei" ([18], 4. 20. 7). The glory of God is the living human and the life of the human is the vision of God. In this perspective of faith, human life is made "holy" and endowed with "dignity" because of its relationship to God. Irenaeus, in speaking of the "glory" of man is not speaking in a secular sense of "self-improvement" for the center is not the human self but the Divine expressed in the human [14].

In the patristic era the reference to dignity had a background and application which was largely ethical or religious. The Christian had to assume a different style in judgment and behavior from the other citizens of the Empire. These themes of the divine image and the place of human beings in creation are themes that recur a number of time in the theological and spiritual writings of the Church [1].

The theme of man's divine image has been developed in different ways in the history of Christian theology and spirituality. In the reflections of the twelfth and thirteenth centuries this theme is integrated into a wider anthropology. the scholastic thinkers sought to identify the characteristics which distinguish human life from all other forms of bodily, created life. The epoch of scholastic philosophy was characterized by an emphasis on the intellectual and rational dimension of God's image impressed onto man. The expression "imago in specula rationis" manifests this truth. The pivotal point here is that human dignity and sanctity consists in man's ability to know himself and God. The faculties of intellect and will, unique to man in the embodied, created world, were seen as the most divine of human attributes.

The language of 'sanctity' and 'dignity' have been shaped by the belief that human life, like all life, is created by God and that it is an icon of God. The constellation of phrases associated with "sanctity of life" take its meaning from the belief that human beings are made in the image of God. In confronting the dilemmas of medical ethics and biomedical research it will be difficult to transpose and interpret this constellation of terms and expressions so that they can function in a secular context which does not believe that human beings mirror the divine. A secular context will lack a canonical anthropology with which to interpret the language of 'sanctity'. Again two conceptual issues confront those who would deploy such language in a secular forum. First there is the difficulty of establishing what the terms mean. Second there is the problem of establishing why we are compelled to accept these terms and principles.

BITS AND SHARES:
THE AUTHORITY OF THE STATE AND THE SANCTITY OF LIFE

Despite the conceptual difficulties which beset the justification of contentfull policy in secular society, biomedicine has been confronted again and again with contentfull moral positions imposed by the state through political force and supported by secular bioethicists. Many of these state interventions have been justified or encouraged under the principle of "sanctity of life". The Hastings Center, for example, in a special issue on organ donation wrote:

The view that the body is intimately tied to our conceptions of personal identity, dignity, and self-worth is reflected in the unique status accorded to the body within our legal tradition as something which cannot and should not be bought or sold. Religious and secular attitudes make it plain just how widespread is the ethical stance maintaining that the body ought to have special moral standing. The powerful desire to accord respect to the *dignity, sanctity,* and identity of the body, as well as the moral attitudes concerning the desirability of policies and practices which encourage altruism and sharing among the members of society produced an emphatic rejection of the attempt to commercialize organ [donation and] recovery and make a commodity of the body (Hastings Center, emphasis added).

The report urged, in effect, that organs and body parts were the *property* of the state and subject to state regulation. The justification for this coercive intervention was cast in appeals to "dignity", and "sanctity". However the meaning of these terms, and their normative force, is far from clear. William Frankena, for example, has summarized six different meanings which are sense of 'sanctity of life' ([16], p. 24). The conceptual dilemmas are twofold. First to determine which meaning should be used and, second, to justify why any particular meaning should have normative force in a secular society.

Earlier it was pointed out that the Missouri Supreme Court had asserted in the *Cruzan* case a *state interest* in the "sanctity of life". The case involved a patient who had been in a persistent vegetative state since an automobile accident. As she was anoxic for twelve to fourteen minutes Mary Beth Cruzan suffered irremediable brain damage. Subsequent to the accident a gastrostomy feeding and hydration tube was placed with the consent of her husband. When it became clear to her parents that their daughter had no chance of recovery, they sought removal of the tube. The employees of the hospital refused to comply with the request. The Supreme Court of Missouri, in a divided opinion, did not agree with the parents' request. The state argued, and the court accepted, that there was a state interest in the sanctity of life under the *parens patriae* doctrine of common law [24]. What is important to note is that the State Supreme Court as well as several of the dissenting opinions of the U.S. Supreme Court upheld "sanctity of life" as a value which trump other values. The Missouri Supreme Court wrote:

The State's interest in life embraces two separate concerns: an interest in the prolongation of the life of the individual patient and an interest in the *sanctity of life itself* [12].

In the view of the majority these general interests are strong enough to foreclose any decision to refuse treatment for an incompetent person unless there exists clear and convincing evidence that the person previously had made such a choice.

While asserting sanctity of life as a value which seems to order other values its meaning is never made clear nor is there an argument given as to why this value should trump other values. The opinions of the justices in these decisions, in fact, reflect the fragmentation of moral vision and language. Indeed Justice Stevens, of the United States Supreme Court, in his dissenting opinion in the appeal of the case, pointed out that: "Life, particularly human life, is not commonly thought of as a merely physiological condition or function. Its sanctity is often thought to derive from the impossibility of any such reduction" ([11], Stevens' dissent, Part III). Stevens's dissent represents a different interpretation of 'sanctity of life' than the one used by the Missouri Supreme Court. Furthermore, the two different interpretations lead to very different outcomes in this case. Some of Stevens's remarks capture the difficulty of speaking of 'sanctity of life'. He points out "[T]he more precise constitutional significance of death is difficult to describe; not much may be said with confidence about death *unless it is said from faith*, and that alone is reason enough to protect the freedom to conform choices about death to individual conscience" ([11], Stevens' dissent, Part III, emphasis added). Stevens shows the difficulty of using particular moral language in a secular society to support an intervention by the state on behalf of the 'sanctity of life'. Indeed one sees in the Cruzan case two different appeals to 'sanctity of life'. One appeal interprets the term to require aggressive medical treatment so that she may be kept alive. The other appeal to sanctity of life concludes that the treatment be withheld. The use of 'sanctity of life' in this case illustrates both the insurmountable difficulties of determining the meanings of moral terms in a secular society and the limited justificatory power of such terms.

In another case, *Bowers v. Hardwick*, which ruled on the constitutionality of statutes which criminalized consensual sodomy, Justice Blackmun made a point similar to Justice Stevens's. He argues "[T]hat certain religious groups, though by no means all religious groups, condemn the behavior at issue give the State no license to impose their judgments on the entire citizenry. The legitimacy of secular legislation depends instead on whether the State can advance some justification for its law beyond its conformity to religious doctrine" ([6], see Justice Blackmun's opinion, Part III).

The remarks of Stevens and Blackmun capture the dilemma and the solution. If contentfull moral claims can only be understood within a particular moral community then the role of the state in enforcing particular moral views, such as 'sanctity of life' and 'Menschenwürde', will have to be rethought. How one resolves moral controversies will depend on which moral language is applied and how it is understood. In the Cruzan case two very different resolutions were argued for yet each resolution was grounded in an appeal to the 'sanctity of life'. As moral terms have many possible meanings the secular state cannot enforce one moral view over others. Such a stance will be arbitrary from the perspective of those who do not share the particular moral view.

The practical implications of this account of the moral authority of the state is illustrated in the practice of critical care medicine in limited democracies. Given the exclaves of privacy in the secular state, men and women can establish limits to their medical care. Such limits are announced in advanced directives to determine which medical care is to be provided and which is to be withheld. At the same time professional groups and health care institutions can establish guidelines for the types of medical care they will offer and how they will deploy their medical resources and expertise. Institutional and professional guidelines may specify the withdrawal of treatment when it is judged to be futile (e.g., [2, 25]) or even stipulate criteria for the exclusion from treatment based on considerations of quality, length of life, and cost. Since there is no canonical, contentfull view of morality which supports the authority of the secular state, individuals, professionals, and institutions will have to articulate their view of appropriate medical action. The state cannot, legitimately, enforce a contentfull view of the moral obligations of medicine.

The state will be called upon in biomedicine, and in all fields, to enforce a morality of procedure rather than a morality of content. Absent a common moral narrative, moral terms such as sanctity of life and Menschenwürde will be incoherent in general secular discussion. The *Cruzan* case gives a very clear illustration of how such terms can be interpreted in contradictory ways. Confronted with moral pluralism the authority of the state to enforce a particular morality evanesces. The role of the state is to ensure the free, peaceable exchanges of its citizens. It functions to enforce agreements, protect unconsented-to violations of privacy, and to distribute commonly held resources. The enforcement of contentfull morality becomes the work of particular communities rather than the work of the state. Under the protection of the exclaves of privacy many understandings of 'sanctity of life' and 'Menschenwürde' can flourish.

Kennedy Institute
Georgetown University
Washington, D.C.
U.S.A.

NOTES

[1] See Justice Steven's dissenting opinion, part III.
[2] The author of this paper holds that objective standards of morality do exist. The conceptual difficulty for secular moral philosophy is epistemological. There is no 'view from nowhere' to enable us to know, by reason alone, which are the correct standards.

BIBLIOGRAPHY

1. Ambrose: 1879, 'De Dignitate Conditionis Humanae', in J. P. Migne (ed.), *Patrologiae cursus completus*, Series II: *Ecclesia latina*, Vol. 17, Garnier, Paris, column 1105–1108.
2. American Thoracic Society: 1991, 'Withholding and withdrawing life-sustaining therapy', *American Review of Respiratory Diseases* **144**, 726–731.
3. Aquinas, T.: 1948, *Summa Theologica*, Christian Classics, Westminster, Maryland.
4. Beauchamp, T. and Childress, J.: 1989, *Principles of Biomedical Ethics* (Third Edition), Oxford University Press, New York.
5. Blackstone, W.: 1969 *Blackstone's Commentaries*, St. George Tucker (ed.), Vol. V., August M. Kelley, New York.
6. Michael J. Bowers v. Michael Hardwick, U.S.L.W., 4919 (U.S., June 30, 1986).
7. Boyle, J.: 1989, 'Sanctity of life and suicide: tensions and developments within common morality', in B. A. Brody (ed.), *Suicide and Euthanasia*, Kluwer Academic Publishers, Dordrecht, The Netherlands, pp. 221–250.
8. Brody, B.A.: 1988, *Life and Death Decisionmaking*, Oxford University Press, New York.
9. Clouser, K.D.: 1973, 'The sanctity of life – an analysis of a concept', *Annals of Internal Medicine* **78**, 119–125.
10. Cook, J.: 1893, *Captain Cook's Journal 1768–71*, Capt. W.S.L. Wharton (ed.), Elliot Stock, London.
11. Nancy Beth Cruzan v. Director of Missouri Department of Health, 111 L.Ed.2d 224, 110 s.Ct. 284, 58 U.S.L.W., 4916 (U.S. June 26, 1990).
12. Cruzan v. Harmon, 760 S.W.2d 408 (MO.Banc 1988).
13. Devlin, P.: 1965, *The Enforcement of Morals*, Oxford University Press, London.
14. Donovan, M.A.: 1988, 'Alive to the glory of God: a key insight into St. Irenaeus', *Theological Studies* **49**, 283–297.
15. Engelhardt, H.T.: 1986, *The Foundations of Bioethics*, Oxford University Press, New York.
16. Frankena, W.: 1977, 'The ethics of respect for life', in O. Temkin, W.K. Frankena and S.H. Kadish (eds.), *Respect for Life*, Johns Hopkins Press, Baltimore, pp. 24–62.
17. Hastings Center: 1985, *Ethical, Legal and Policy Issues Pertaining to Solid Organ Procurement: A Report of the Project on Organ Transplantation*, The Hastings Center, Hastings-on-the-Hudson, New York.
18. Irenaeus: 1979, 'Against heresies', in A. Roberts and J. Donaldson (eds.), *The Ante-Nicene Fathers*, Vol. 1, W. Eerdmans, Grand Rapids, Michigan, pp. 26–29.
19. Jonsen, A. and Toulmin, S.: 1988, *The Abuse of Casuistry*, University of California Press, Berkeley.

20. Lyotard, Jean-François: 1984, *The Postmodern Condition*, Manchester University Press, Manchester.
21. MacIntyre, A.: 1981, *After Virtue*, University of Notre Dame Press, Notre Dame.
22. MacIntyre, A.: 1988, *Whose Justice? Which Rationality?*, University of Notre Dame Press, Notre Dame.
23. The National Commission for the Protection of Human Subjects of Biomedical and Behavioral Research: 1978, *The Belmont Report*, Department of Health, Education and Welfare, No. (OS) 78–0012 Washington, D.C.
24. Payton, S.: 1992, 'The concept of the person in the *Parens Patria* jurisdiction over previously competent persons', *The Journal of Medicine and Philosophy* **17**, 605–645.
25. Society of Critical Care Medicine: 1990, 'Consensus report on the ethics of foregoing life-sustaining treatments in the critically ill', *Critical Care Medicine* **18**, 1435–1439.
26. Wildes, S.J., K.: 1993, 'The priesthood of bioethics and the return of casuistry', *The Journal of Medicine and Philosophy* **18**, 33–49.

MARTIN HONECKER

ON THE APPEAL FOR THE RECOGNITION OF
HUMAN DIGNITY IN LAW AND MORALITY

I.

In his paper on "The Sanctity of Human Life: Secular Moral Authority,
Biomedicine, and the Role of the State" [11], Kevin Wm. Wildes operates
on the assumption of moral pluralism as a given fact. Pluralism results
from differing interpretations of a particular matter: What, e.g., does "life"
mean? Does it refer simply to biological existence or to something
qualitative, personal, spiritual, fulfilled? In what sense life should be considered sacred and inviolable depends upon how it is understood.
Furthermore, Wildes points to the fuzziness of such terms as "human
dignity" and "sanctity of life". He refers to a "fragmentation of moral
language" and especially sees the loss of a uniform moral language as
one of the essential causes of pluralism. "The particularity of moral
language" raises the question, whether rationally based common morality
exists at all. Unlike medieval natural law, whose fundamental principles were accessible to reason, and unlike the Enlightenment, for
which reason was the measure, Wildes sees moral arguments and stand-
points as being employed in a pluralistic fashion and as therefore being
relative.

In addition to moral pluralism, religious and ideological pluralism
makes respect for an individual's freedom of religion and conscience
imperative. Given this situation, is it even possible to appeal to common
fundamental values? On the basis of what criteria are generally binding
decisions, e.g. in bioethics and medicine, to be made? Wildes' description of a highly problematic situation is extensive and vivid. He therefore
recommends respect for the will of the patient in order to arrive at

257

K. Bayertz (ed.), Sanctity of Life and Human Dignity, 257–273.
© 1996 *Kluwer Academic Publishers. Printed in the Netherlands.*

anything like "informed consent". Medical doctors, patients, citizens and scientists encounter each other as "moral strangers: without a common starting point or shared moral vision". Only free agreement among those involved as moral subjects can legitimate medical action and intervention. Moral authority is placed entirely in the hands of the patient concerned. In light of this State of affairs, procedures as well as their regulation and fairness become decisive criteria.

Against this backdrop, Wildes reflects the role the State and the legal system should play in the decision-making process. Should the law guarantee the inviolability of human dignity? Should it serve as advocate and guardian of the sanctity of life and therefore regulate euthanasia or the discontinuance of medical treatment? Or is the decision concerning life and death not ultimately a private matter solely concerning the patient? Wildes demands that the State remain morally neutral and only guarantee and protect the individual's freedom of choice. Especially with respect to the Missouri Supreme Court in the Cruzan Case, Wildes discusses the question, whether and to what extent a State court is actually in a position to make and enforce moral decisions. Wildes is of course not arguing for random decisions, but rather for "institutional and professional guidelines". "The enforcement of contentful morality becomes the work of particular communities rather than the work of the State. Under the protection of the exclaves of privacy many understandings of 'sanctity of life' and 'Menschenwürde' can flourish."

The difference with respect to the cultural context immediately strikes the German commentator of Wildes' thoughts. The American and German legal systems are different. In the U.S. case law prevails, whereas in the Federal Republic of Germany legal regulation is the business of parliament as law-giver. Naturally, differing legal systems do not necessarily mean a difference in moral valuation. Within the cultural context of the U.S., pluralism is apparently natural and therefore the moral consequence is a decision-making process geared toward the individual nature of a given case. The autonomy and will of the patient form the basis for medical action (*voluntas aegroti suprema lex esto*), whereas the German tradition seeks objective standards (*salus aegroti suprema lex esto*).

1. Does ethical relativism necessarily follow from acknowledging the plurality of moral notions and language? Certainly individual situations, values, and preference must be regarded, but must this

necessarily lead to a theory of complete value-relativism? Doesn't even a pluralistic society need to reach agreement concerning the protection of central values, of fundamental human rights? Is there not such a thing as a common point of reference even among different ways of life? The issue here is basic moral consensus: Can moral legitimation be solely derived from the will of an individual? Put more drastically: Is it possible, and if so, under what circumstances, to protect human dignity or life against an individual's will?

2. Moral pluralism is indisputably a fact, but does this fact in itself represent a norm? How can a pluralistic society survive in the long run without a minimum of moral consensus? Doesn't the debate over moral issues necessarily presuppose the possibility of reaching an agreement? Wildes' reference to pluralism must be examined critically. Is it meant descriptively or prescriptively, in a normative sense?

3. I agree with Wildes in emphasizing the secular nature of the State. In a religiously pluralistic society, the State must maintain its neutrality. Especially as a Protestant theologian, I expressly agree with this postulate. It is the duty of the State to secure the peace for its citizens, not to assert a mandatory ideology. But a distinction must be made between religious and ideological neutrality on the one hand and moral neutrality on the other. An ideologically neutral State cannot totally ignore moral issues, since it not only regulates procedures, but also must deal with issues of a substantially moral nature. I shall discuss this problem in terms of two questions: (1) What does human dignity as a fundamental norm mean? (2) What is the relationship between law and morality, and what is the duty of the State?

II.

Respect for human dignity stands at the very beginning of the Constitution of the Federal Republic of Germany. Article I,1 states: "The dignity of man is inviolable. All the powers of State must respect and protect it." The consequence thereof is stated in Article I,2: "The German People therefore commit themselves to inviolable and inalienable human rights as the basis of every human community, of freedom and of justice in the world." Furthermore, constitutional control over the powers of State also results from the recognition of human rights. Is this commitment to human dignity to be understood as an absolute moral principle? Put differently: Is the appeal to human dignity meant deontologically, thereby leaving no room for teleological considerations or a weighing of options? In order to answer this question, the history of Article I must be called to attention. With regard to National Socialism's dictatorship and its contempt for humanity, the constitutional convention initially intended to make the protection of the individual the basis for State action. Therefore the Herrenchiemsee Draft begins by stating: "The State exists for the sake of man, not man for the sake of the State." This formulation was rightfully criticized on the grounds that the constitution of a newly forming State cannot open with a critique of the State. The General Statement on Human Rights, dated December 10, 1948, states in its Article 1: "All human beings are free and born with equal dignity and rights. They are endowed with reason and conscience and should encounter each other in the spirit of brotherhood." The emphasis on human dignity was aimed at defining the State's obligation to recognize and protect the individual. Later, President Theodor Heuss was to note that already during the constitutional convention the appeal to human dignity was an "uninterpreted formula". What are the interpretational problems of this term [6]? The conviction that absolute dignity is inherent in man as such is a modern development. The term "human dignity" stands for a confluence of differing views of man. Human dignity first achieved constitutional status in Ireland's Constitution of 1937. Antiquity already knew the term human dignity (*dignitas hominis*), e.g. Cicero. Within this understanding of dignity, two aspects must be distinguished. First of all, dignity refers to rank, to a person's honor within society, to his public prestige. Thus it is apparent that the modern understanding of dignity has replaced the traditional class-oriented idea of honor. Therefore acknowledging a person's dignity means to honor him, to respect his public position and prestige. To offend a person's dignity is to offend his honor, which, according to his social position (man, woman, nobility, common citizen, etc.), can be defined in different ways. Secondly, Cicero sees dignity as a characteristic of human nature ([2], I, 106). Man as a rational being is accorded a predominant position (*praestantia*) within the cosmos. Behind this understanding of human dignity lies the stoic idea that all human beings have a common rational nature.

The stoic idea of the inherent value of man based on his rational nature was combined in the early Church with the biblical and Christian teaching concerning man's particular position with respect to God, his *imago dei* (Gen.1,27; 9,6; Ps. 8,5ff, etc.). Of course the conviction that man was created in the image of God is a statement of faith. Recognition of human dignity is based on God's creation of man as a rationally endowed creature.

Furthermore, the Italian humanists used the term dignity to emphasize man's exceptional status. In "De dignitate hominis" (1496), Picio della Mirandola uses the term programatically. He utilizes the stoic notion of man as a microcosmic representation of universal reason. Thus the stoic idea of man's equality with respect to dignity was renewed. During the ensuing years, the term dignity served to pave the way for a secular understanding of man. Accordingly, dignity distinguishes man as a rational being. Dignity to Kant was something beyond valuation and without equal ([8], p. 439). Man should be an end unto himself, never the means to an end. This constitutes the dignity of man, his inner value. Thus dignity becomes synonymous with autonomy. Dignity is man's "prerogative" as a rational and personal being and distinguishes him from all other living beings. Therefore a human being should be respected as such and not subjected to serving as a means for the accomplishment of other goals.

In the nineteenth century human dignity became a political catchword. Demands for "dignified existence" and "humane conditions" are among the major slogans of the early socialist movement. The rediscovery of the term human dignity in the twentieth century was a reaction against totalitarianism's reign of terror. It gained admission into constitutional texts. Finally, Ernst Bloch [1] understood dignity as something not yet achieved, the utopian expectation of a life without affliction, subserviance and dependence.

This historical sketch of the term human dignity gives some idea of the range of its understanding and theoretical underpinnings. The stoic conviction of the equal rational nature of all men, the biblical view of man as an image of God, Kant's understanding of autonomy and the longing for dignified human existence all come together in this one term. With respect to its origin it is not Christian, but capable of Christian interpretation from the vantage point of creation. Initially, the term only gained admission into Christian dogmatics with some degree of difficulty, since leading dogmatic terms, such as creation and salvation, were not, and are still not, anthropologically oriented. Within the juridical context the term is equally unclear. There are attempts to explain human dignity in terms of Kant's notions, but also those which view a person's value as something belonging to man's natural constitution.

In the meantime, the notion of human dignity has been expanded to include social demands and rights [7]: What are the material prerequisites for a life of dignity? Thus human dignity becomes a term closely related to the value "quality of life".

On the basis of its origin, no religious or other group can claim a monopoly on the interpretation of the term human dignity. Depending on the contextual and ideological background, the word takes on a different color. The origin of this fundamental norm lies outside the area of jurisprudence. Nevertheless, it would be a mistake to conclude that the term human dignity is a hollow one. This term was taken up by constitutions as a constitutive principle in reaction to crimes against human dignity. Extermination, slavery, abduction, torture, deprivation of personal liberty, and persecution on political, racial or religious grounds are offences against human dignity. First of all, reference to human dignity has a negative function, namely of identifying acts of degradation. The simple appeal to the term alone is of course not sufficient for a positive understanding of it. The formal comment that "therefore", because the dignity of man is inviolable, the German People commit themselves to inviolable and inalienable human rights, demonstrates clearly that a more specific interpretation of the legal norm human dignity is to be achieved with the help of individually specified human rights. Among these are the ban on discrimination, the principle of equality, protection of life and the guarantee of personal freedoms. All action and legislation of the State is thus obligated to the principle of the inviolability of human dignity. The State must refrain from all action infringing on human dignity. Furthermore, no one must tolerate violation of his dignity by another or being subjected to the discretion and whim of another without the right of legal contestation. A human being, as person, has legally guaranteed rights and can demand of the State that it deter infringement upon his dignity. The positive guarantee of human dignity admittedly requires a corresponding order of government. If all the powers of State are obligated to respect and protect human dignity, then only that State can fulfil this obligation which is founded upon democratic and social welfare principles. The principles of a State founded upon respect for the law, social welfare and democracy can be derived from the fundamental norm of human dignity. For only a State based on respect for the law can guarantee its citizens legal protection. The deterrence of humanly degrading affliction can only be expected from a State which acknowledges its responsibility to take precautionary measures to ensure the welfare of its citizens, i.e. from a welfare-oriented State. The principle of democracy depends on the freedom and participation of citizens. Finally, human dignity cannot be limited: Even prisoners, the legally incapacitated or those incapable of making decisions do not forfeit their human dignity and have the right to be recognized as persons.

If human dignity is thus understood as a foundation of individual human rights and furthermore requires interpretation in the form of specific human rights, then it must be clear that human dignity is not a principle, from which individual demands can be positively deduced. The appeal to human dignity sets out certain limits. When human dignity is infringed upon, these limits are overstepped. Torture, the death penalty (according to German opinion), and all cruel punishments are incompatible with human dignity. Within the area of personal existence falling under the protection of human dignity, there is room for individual self-determination.

What conclusions may be drawn from these considerations concerning human dignity in view of bioethics? In the first place, human dignity as a reference point sets limits to arbitrary behavior. At the same time it allows for a certain amount of leeway both in terms of realizing human dignity as well as in terms of specifying individual human rights. That the interpretation and application of human rights are, despite their generally accepted universality, dependent upon the cultural and religious context is common knowledge. Western secular views, Islam, Hinduism and Confucian influenced Eastern thought each interpret human rights differently. But a relativism in principle and a postmodern principle of arbitrariness cannot be derived from the plurality of views concerning human rights. There is, after all, a hard core of human rights principles. Recognition of a person's bodily rights is fundamental. Since life is the vital basis for personal existence, the protection of life is fundamental in bioethics. Only those living can claim further human rights. In addition to the rights of the physical person, those of the spiritual person must be regarded as well. Rights of spiritual persons include freedom of conscience, thought and religion. In order to effectively make use of the rights of the physical and spiritual person, the rights of the political person must be protected. The rights of the political person include rights of cooperation and participation in public life. In bioethics these rights can also be referred to as rights of self-determination. Given the multitude of diverging views concerning human rights, this hard core of human rights should be acknowledged worldwide. For purposes of emphasis, all these may be summarized in a single word: "dignity". Regard for dignity implies respect for every individual's right to life, freedom of thought and conscience and demonstrates serious concern for his political and public freedom.

Is such argumentation from the vantage point of human dignity deontological? This is to a large degree a question of semantics. Indisputably it is deontological to accept a non-derivable axiom, according to which each human life has the same inviolable dignity. This is why Article I of the Constitution of the Federal Republic of Germany speaks of an obligation, a commitment and a binding principle. The deontological basis of the German constitution does not, however, exclude teleological argumentation within the course of its application: The right of the physical person must be balanced with the political person's right of self-determination. The old principle forbidding harm (*nil nocere*) necessitates agreement over what is harmful for a patient and what "harm" means. Thus discussion concerning the interests of patients, their relatives, doctors, perhaps even of society, is not severed or excluded from appeal to the principle of human dignity. Assessing the consequences and weighing alternatives on the one hand and orientation by the principle of human dignity on the other hand are not principally alternatives. Granted, the sense in which appeal is made to the inviolability of human dignity must be clarified. Does it involve a perspective intended to provide fundamental orientation or an ideal norm, against which every individual decision is to be measured? If human dignity were an ideal norm, its reconciliation with empirical reality or the concrete situation would remain an unsolved problem. At least in terms of human rights, the right to have rights is decisive. If a human being does not have the fundamental right to be a member of a community and to be recognized as a person, then a humane life is impossible. If a human being is not recognized as a bearer of rights, then his inalienable dignity will in no way either be recognized or respected. It is not my intention to pursue this fundamental problem of moral theory any further, but rather to consider the State's role in protecting human dignity.

III.

Recognition of human dignity is to be understood both as a fundamental moral norm and as a legally guaranteed fundamental right. Thus both the question of the relationship between law and morality as well as the question of the responsibility of the State for protecting human rights arise. First, two inapplicable models concerning the relationship of law and morality shall be described. No distinction will be made between different forms of law (constitutional, criminal, civil). Granted, such a distinction is essential with respect to a particular case. It is necessary to be clear on the fact that law is not to be equated with criminal law. The kind of law appropriate to a particular case is, however, a legal technicality and a matter concerning legal policy.

The identity model sees law and morality as being closely related. Accordingly, it is the duty of the law to guarantee and protect a minimum of morality with the aid of legal force. The law puts sanctions upon certain important moral norms. Morality, on the other hand, legitimates the law. For bioethics this would mean that decisions are to be made and action taken in accordance with given moral norms, but without any regard for the will of the patient. The identity model of law and morality can be objected to both from the moral and legal vantage point. Fundamentally moral forms of behavior, such as confidence and care, cannot be induced by means of legal sanctions. Thus spiritual cruelty is not a criminal offence, even if it is prima facie morally reprehensible. Duties out of love are not legally binding. From a legal perspective, an appeal to moral authority does not constitute a right. The law rests upon moral presupposition. Furthermore, not every law as such has a moral content. Within construction law, traffic law, etc., very different laws can have a morally defensible basis. Whether it is in the interest of legal security that the State regulate at all is a controversial matter. This is especially true in the area of medical law. In the Federal Republic of Germany, nearly twenty years of debate have passed without resolving the question of whether or not a transplantation law is necessary. Is the law pertaining to the medical profession adequate for the regulation of circumstances and conditions under which organ transplants are to be permitted? Furthermore, will it sufficiently provide for a corresponding decision-making procedure which is both clear and controllable?

For this reason the separation model clearly distinguishes between law and morality: What makes a law legal is that it is the result of a legal procedure.Thus the law is understood as positive, established law. Not the moral content of a law is important, but rather the fact that it is in force. Such a legally positivistic interpretation can draw on common attempts to separate law and morality. Thus, e.g., a distinction is made between law, as something merely concerned with outer behavior, and morality, as something solely concerned with inner convictions. However, this distinction does not apply, since intent and motive must on occasion be considered in criminal law.The same can be said of the distinction between legality and morality. This can be illustrated by pointing to the phenomenon of a wrongful law. The legislation on euthanasia during the "Third Reich", which legalized killing unwanted human life, which was considered "unworthy to live", while, at the same time, was both willing and able to live, is a wrong. Correct legislative procedure involves legalization of what is morally legitimate. Insight into the positive nature of law is not sufficient grounds for its binding quality. On the contrary, the binding quality of law as such is incapable of positivistic justification.

Thus, if law and morality can neither be identified nor separated, a discrepancy or mediating model is required. Law is, then, more than just regulation of the minimum required for human co-existence and peace under the law. Furthermore, law is more than just something which is posited along the lines of Hobbes' dictum that "auctoritas, non veritas facit legem", more than a coercive order. But law is also not just the application of moral precepts to legislation and decision-making. Above all, positive law comes about by means of legal policy. Admittedly, however, legal policy is not free with respect to the laws it creates, but must pay attention to the compatibility of positive law and morality. Such an examination of the compatability of positive law and generally accepted morality is only possible in an atmosphere of free and open debate. This is why legislative procedure necessarily involves parliamentary debate, expert hearings and public discussion. The public dimension of legislation as well as of the pronouncement of judgement has a controlling function and allows for corrections to be made. Law is formed and established within a process open to scrutiny. Naturally the legislation process presupposes a culture's understanding of law. At this point, the differences existing between the legal traditions in the U.S. and Germany become apparent. Different cultural attitudes to law lead to different kinds of legal regulation. The German tradition of applying the rule of human dignity to all actions touching upon the life and rights of an individual results from the experience of tyranny, arbitrary justice and inhumanity under the reign of National Socialism. American confidence in the reasonability of those concerned, as well as the optimistic assumption that the autonomous decision of those concerned or involved will always yield what is morally right, is rooted in a different cultural tradition and history.

Indeed, this issue is not new. In the debate centering around the reform of legislation concerning sexual crimes, the same arguments used in medical legal ethics today emerged. Thus the issue can be clarified by examining the debate concerning the relationship between law and morality in England in the late 1950's.

The debate over the relationship between law and morality was carried out after 1958 in England between Judge and Lord Patrick Devlin and the legal philosopher H.L.A. Hart. At issue was "the legal enforcement of morals". The dispute came about following the 1957 publication of the Wolfenden-Report which recommended exempting homosexual acts between adults from legal prosecution. Both Devlin and Hart supported such a reform. Both men, then, took a liberal stance. Nevertheless, they disagreed in their individual reasoning. Lord Devlin was opposed to Bentham's and Mill's utilitarian justification of punishment, according to which only doing harm to another justifies punishment. In Devlin's opinion, morality and religion are the foundations of society and should therefore be protected. It is "reasonable man", i.e. "right-minded man", not "rational man", and certainly not the expert, who determines the limits of personal freedom. He demanded that heed be given to "toleration of the maximum individual freedom that is consistent with the integrity of society" ([3], p. VIII). Thus Lord Devlin was in favor of an interpretation closely linking moral and legal precepts. Law and morality, as he viewed them, did not represent different realms and could therefore not be separated. For this reason morality can demand legal protection of an individual's rights. The morals of a society are a unity and comparable to a seamless garment. "Every society has a moral structure as well as a political one." "Society cannot tolerate rebellion; it will not allow argument about the rightness of the cause" ([3], p. 9). Just as a society cannot tolerate rebellion, it also cannot tolerate being unfaithful to its moral precepts. When the integrity of society is at stake, intolerance, rejection and disapproval are called for. Therefore a society's tolerance has its limits where the foundations of its value system are called into question. In this case morality must be protected by means of the law or "legal enforcement of morals." According to Devlin, morality is the foundation of every society. Therefore legal changes risk the danger of moral decay, "moral disestablishment". The moral standards, with which a reasonable man agrees, must be safeguarded by the law. "The morality which the law enforces must be popular morality" ([3], p. IX). Especially the church should speak up in defense of the moral foundations of society. "A State which refuses to enforce Christian beliefs has lost the right to enforce Christian morals" ([3], p. 5).

In his critique of Lord Devlin, Hart takes a contrary position. Hart accuses Devlin of wanting to maintain existing moral norms and, thus, the social status quo by legal means. According to Hart, only such actions are worthy of punishment which, after considering the interests involved, are judged clearly harmful. In addition, Hart actually disputes that a society requires an homogenous, stable morality. By means of this argument he rejects a "legal moralism". Limitations upon freedom are only permissable if they are imperative to social co-existence. According to Mill, the law should only prevent harm being done to others. Harm done to oneself does not entail prosecution. Therefore the law must principally maintain moral neutrality. Seen rationally, criminal law should be limited to protecting the interests of the third party. Only in exceptional cases should this not apply. This especially pertains to laws concerning sexual crimes. In this case, Hart deviates from Mill's position. The criterium for determining the punishable character of sexual acts is, in the case that both people involved agree, harm done to a third party. According to Hart, this also justifies a moral paternalism which protects children and those under-age from seduction. But only insofar as the moral well-being of children is concerned is a limitation upon the freedom of sexual behavior permissable. In this connection the scope of Hart's definition of harm remains unclear. A broad definition would lead to Devlin's thesis that the ethical foundations of society can and must be legally protected. This approach, however, would then include Devlin's thesis, according to which law is based upon a particular culture. Law is not only determined by the demands of autonomy or an individual's self-realization alone, but also by considerations of cultural survival.

Some of Devlin's arguments are admittedly problematic. Devlin is certainly right when he states that the law rests upon given ethical foundations and values. The institution of monogamy can serve as an example for Western culture, as well as notions of human dignity and other legally protected rights. But the question remains, whether the morality of a society is really comparable to a seamless garment and, thus, whether deviation in a particular instance necessarily destroys the foundations of society. Is, therefore, every deviation from the prevailing moral code "treason"? Is recognition of the prevailing moral code as important as recognition of the government? This is the question posed by social and moral pluralism.

Furthermore, Devlin's thesis that morality always requires religion and that, without recognition of an accepted religion, morality decays ([3], p. 23), is also controversial. The question of ideological pluralism and the State's duty to remain neutral, which is important for the formation of the legal order, is relevant here.

Finally, orientation according to the judgement of the "reasonable man" must be considered ([3], pp. VIII–IX and pp. 15–16). Can the judgement of a reasonable man be equated with the will of the majority? Is the morality of the majority, the right, valid morality. The danger here, as Hart correctly points out, is that of "moral populism" ([4], pp. 77ff). Devlin's intention of guarding against the moral dictates of a self-elected elite is without doubt justified. However, this should not rule out the search for a criterion of judging existing morality. In addition to such a criterium, Devlin also leaves the question concerning the necessity and the limits of tolerance against dissenting opinion open.

The debate between Devlin and Hart was carried on by Basil Mitchell. Accepting Hart's critical objections, Mitchell nevertheless takes Devlin's basic position and asks what the moral "essentials" of a society are ([9], p. 35). However, it is a false alternative to demand either protection of existing morality in its entirety or of only those values whose disrespect every society punishes. When Devlin argues for the protection of the "essentials of a society", then this can only mean protection of "the essential institutions of a society" ([9], p. 121). The law not only has the duty to protect individuals from harm but also to maintain a society's fundamental institutions. Therefore the law cannot principally be neutral or value-free. It must ward off behavior undermining the ethos of a society ([9], p. 134 and p. 139). Mitchell names cruelty, cruelty to animals, violence, racism, and prostitution as examples of such behavior. But existing morality itself does not stand above criticism. In view of the permissibility of a "legal enforcement of morals", Mitchell concludes that the law must not only protect the individual but also society's fundamental institutions against harm. Therefore the law cannot be "morally neutral". Nevertheless, morality itself requires critical scrutiny. For all these reasons, however, a clear distinction between public and private morality cannot be made. This has consequences for legislation. It should demonstrate regard for privacy, uphold the principle of equality before the law, be guided by consensus among the majority of the "knowledgeable", not cause disadvantages or suffering by, e.g., enabling the practice of extortion. Furthermore, the law should continue to leave off punishing where means of help do not exist.[1]

These principles and criteria, thus the conclusion, oppose the prosecution of homosexual acts between adults.

The debate between Devlin and Hart demonstrates the fundamental significance of determining the relationship between law and morality (see [5]). Legal prosecution by itself is not enough. The value a society attaches to its legal institutions must also be considered. In addition to sanctions in the event of harm, it must be considered that questionable ethical behavior can serve as an example to others. Thus Hart, unlike Mill, affirmed the legitimacy of moral paternalism and the permissability of protecting against self-inflicted injury. Furthermore, the effects legal changes can have on the moral values of a society must also be considered. Such considerations are not only necessary with respect to laws concerning sexual crimes, but also to the legalization of certain drugs, euthanasia or genetic manipulation. Thus the ultimate question is whether the law can set limits even to individual freedom and tolerance where fundamental ethical principles, the basic value system of a society, are at stake.

Thus two fundamental questions arise. What effect do individual decisions have on the morality of a society and to what extent are case decisions capable of generalization. Furthermore, there is a duty to protect those incapable of making decisions from self-inflicted harm or damage. That is the problem with moral paternalism; it is repeatedly required to protect the interests of the weak. Often a medical doctor will have to treat a patient in accordance with his best knowledge and conscience because a clear expression of the patient's will is impossible or a patient's will can lead to self-inflicted injury, e.g. due to depression or psychiatric illnesses. Therefore treatment according to therapeutic standards makes sense. In addition, the German legal tradition also requires observance of relevant laws in order to guard against malpractice. However, the differences between legal systems and cultural interpretations of the law do not preclude general observations concerning the relationship between law and morality.

One must, of couse, distinguish between the moral legitimacy of individual laws and the recognition of their being in force as such. General failure to provide for legal regulation does not follow from moral pluralism. "The pluralism of moral viewpoints does not preclude the (moral) general validity of the law" ([10], p. 21). In order to arrive at acceptable laws, legislative criteria must be observed.

(a) Legislative procedure must recognize and consider the interests of those concerned. Providing for hearings and fair procedure are pre-requisites for ensuring the general recognition of legal regulations.
(b) Those to whom a law pertains must be treated equally. Dis-criminating as well as privileging individual groups of people must be avoided.
(c) Laws must be equally binding upon those subject to them as well as those administering them.

These formal legislative criteria are prerequisites for moral recogni-tion of legal regulation. But as important as such formal criteria are, they alone do not guarantee moral recognition of the law. The law must guarantee and protect fundamental rights of human existence, such as life, bodily integrity (forbidding torture) and the fundamental freedoms of the person. Since these are legal rights valuable to all people and their co-existence, there can be no such thing as a value-free legal system. The distinction and coordination of law and morality can only be achieved within an on-going legislative process. Within this process, respect for human dignity serves as an orientation point and a critical norm insofar as it provides fundamental orientation and lends purpose to moral and legal debate.

IV.

In conclusion, I would like to make a general remark. In this paper I have discussed the interpretation of human dignity and offered some thoughts concerning the relationship between law and morality with respect to the State's role in protecting human dignity. Thereby I intentionally avoided the expression "sanctity of life". As a theologian, I have trouble using the word "sacred" (heilig) within a secular context. The sacred as *"mysterium fascinosum at tremendum"* is a synonym for the word religion among scientists of religion. Religion is concerned with man's relationship to God. Similarly, "sacred" means "belonging to God". Certainly a Christian or religious person sees life as a gift from God, sees life in connection with God and thus can say that he owes his life (created life) to God and that it is therefore sacred. This, however, is a statement of faith, of one's creed. The question for bioethics concerns the necessity and limits of protecting human life. The goal is not a religious interpretation of life but an ethical consensus, to which both the religious and non-religious can submit. For this reason, the expression "sanctity of life" is inappropriate in ethical discourse.

Furthermore, emphatic reference to "the sanctity of life" within ethical discourse is formally an exhortatory appeal to respect for life, not an argument capable of solving conflictual and definitional problems. Not even those who speak of "the sanctity of life" can, in a particular case, presuppose and demand that, in every event, and without consideration of the consequences, biological human life must be sustained. In conflicting situations and borderline cases, purely emphatic appeal to the "sanctity of life" cannot replace the complex task of analyzing the situation and weighing alternative consequences.

Therefore reflection on the fundamental norm of inviolate human dignity should be connected with considerations of protecting the life of a human being.

(a) The term "human dignity" involves concentration upon the distinctive nature of human personal life. Human dignity is the very essence of humanity.
(b) Human dignity demands recognition of the equality of all human beings. All people have the same dignity. They are to be respected equally as persons.
(c) Mutual recognition of dignity leads to the recognition of equal rights, i.e. generally acknowledged human rights. Mutual recognition of human dignity includes the immorality and illegal nature of evident disregard for dignity and fundamental rights.

Even if an epistemological pluralism must be accepted as a fact not to be overlooked, it does not as such disprove the necessity and possibility that, despite varying interpretations and positions, there can and must be agreement concerning fundamental moral norms. Admittedly, fundamental norms always require and are capable of interpretation. Furthermore they must be interpreted and applied to different situations and living conditions. Uniformity of moral decisions does not necessarily logically follow from uniform recognition of a fundamental norm. But common reference to a fundamental norm encourages dialogue where positions and opinions differ and is therefore a prerequisite to understanding.

Evangelisch-theologisches Seminar
University of Bonn
Germany

NOTE

[1] See [9], p. 135: "The respect of the most reasonable man".

BIBLIOGRAPHY

1. Bloch, E.: 1961, *Naturrecht und menschliche Würde*, Suhrkamp, Frankfurt/Main.
2. Cicero: 1975, *De officiis*, ed. G. P. Goold, transl. W. Miller, Loeb Classical Library, Harvard University Press and William Heinemann, Cambridge (MA) and London, I, 105–106, p. 91.
3. Devlin, P.: 1965, *The Enforcement of Morals*, Oxford University Press, Oxford.
4. Hart, H.L.A.: 1963, *Law, Liberty and Morality*, Oxford University Press, Oxford.
5. Hoerster, N.: 1970, 'Strafwürdigkeit und Moral in der Angelsächsischen Rechtsphilosophie', *Zeitschrift für die gesamte Strafrechtswissenschaft* 20, pp. 538–570.
6. Horstmann, R.P.: 1980, 'Menschenwürde', in J. Ritter *et al.* (eds.), *Historisches Wörterbuch der Philosophie*, vol. 5, pp. 1124–1127.
7. Huber, W.: 1992, 'Menschenrechte/Menschenwürde', *Theologische Realenzyklopädie* 22, pp. 577–602.
8. Kant, I.: 1911, *Grundlegung zur Metaphysik der Sitten*, Academy-Edition, vol. 4, Reimer, Berlin, pp. 385–463.
9. Mitchell, B.: 1970, *Law, Morality and Religion in a Secular Society*, Oxford University Press, Oxford.
10. Otte, G.: 1981, 'Recht und Moral', *Christlicher Glaube und moderne Gesellschaft*, vol. 12, pp. 7–36.
11. Wildes, K.W., S.J.: 1996, 'The sanctity of human life: secular moral authority, bio-medicine, and the role of the state', this volume, pp. 241–256.

BJÖRN HAFERKAMP

THE CONCEPT OF HUMAN DIGNITY:
AN ANNOTATED BIBLIOGRAPHY*

This bibliography contains a selection of representative titles on the topic "human dignity". The most important titles are annotated, whereby the views given are those of the respective authors and do not amount to a critical evaluation. Wherever possible, English editions are listed. Special attention was given to use of the term "human dignity" within the field of medicine.

1. History of the concept "human dignity"
   1.1 Sources
   1.2 General historical accounts
   1.3 Ancient Times and the Renaissance
   1.4 Contemporary approaches

2. The concept of human dignity within the field of law
   2.1 International
   2.2 Germany
      2.2.1 Basic Law: commentaries and handbooks
      2.2.2 Miscellaneous

3. The concept of human dignity within the field of medicine
   3.1 General
   3.2 Reproduction / Gene technology
   3.3 Life and death
   3.4 Psychiatry

K. Bayertz (ed.), Sanctity of Life and Human Dignity, 275–291.
© 1996 *Kluwer Academic Publishers. Printed in the Netherlands.*

1. HISTORY OF THE CONCEPT "HUMAN DIGNITY"

1.1 *Sources*

1. Ambrose: 1879, '*De dignitate conditionis humanae*', in J. P. Migne (ed.), *Patrologiae cursus completus*, Series II: *Ecclesia latina*, Vol. 17, Garnier, Paris, column 1105–1108.
2. Bernard of Clairvaux: 1854, '*De cognitione humanae conditionis*', in J. P. Migne (ed.), *Patrologiae cursus completus*, Series II: *Ecclesia latina*, Vol. 184, Garnier, Paris, column 485–508.
3. Cicero: 1975, *De officiis*, ed. G. P. Goold, transl. W. Miller, Loeb Classical Library, Harvard University Press and William Heinemann, Cambridge (MA) and London, I, 105–106, p. 91.
4. Manetti, G.: 1452, *De dignitate et excellentia hominis libri IV*, in *Janocius de Manectis, idem*, Basel 1532. Republ. by E. R. Leonard, Padua 1975.
5. Pico della Mirandola, G.: 1486, *Oratio de hominis dignitate*, in *Opera omnia*, Basileae, 1572; 'Oration on the dignity of man', trans. by E. L. Forbes in Cassirer *et al.* (eds.), 1948, *The Renaissance Philosophy of Man*, Chicago University Press, Chicago, pp. 223–254.
6. Hume, D.: 1987, 'Of the dignity or meanness of human nature', in *Essays Moral, Political and Literary*, ed. with foreword, notes and glossary E.F. Miller, Liberty Classics, Indianapolis (revised edition).
7. Kant, I.: 1964, *Groundwork of the Metaphysic of Morals*, translated and analysed by H.J. Paton, Harper Torchbooks/ The Academic Library, Harper & Row, New York/ Hagerstown/ San Francisco/ London, pp. 95–112.
8. Bein, E.: 1986, *Menschenwürde – Ein Arbeitsbuch*, Diesterweg, Frankfurt/Berlin/ München.

This book is written for the teaching of philosophy and contains a collection of more than twenty source texts, from Plato to Hans Jonas, via Cicero, Pico and Kant, and finally a few announcements from the German Federal Constitutional Court. It is one of the few existing historical anthologies on the subject of human dignity. The author has included introductions to and links between the sources, ordering them systematically, according to periods in history.

1.2 *General Historical Accounts*

9. Baker, H.: 1961, *The Image of Man: A Study of the Idea of Human Dignity in Classical Antiquity, the Middle Ages and the Renaissance*, Harper and Row, New York.

The book examines the classical, the Christian and the renaissance view of man from the presocratics to the reformation. It concludes that from Socrates to protestantism it was generally agreed that man is set apart from all other mortal creatures. The study intends to isolate some of the basic motifs that gave continuity to the belief in human dignity.

10. Cancik, H.: 1987, ' "Die Würde des Menschen ist unantastbar." Religions- und philosophiegeschichtliche Bemerkungen zu Artikel 1, Satz 1, Grundgesetz', in H.

Funke (ed.), *Utopie und Tradition, Platons Lehre vom Staat in der Moderne*, Königshausen & Neumann, Würzburg, pp. 73–107.

The author delivers an historically critical analysis of the concept of human dignity. Humanists during the Renaissance did not secularise the ideas behind Christian orthodoxy, instead renewing ancient traditions. Human dignity was always characterised by power over the world, over animals and over human beings. Attempts to back up Article 1 of the German Constitution with religion or by linking it to the *Quattrocento* are untenable.

11. Horstmann, R.-P.: 1980, 'Menschenwürde', in J. Ritter and K. Gründer (eds.), *Historisches Wörterbuch der Philosophie*, Vol. 5, Wissenschaftliche Buchgesellschaft, Darmstadt, column 1124–1127.

This article provides an historical overview of the content and function of the concept of human dignity, as well as the changes it has undergone. With the aid of relevant comments by Cicero, Pico, Pascal, Kant and others, it presents the components of this concept's history stemming from natural law, theology and philosophy, such as endowment with reason, *imago dei* and capacity for self-determination. Attention is paid to the significance of human dignity within political and legal fields, as well as the present state of the debate between substantialist and sociological viewpoints. The article ends with a bibliography of essential texts.

1.3 *Ancient Times and the Renaissance*

"*Dignitas*" originally stems from "*decus*" (decoration), in Rome's political and social circles meaning "public position and status". In the late Imperial "*notitia dignitatum*" it is full beaurocratic terminology of status. Philosophical use of the term, for example by Cicero, Seneca and early Christian writers, comes closer to the term dignity as it is understood today. During the Renaissance, the image of humanity shifted its accentuation from "*miseria hominis*", a conception popular with Christians in the Middle Ages, to "*dignitas hominis*". This was partly due to authors such as Pico and Manetti.

12. Auer, A.: 1956, 'G. Manetti und Pico della Mirandola: *De hominis dignitate*', in Festschrift für K. Adam, *Vitae et veritate*, Patmos, Düsseldorf, pp. 83–103.

13. Buck, A.: 1990a, 'Der Begriff der Menschenwürde im Denken der Renaissance, unter besonderer Berücksichtigung von Gianozzo Manetti', in G. Manetti, *Über die Würde und Erhabenheit des Menschen*, Meiner, Hamburg, pp. VII–XXXIV.

14. Buck, A.: 1990b, 'Giovanni Pico della Mirandola und seine "Rede über die Würde des Menschen" ', in G. Pico della Mirandola, *Über die Würde des Menschen*, Meiner, Hamburg, pp. VII–XXVII.

15. Cassirer, E.: 1959, ' "Über die Würde des Menschen" von Pico della Mirandola', *Agorà* **12**, 48–61.

16. Drexler, H.: 1943, '*Dignitas*', in R. Klein (ed.), 1966, *Das Staatsdenken der Römer*, Wissenschaftliche Buchgesellschaft, Darmstadt, pp. 231–254.

17. Dürig, W.: 1957, *'Dignitas'*, in T. Klauser (ed.), *Reallexikon für Antike und Christentum*, Vol. 3, Hiersemann, Stuttgart, column 1024–1035.
18. Kristeller, P.O.: 1972, 'The dignity of man', in Kristeller, *Renaissance Concepts of Man and Other Essays*, Harper and Row, New York; repr. in Kristeller, 1985, *Renaissance Thought and Its Sources*, ed. by M. Mooney, Columbia University Press, New York, pp. 169–181.
19. Trinkaus, C.: 1970, *In Our Image and Likeness. Humanity and Divinity in Italian Humanist Thought*, 2 Vols., University of Chicago Press, Chicago.
20. Wegehaupt, H.: 1932, *Die Bedeutung und Anwendung von dignitas in den Schriften der republikanischen Zeit*, Eschenhagen, Ohlau in Schlesien.

1.4 *Contemporary Approaches*

The following selection, entitled "contemporary approaches", is intended to shed light upon the controversial ways in which the concept of human dignity is applied. For this reason, the selection also includes works which fundamentally criticise the concept.

21. Bloch, E.: 1961, *Natural Law and Human Dignity*, transl. D. J. Schmidt, MIT Press.

In opposition to rejections or misinterpretations of natural law in civil, clerical, Fascist and Socialist writings, the author attempts to reveal the "true intentions behind the original natural law". The goal of natural law is human dignity, and it is reliant upon the social Utopia of economic freedom. The establishment of upright gait is a postulate beyond natural law.
22. Ginters, R.: 1982, 'Menschenwürde und Menschenrechte: Berufung auf sie – Appell oder Argumentation?' in R. Ginters, *Werte und Normen. Einführung in die philosophische und theologische Ethik*, Vandenhoeck & Ruprecht, Patmos, Göttingen and Düsseldorf, pp. 116–147.

Starting with the reason why each essential, ethical question may be answered without resorting to "human rights" or "human dignity", the author goes on to explain the appellative, paranetic function of the concept of human dignity, distinguishing between descriptive and normative components. Human dignity signifies the destiny of the human being to realise non-moral goods (happiness) on the one hand, and – more importantly – morality, on the other.
23. Gotesky, R. and E. Laszlo (eds.): 1970, *Human Dignity. This Century and the Next*, Gordon & Breach, New York.

This volume is a collection of papers concerning human dignity as a catch-all word for the unresolved problems of individuals and societies. Though the concept usually is vague, it is generally acknowledged and of universal relevance. The contributions consider its meaning in the context of human rights, technology, war, and the ideal society.
24. Hill, T.E.: 1992, *Dignity and Practical Reason in Kant's Moral Theory*, Cornell University Press, Ithaca and London.

A collection of essays on Kant's ethics, in which special attention is paid to Kant's concept of dignity.

25. Honnefelder, L.: 1991, 'Person und Menschenwürde. Zum Verhältnis von Metaphysik und Ethik bei der Begründung sittlicher Werte', in W. Pöldinger and W. Wagner (eds.), *Ethik in der Psychiatrie. Wertebegründung – Wertedurchsetzung*, Springer, Berlin, pp. 22–39.

The author argues in favour of a dualistic view of the human personality. Human dignity results from humans being moral subjects. It is a self-attribute, contained within a practical relationship to the self. It underlies all other values and calls for particular metaphysical assumptions. He rejects a metaphysics as favoured by authors ranging from Hume to Singer, because it gives rise to certain problems of justification regarding the worth of protecting human life.

26. Luhmann, N.: 1965, *Grundrechte als Institution*, Duncker & Humbldot, Berlin.

The intellectual conditions ontologically essential for natural law and dogmatic concepts of worth have distanced themselves considerably from the controllable truths of the empirical sciences. The author criticises "very elastic methods" of legal interpretation which merely delegate the weighing up of interests to the residing judge. Human dignity is on no account naturally given, but the result of successful self-portrayal. The human being itself is primarily responsible for its dignity. Through the latter's exposure as an attendant institution of German Basic Law, however, it has gained an important corrective function. The assumption that this catalogue of basic rights is a "system" of values stemming from the essence of humanity is illusionary.

27. Meyer, M.J. and W.A. Parent (eds.): 1992, *The Constitution of Rights – Human Dignity and American Values*, Cornell University Press, Ithaca and London.

Contains essays of twelve authors who want to provide a richer understanding of both the moral commitments that justify democratic institutions and the place of the idea of human dignity within the broader moral, political and legal scheme. Their essays cover a wide variety of topics and points of view and concern human dignity in moral theory in general as well as the role of human dignity in particular American constitutional rights.

28. Skinner, B.F.: 1971, *Beyond Freedom and Dignity*, Knopf, N.Y.

The author is concerned with matters of behavioural control, concentrating on the concept of dignity with regard to performance. If a person's behaviour is attributed to external circumstances and not to that person's autonomy, then that person's dignity is questioned. Scientists increasingly attribute the carrying out of, and responsibility for a certain type of behaviour to the environment, and not to the individual person. "Defenders of dignity", in Skinner's view, help to mask abuse of behavioural controls and to block more effective behavioural technologies.

29. Wagner, H.: 1992, *Die Würde des Menschen – Wesen und Normfunktion*, Königshausen und Neumann, Würzburg.

The determined positivism/empiricism of today is an obstacle to human dignity and its unequivocal justification. At most, the empirical sciences are able to establish a human superiority in relation to other living creatures. And yet human dignity is rooted in the fact that human beings exist as subjects. There is a difference between the principle of dignity and its concrete application, meaning that "humanity is called upon to give its opinion, alert and appropriately evaluating, to each instance of the concrete".

30. Bljumkin, I. A.: 1972, 'Die menschliche Würde', in F. V. Konstantinov, *Die Persönlichkeit im Sozialismus*, ed. Akad. d. Wiss. d. UdSSR, Akademie-Verlag, Berlin, pp. 209–220.
31. Gewirth, A.: 1992, 'Human dignity as the basis of rights', in M.J. Meyer and W.A. Parent (eds.), *The Constitution of Rights*, Human Dignity and American Values, Cornell University Press, Ithaca, N.Y., pp. 10–28.
32. Holderegger, A. , R. Imbach and R. Suarez de Miguel (eds.): 1987, *De Dignitate Hominis* – Mélanges offerts à C.-J. Pinto de Oliveira, Universitätsverlag, Freiburg (Ch).
33. Marcel, G.: 1965, *The Existential Background of Human Dignity*, Harvard University Press, Cambridge.
34. Meyer, M.J.: 1987, 'Kant's concept of dignity and modern political thought', *History of European Ideas* **8**, 319–332.
35. Meyer, M.J.: 1989, 'Dignity, rights, and self-control', *Ethics* **99**, 520–534.
36. Pinto de Oliveira, J.-C.: 1992, *Ethique chrétienne et dignité de l'homme*, Universitätsverlag, Freiburg (Ch).
37. Schüller, B.: 1978, 'Die Personenwürde des Menschen als Beweisgrund in der normativen Ethik', *Theologie und Philosophie* **53**, 538–555.
38. Spaemann, R.: 1985, 'Über den Begriff der Menschenwürde', *Scheidewege* **15**, 20–36; also in E.-W. Böckenförde and R. Spaemann (eds.): 1987, *Menschenrechte und Menschenwürde*, Klett-Cotta, Stuttgart, pp. 295–313.
39. Wolbert, W.: 1987, *Der Mensch als Mittel und Zweck. Die Idee der Menschenwürde in normativer Ethik und Metaethik*, Aschendorff, Münster.

2. THE CONCEPT OF HUMAN DIGNITY WITHIN THE FIELD OF LAW

In the course of the 20th Century, the concept of human dignity was integrated into the Constitutions of several countries. In addition to the philosophical significance it already had, it hereby gained additional legal significance.

2.1 *International*

40. Bielefeldt, H., W. Brugger and K. Dicke (eds.): 1992, *Würde und Recht des Menschen. Festschrift für J. Schwartländer*, Königshausen & Neumann, Würzburg.
    This volume contains several essays on the subject of human dignity, concentrating on its relationship to pluralism and pragmatism, to the universality of human rights, and to Islamic views.
41. Cohn, H.H.: 1983, 'On the meaning of human dignity', *Israel Yearbook on Human Rights* **13**, 226–251.
    The author, Deputy President Emeritus of the Supreme Court of Israel, reviews

the function possessesed by the human dignity concept within approximately twenty international announcements, and comes to the conclusion that human dignity is the source from which human rights are derived. He analyses texts from German authors and German law courts because German law has set a legislative and jurisprudential precedent that cannot be disregarded. In a third part he proposes that human dignity must also encompass the dignity of human death. Finally he examines the Jewish tradition, in which there is no counterpart for the term "dignity", but the word "*kavod*" has a similar tradition and meaning.

42. Newell, J.D.: 1982, 'The concept of human dignity in some contemporary legal contexts', *Archiv für Rechts- und Sozialphilosophie*, Supplementa, vol. 1, 411–417.

   Though the concept of human dignity has acquired a new importance in biomedical ethical contexts as euthanasia or suicide, the literature reveals an alarming lack of agreement concerning the meaning of "human dignity". The author analyses four usually employed concepts of human dignity, each of which suffers from the same defect: the fragmentation of the personality of the individual. He holds that the person is a subject-object entity whose dignity is many-faceted. If the latter is attacked, e.g. by a terminal disease, a right to protect the individual's dignity, i.e. a legal right to die with dignity should be granted.

43. Nino, C.S.: 1991, 'The principle of the dignity of the person', in Nino (ed.), *The Ethics of Human Rights*, Clarendon, Oxford, pp. 164–185.

   The author seeks to disqualify ethical dogmatism and ethical scepticism and to show that morality is a human creation. In the light of a constructivist meta-ethical conception, following Kant, Rawls and others, he discusses principles like human dignity, from which fundamental individual rights could be derived. The principle of the dignity of the person is opposed to normative determinism in which the volition of an individual does not count and a comprehensible conception of society cannot be based. It underlies the possibility of a dynamic treatment of the other liberal principles.

44. Schachter, O.: 1983, 'Human dignity as a normative concept', *American Journal of International Law* 77, 848–854.

   No explicit definition of "human dignity" is accepted. It has been left to intuitive understanding dependent on cultural factors. The Kantian proposition that individuals are not to be perceived or treated merely as instruments or objects is generally acceptable and has further implications as corollaries. In a philosophical sense human rights are derived from human dignity, not from the state or any other external authority. Drawing upon human dignity even new rights can be formulated for application to new situations. Not only by asserting claims of right, but also through informal channels respect for dignity can be fostered.

45. Verdross, A.: 1977, 'Die Würde des Menschen als Grundlage der Menschenrechte', *Europäische Grundrechte-Zeitschrift* 4, 207–208.

and

46. Verdross, A.: 1979, 'La dignité de la personne humaine comme base des droits de l'homme', in U. Häfelin, W. Haller and D. Schindler (eds.), *Menschenrechte, Föderalismus, Demokratie*, Schulthess, Zürich, pp. 415–421.

   On several occasions, the member states of the United Nations have acknowledged human dignity and the fact that it has exceedingly positive foundations. Evidence of historical roots in Near Eastern, European and American cultural circles serves to support this exceedingly positive view. Sources cited include the Bible, Greek philosophy, Kant and Hegel.

47. Archiv für Rechts- und Sozialphilosophie: 1988, *Menschen- und Bürgerrechte*, Beiheft 33, ed. U. Klug and M. Kriele, Steiner, Stuttgart.
48. Häberle, P.: 1980, 'Menschenwürde und Verfassung am Beispiel von Art. 2 Abs. 1 Verf. Griechenland 1975', *Rechtstheorie* **11**, 389–426.
49. Howard, R.E. and J. Donelly: 1986, 'Human dignity, human rights, and political regimes', *American Political Quarterly* **80**, 802–817.
50. Khushalani, Y.: 1982, *Dignity and Honour of Women as Basic and Fundamental Human Right*, Nijhoff, The Hague.
51. Mandy, P.: 1991, 'Besoins élémentaires et dignité humaine. L'application économique de "noyau intangible" des droits de l'homme', in P. Meyer-Bisch (ed.), *Le noyau intangible des droits de l'homme*. Actes du VIIe colloque interdisciplinaire sur les droits de l'homme, Editions Universitaires, Fribourg/Suisse, pp. 149–170.
52. Mastronardi, P.A.: 1978, *Der Verfassungsgrundsatz der Menschenwürde in der Schweiz*, Duncker & Humblot, Berlin.
53. Mastronardi, P.A.: 1979, 'Die Menschenwürde als Verfassungsgrundsatz in der Schweiz', *Jahrbuch des Öffentlichen Rechts der Gegenwart*. N.F., Vol. 28, 469–485.
54. Paust, J.: 1984, 'Human dignity as a constitutional right: a jurisprudentially based inquiry into criteria and content', *Howard Law Journal* **27**, 150–158.
55. Recaséns-Siches, L.: 1977, 'Dignity, liberty and equality', in G. Dorsey (ed.), *Equality and Freedom*. Papers on the World Congress on Philosophy of Law and Social Philosophy 1975, Vol.1, Oceana Publ., Dobbs Ferry, N.Y.; Sijthoff, Leiden, pp. 3–25.
56. Rodrígez Molinero, M.: 1977, '*Libertad, igualdad y dignidad de la persona*', in G. Dorsey (ed.), *Equality and Freedom*. Papers on the World Congress on Philosophy of Law and Social Philosophy 1975, Vol.1, Oceana Publ., Dobbs Ferry, N.Y.; Sijthoff, Leiden, pp. 143–149.
57. Wetlesen, J.: 1990, 'Inherent dignity as a ground of human rights – a dialogical approach', in W. Maihofer and G. Sprenger (eds.), *Revolution and Human Rights*, Proceedings of the 14th IVR World Congress 1989, Series Archiv für Rechts- und Sozialphilosophie, Beiheft 41, Steiner, Stuttgart, pp. 98–114.

2.2 Germany

2.2.1 Basic Law: Commentaries and Handbooks

The commentaries and articles from handbooks on Basic Law included here contain basic texts to aid interpretation of the human dignity norm. They are significant in Germany in both legal studies and practice, describing the scope and legal function of the dignity norm.

58. Benda, E.: 1983, 'Die Menschenwürde', in E. Benda, W. Maihofer and H. J. Vogel (eds.), *Handbuch des Verfassungsrechts*, de Gruyter, Berlin/New York, pp. 107–128.
59. Häberle, P.: 1987, 'Die Menschenwürde als Grundlage der staatlichen Gemeinschaft', in J. Isensee and P. Kirchhof (eds.), *Handbuch des Staatsrechts*, Vol. 1: Grundlagen von Staat und Verfassung, Juristischer Verlag Müller, Heidelberg, pp. 815–861.
60. Herzog, R.: 1966, 'Menschenwürde', in H. Kunst and S. Grundmann (eds.), *Evangelisches Staatslexikon*, Stuttgart, XXIff.
61. Klein, F.: 1990, 'Artikel 1, Menschenwürde – Menschenrechte', in B. Schmidt-Bleibtreu (ed.), *Kommentar zum Grundgesetz*, Luchterhand, Neuwied, 7th ed., pp. 122–135.
62. Kunig, P.: 1992, 'Kommentar zu Artikel 1', in I. von Münch and Ph. Kunig (eds.), *Grundgesetz-Kommentar*, Vol. 1, Beck, München, 4th ed., pp. 77–141.
63. Maunz, T. and G. Dürig: 1990, *Grundgesetz-Loseblatt-Kommentar*, 7th ed., Beck, München.
64. Münch, I. von: 1981, 'Kommentar zu Art. 1', in Münch (ed.), *Grundgesetz-Kommentar*, vol. 1, Beck, München, 2nd ed., pp. 65–104.
65. Podlech, A.: 1984, 'Kommentar zu Artikel 1 Absatz 1 GG', in Azzola *et al.*, *Kommentar zum Grundgesetz für die Bundesrepublik Deutschland*, vol.1, Luchterhand, Neuwied, 2nd ed. 1989, pp. 199–225.
66. Starck, C.: 1985, 'Artikel 1', in H. von Mangoldt, F. Klein and C. Starck, *Das Bonner Grundgesetz*, vol. 1, Franz Vahlen, München, 3rd ed., pp. 24–149.

2.2.2 *Miscellaneous*

67. Dürig, G.: 1956, 'Der Grundrechtssatz von der Menschenwürde', *Archiv des Öffentlichen Rechts* 81, no. 2, 117–157.

This essay is one of the sources most frequently cited within legal literature in the German language on the problems raised by human dignity. It contains the *"object formula"* which was canonised by the German Federal Constitutional Court to explain Article 1, Paragraph 1 of German Basic Law. This formula states that the human being may not be "degraded to an object, to a mere means, to a tenable parameter" and therefore not to "the object of State proceedings" either. Dürig goes on to say that human dignity has been acknowledged by Basic legislation as an objective and fundamental value as reference to God is no longer generally accepted. This value calls for forbearance and breaks with liberalistic optimism that human dignity should be no concern of the State. As a principle of Basic Law, it amounts to an "axiomatic decision for eternity" which places the State under an obligation to defend human dignity from third-party attacks. An as yet unborn or already dead human being also possesses human dignity.

68. Eger, T., B. Nagel and P. Weise: 1991, 'Effizienz und Menschenwürde – ein Gegensatz? Anmerkungen zu einer Kontroverse zwischen Ott / Schäfer und Fezer', in C. Ott and H.-B. Schäfer (eds.), 1991, *Ökonomische Probleme des Zivilrechts*, Springer, Berlin *et al.*, pp. 18–34.

It is useful to examine each phenomenon with regard to what the alternative would cost. Human dignity may be described and priced as a good, whereby the popular criteria put forward by Pareto and Kaldor/Hicks are unsatisfactory, both

in theory and in practice. If we do away with the absolution of efficient allocation, this point of view is extremely well suited for analysing social phenomena.
What the lawyers term the weighing up of goods, the economists term "trade-off".
An economic analysis enables the prices implied by legal decisions to be made
visible, including those for human dignity or freedom.

69. Geddert-Steinacher, T.: 1990, *Menschenwürde als Verfassungsbegriff, Aspekte der
    Rechtsprechung des Bundesverfassungsgerichts zu Artikel 1 Absatz 1 Grundgesetz*,
    Duncker & Humblodt, Berlin.

    This text delivers a legislative analysis on the basis of decisions by the German
    Federal Constitutional Court. Paraphrases of human dignity, for example the "object
    formula" borrowed from Kant's philosophy (see [67]), are examined for their suit-
    ability in deciding concrete cases. In addition, this text explains the position held
    by Article 1, Paragraph 1 in German Basic Law, its "radiating effect" on the basic
    rights which follow it, as well as individual functions and types of argumentation.
    Its goal is to render the dignity norm, as used to justify German Federal Constitutional
    Court decisions, more transparent, and thus controllable.

70. Hoerster, N.: 1983, 'Zur Bedeutung des Prinzips der Menschenwürde', *Juristische
    Schulung* **23**, 93–96.

    The author adopts a critical stance with regard to the concept of human dignity.
    He considers the interpretation common within Basic Law to be problematic, since
    the decisive question concerning which kinds of human self-determination are
    legitimate cannot always be answered by intersubjectively accepted criteria. When
    applying the principle of human dignity, evaluation is unavoidable, since human
    dignity is not something which is given and recognisable. In addition, the end
    result arrived at through human dignity is, in nearly every case, anchored within
    current German Law anyway. If the legislator did not take the necessary evalua-
    tion upon himself, instead delegating it to the residing judge by means of the
    human dignity principle, then this would amount to giving the judge a blank cheque
    – free to decide according to his own personal values.

71. Krawietz, W.: 1977, 'Gewährt Art. 1 Abs. 1 GG dem Menschen ein Grundrecht
    auf Achtung und Schutz seiner Würde?', in D. Wilke and H. Weber (eds.),
    Gedächtnisschrift für F. Klein, Vahlen, München, pp. 245–287.

    Human dignity is neither a platonic idea nor an eternally binding value, but a
    legal value anchored within German Basic Law, as the result of a normative decision.
    Dürig's humanistic interpretation (see [67]) is scientifically risky and methodo-
    logically questionable. Human immediate self-interest has the utmost priority, and
    Article 1, Paragraph 1 of German Basic Law guarantees its protection as a basic
    right, impressively and unambiguously. Any other view would contradict the system.
    The vagueness of the regulation is non-problematic insofar as it is the task of
    legislation to convert it into concrete terms when deciding concrete problems.

72. Maihofer, W.: 1968, *Rechtsstaat und menschliche Würde*, Klostermann,
    Frankfurt/Main.

    According to Maihofer, former German Minister of the Interior, human dignity
    is not the product of natural determination, but of human self-determination. It is
    threatened by behaviour and by relationships, and thus justifies not only a call
    for forbearance, but also for action. Basic Law principles regarding the legal
    State, the social State and democracy are founded in human dignity as State

purposes. Human dignity protects humanity (Kant), sociality (Marx) and singularity (Sartre). It is not a given, objective value. The legal field cannot escape evaluation, however, since human dignity is intersubjectively part of "the nature of things".

73. Schreiber, H.-L.: 1989, 'Die Würde des Menschen', in G. Borsi (ed.), *Die Würde des Menschen im psychiatrischen Alltag*, Vandenhoeck & Ruprecht, Göttingen, pp. 15–23.

This essay contains a history of human dignity and refers to Luhmann and Skinner's interpretation of the concept. The author also establishes that the claim to respect – guaranteed by Article 1, Paragraph 1 of German Basic Law – is inviolable. He warns against simple appeals to "dignity" in order to avoid essential and concrete debates and decisions. The dignity precept is a kind of "regulating principle"; Spaemann's use of it, for example, is unacceptable.

74. Vitzthum, W.: 1985, 'Die Menschenwürde als Verfassungsbegriff', *Juristenzeitung* 40, 201–209.

Protecting human dignity is essential for the legitimation of State power. The author acknowledges Dürig's "object formula" (see [67]) and a negative definition of human dignity. Contrary to Luhmann's performance-orientated concept, he offers a reminder that protection of human dignity is still important when active self-portrayal is unsuccessful. Individual basic rights are not sufficient to prevent humanity from being biomedically endangered. Only the principle of dignity itself fully covers the scope of technologically viable legal violations and can provide adequate opposition.

75. Alexy, R.: 1986, *Theorie der Grundrechte*, Suhrkamp, Frankfurt/Main, pp. 94–97 and pp. 321–326.

76. Starck, C.: 1981, 'Menschenwürde als Verfassungsgarantie im modernen Staat', *Juristenzeitung* 36, 457–464.

77. Stern, K.: 1983, 'Menschenwürde als Wurzel der Menschen- und Grundrechte', in N. Achterberg *et al.* (eds.), *Recht und Staat im sozialen Wandel*, Festschrift für H. Scupin, Berlin, pp. 627–642.

78. Stöcker, H.A.: 1968, 'Menschliche Würde und kritische Jurisprudenz', *Juristenzeitung* 23, 685–691.

3. THE CONCEPT OF HUMAN DIGNITY WITHIN THE FIELD OF MEDICINE

Since the beginning of the 1980s, the concept of human dignity has been experiencing a boom within the field of bioethics. This is especially true for the German-speaking countries. The concept notably comes into play in connection with the ethical problems surrounding gene and reproduction technology and euthanasia.

3.1 *General*

79. Hansson, M.G.: 1991, *Human Dignity and Animal Well-Being. A Kantian Contribution to Biomedical Ethics*, Almqvist & Wiksell, Stockholm.

The conclusion of this study is that Kantian ethics provides a fruitful ethical approach which gives differentiated answers to questions concerning biomedical and ecological ethical problems. The Kantian account assigns a special value to man because he is the only being with the capacity to be the author of universal moral laws.

80. Hoerster, N.: 1991, 'Der Fötus als "Ebenbild Gottes"?', in Hoerster, *Abtreibung im säkularen Staat. Argumente gegen den §218*, Suhrkamp, Frankfurt/Main, pp. 114–127.

A religious view of the human race comprehends the right to live of foeti as a direct consequence of *imago dei*. Making the beliefs prerequisite for this view the basis of a criminally punishable act would not be compatible with the tolerance demanded of a Western secular State. The concept of human dignity is therefore used as a secular veil to hide a religious view of humanity. And yet, appeals to human dignity and the propagation of a certain normative requirement exist incoherently side-by-side.

81. Holzhüter, W.: 1989, *Konkretisierung und Bedeutungswandel der Menschenwürdenorm des Artikel 1 Absatz 1 des Grundgesetzes*, Dissertations Druck, Darmstadt.

This dissertation examines the idea of human dignity from an historical point of view, as well as international use of the norm, subsequently turning to problems such as abortion, imprisonment, data protection, etc. It turns out that violations of human dignity in Germany are usually treated in connection with other elements of Basic Law, such as Article 2, Paragraph 1. In addition, the dignity Article is elastic in its interpretation, so that a change in content may not be prevented. On the basis of this, the author dedicates the last few chapters to an attempted evaluation of reproduction medicine, gene technology, euthanasia and organ removal.

82. Inoue, T.: 1988, 'Dignity of life', in J. Bernard *et al.* (eds.), *Human Dignity and Medicine. Proceedings of the Fukui Bioethics Seminar*, Japan 1987, Elsevier, Amsterdam, pp. 11–16.

In Buddhism not only all sentient beings, but all living, i.e. existing things represent the virtue and wisdom of Buddha. Dohgen said: "All existence has Buddhahood." Buddhahood or Buddha nature is the rationale for the value of all things. Underlying this view is the notion of causal relationship through which all beings are connected. The Buddhist holds the precept not to indulge or engage in unnecessary, wasteful killing of human beings or even nonbiological beings. The Buddhist concept of *Innen*, i.e. of indirect and direct causes, and the cycle of reincarnation are essential for the dignity of human as well as of all life.

83. Ziegler, J.G,: 1977, 'Organübertragung – Medizinische, moraltheologische und juristische Aspekte', in J.G. Ziegler (ed.), *Organverpflanzung. Medizinische, rechtliche und ethische Probleme*, Patmos, Düsseldorf, pp. 52–127.

Human dignity does not grant an unlimited right of access to oneself or others, but merely a usufructuary right to one's own physical or spiritual powers. Several ideal values are higher than corporeal integrity, meaning that organ donation is morally justifiable when intended to help. Functionalisation of personal existence for social purposes, e.g. in the 1789 Declaration of Human Rights or in Socialism, merits rejection; absolute justification, e.g. in the 1776 Declaration of Independence, merits priority. Organ removal and donation are not to be decided at will.

84. Auer, A.: 1989, 'Bioethische Argumentation mit der Menschenwürde?', in D. Schwab (ed.), *Staat, Kirche, Wissenschaft in einer pluralistischen Gesellschaft, Festschrift für P. Miktat*, Berlin, pp. 13–28.
85. Böhler, D.: 1992, 'Diskursethik und Menschenwürdegrundsatz zwischen Idealisierung und Erfolgsverantwortung', in K.-O. Apel and M. Kettner (eds.), *Zur Anwendung der Diskursethik in Politik, Recht und Wissenschaft*, Suhrkamp, Frankfurt/Main, pp. 201–231.
86. Eibach, U.: 1976, *Medizin und Menschenwürde. Ethische Probleme der Medizin aus christlicher Sicht*, Brockhaus, Wuppertal.
87. Herzog, R.: 1987, 'Die Menschenwürde als Maßstab der Rechtspolitik', in H. Seesing (ed.), *Technologischer Fortschritt und menschliches Leben*, Schweitzer, München, pp. 23–32.
88. Lauenstein, D.: 1976, 'Die "Würde des Menschen" bei Arzt und Patient. Die Deklarationen des Weltärztebundes von Tokio gehen ebenso wie unser Grundgesetz auf Kants Philosophie zurück', *Deutsches Ärzteblatt*, 1976, 2173–2178.
89. Pap, M.: 1986, 'Die Würde des werdenden Lebens', *Medizinrecht* 4, 229–236.
90. Szawaski, Z.: 1989, 'Dignity and Technology', *Journal of Medicine and Philosophy* 14, 1989, 243–249.
91. Trockel, H.: 1971, 'Menschenwürde und medizinisch-biologische Forschung', *Neue Juristische Wochenschrift* 24, 217–220.

3.2 Reproduction and Gene Technology

92. Benda, E.: 1984, 'Erprobung der Menschenwürde am Beispiel der Humangenetik', in R. Flöhl (ed.) (1985), *Genforschung – Fluch oder Segen*, Schweitzer, München, pp. 205–231.

Benda, former German Minister of the Interior and former President of the German Federal Constitutional Court, is of the opinion that the human dignity principle may, as an empty formula, underly "metaphysical notions of encumberance" or an ideologisation. Even if Article 1 of German Basic Law does primarily deal with the individual human being, irreparable manipulations of the human image must be forbidden. If the human being were to be mass-produced, for example, then this must surely lead to its loss as a creature. Benda rejects notions of perfectibility with the observation that human imperfection is the very definition of human essence. This is why Basic Law prohibits all intentions which extend beyond operations intended to heal.

93. Birnbacher, D.: 1987, 'Gefährdet die moderne Reproduktionsmedizin die menschliche Würde?', in V. Braun, D. Mieth and K. Steigleder (eds.), *Ethische und rechtliche Fragen der Gentechnologie und der Reproduktionsmedizin*, Schweitzer, München; repr. in A. Leist (ed.): 1990, *Um Leben und Tod. Moralische Probleme bei Abtreibung, künstlicher Befruchtung, Euthanasie und Selbstmord*, Suhrkamp, Frankfurt/Main, pp. 266–281.

The concept of human dignity is often uncertain and ambiguous. It does not always prevent heteronomy, identity change or intervention in the natural course of things. The author distinguishes between an individual concept of human dignity and one based on the entire genus, and qualifies some aspects of modern medicine as violations of human dignity in this second sense. And yet it is not always dangers to individual human dignity which render these new procedures morally unsound; often it is the increased risk of other dangers. In contrast to the principle of individual human dignity, dignity of the human genus requires that a weighing-up process take place. The Law needs to draw very definite lines, whereas the ethicist is faced with the often impossible task of finding universal norms capable of inspiring consensus.

94. Enders, C.: 1986, 'Die Menschenwürde und ihr Schutz vor gentechnologischer Gefährdung', *Europäische Grundrechte-Zeitschrift* **13**, 241–252.

The author believes that the concept of human dignity – in addition – essentially protects the image of humanity and thus the future of the entire human race. The debates often deal with conventional situations of danger in which resorting to human dignity is hardly suitable. Human dignity is unable to live up to the expectations placed in it, and yet its anchored position within Basic Law keeps a certainty within Western tradition alive.

95. Fechner, E.: 1986, 'Menschenwürde und generative Forschung und Technik', *Juristenzeitung* **41**, 653–664.

Human dignity is a humanitarian success within a world full of violence and cruelty. The image of humanity within German Basic Law is not so much Kant's conception as that of Catholic social doctrine. In addition, human dignity and human rights have an unstable position internationally due to varying religious and social views, meaning that States "burdened with human dignity" could be at a fateful disadvantage. A voluntarist justification should make it possible, independent of naturalistic and religious world views, and in the light of violation of the human image, e.g. by the National Socialists, to encourage spontaneous respect for human dignity. A weighing-up process is necessary when the dignity norm is applied to research and technology.

96. Kluxen, W.: 1986, 'Fortpflanzungstechnologien und Menschenwürde', *Allgemeine Zeitschrift für Philosophie* **11**, 1–15.

The term "human dignity" may be ideologically threadbare, but philosophically it is clear: it denotes the existence of human beings as moral subjects, about the acknowledgement of which consensus must not only be capable of being achieved, but must also be achieved. The human being must never be treated just as an object. All members of the species *homo sapiens* possess human dignity. Any kind of intervention in the genotypes of fertilised egg cells amounts to manifest heteronomy for those involved. Reproduction technologies are only justifiable in cases of morally tenable denaturalisation and if the resulting social situation is favourable for the child. Advancing sexual permissiveness is a danger to human dignity and the vital basis of society.

97. Köbl, U.: 1985, 'Gentechnologie zu eugenischen Zwecken – Niedergang oder Steigerung der Menschenwürde?', in H. Forkel and A. Kraft (eds.), *Beiträge zum Schutz der Persönlichkeit und ihrer schöpferischen Leistungen. Festschrift für H. Hubmann*, Metzner, Frankfurt/Main, pp. 161–192.

The author analyses the status of the human dignity argument. She criticises the religious and metaphysical views of Christian writers. In order to preserve human dignity in the light of gene technological manipulation, she demands that human powers of self-determination and self-education, of autonomous recognition and realisation of values, be neither removed nor limited. She objects rigorous moral dogmatism and the normative dominance of individual persons and groups. Reticence concerning trenchant restriction is recommended in order to prevent aid from being carelessly rejected.

98. Lerche, P.: 1986, 'Verfassungsrechtliche Aspekte der Gentechnologie', in R. Lukes and R. Scholz (eds.), *Rechtsfragen der Gentechnologie*, Heymanns, Köln *et al.*, pp. 88–111.

The level of significance which is attributed to "human dignity" in the evaluation of gene technology should undergo a drastic reduction. Human dignity only merits defence within a limited area "where legal contemporaries give their assent as a matter of course", namely the prohibition of human degradation, respect for individuality and the guarantee of humane living conditions. Areas of protection beyond these merely serve to weaken the principle of human dignity, as in the case of abortion, nuclear energy technology or human genetics. They are all sufficiently covered by other, independent basic laws. Transferring these problems to the protection of human dignity track is not a solution, instead leading to a fragmentation of our consciousness of what is right and wrong.

99. Neumann, U.: 1988, 'Die "Würde des Menschen" in der Diskussion um Gentechnologie und Befruchtungstechnologien', in *Archiv für Rechts- und Sozialphilosophie*, Beiheft 33, ed. U. Klug and M. Kriele, Steiner, Stuttgart, pp. 139–152.

The principle of human dignity is so irritatingly ambiguous that it raises considerable subsumption problems. Extending its area of application leads to a normative dilution, which in turn endangers consensus. Theological debates and certain areas of debate on Basic Law are plagued by an inconsistent, naturalistic point of view which is incapable of separating biological Nature sufficiently from moral and reasonable human nature, and which implies metaphysical assumptions that are factually unsuitable for intersubjective acknowledgement. Finding normative alternatives to the precious human dignity principle is primarily the task of a philosophical ethics, the significance of which is emphasised by the author.

100. Schöne-Seifert, B.: 1990, 'Philosophische Überlegungen zu "Menschenwürde" und Fortpflanzungs-Medizin', *Zeitschrift für philosophische Forschung* **44**, 442–473.

The concept of human dignity serves to denote *ex post* a norm violation deeming independent justification, and it is a specifically German debating characteristic. The author identifies a descriptive and a normative component within the concept. By "coupling a quality 'human dignity' and a claim to human dignity which it seems to justify", they suggest a naturalistic fallacy within the term itself, so to speak. The concept touches on the impersonal realisation of values, self-determination, charity, the welfare of children, parents and human beings in general, as well as the rules and principles characterising norms. Respect for human dignity is based on a bundle of *prima facie* norms and represents a weighing-up rule, whereby due consideration for 'personal' interests is demanded above that of all other interests.

101. Vitzthum, W.: 1985, 'Gentechnologie und Menschenwürde', *Medizinrecht* **3**, 249–257.

Because of abstract and concrete dangers, the author believes it necessary to define limits for gene technology in the area of tension between human dignity and the freedom to research. Discussing philosophical and theological justifications of the dignity norm and the concept of human dignity within German Basic Law, he weighs up the benefits and dangers represented by applications of gene technology. He attributes dignity to the fertilised egg cell, and yet the survival of the latter is not to enjoy absolute priority over other Basic Law values. He distinguishes between concrete and abstract dangers with regard to the Law.

102. Leist, A.: 1990, 'Die normative Schwäche des Würdebegriffs', in Leist, *Eine Frage des Lebens, Ethik der Abtreibung und künstlichen Befruchtung*, Campus, Frankfurt/Main and N.Y., pp. 210–216.

103. Knoppers, B.M.: 1991, *Human Dignity and Genetic Heritage*, Law Reform Comission of Canada, Ottawa.

104. Reiter, J.: 1988, 'Menschenwürde und Gentechnologie', in H. Seesing (ed.), *Technologischer Fortschritt und menschliches Leben*, Part 2, Schweitzer, München, pp. 16–34.

105. Struck, G.: 1988, 'Die "Würde des Menschen" als Argument und Tabu in der Debatte zur Fertilisations- und Gentechnologie', in *Archiv für Rechts- und Sozialphilosophie*, Beiheft 33, pp. 110–118.

106. Vitzthum, W.: 1987, 'Gentechnologie und Menschenwürdeargument', *Zeitschrift für Rechtspolitik*, 33–37; repr. in: ARSP 1988, Beiheft 33, pp. 119–138.

3.3 *Life and Death*

107. Burkart, M.: 1983, *Das Recht, in Würde zu sterben – ein Menschenrecht. Eine verfassungsrechtliche Studie zur Frage der menschenwürdigen Grenze zwischen Leben und Tod*, Schulthess, Zürich.

This study examines the significance attributed to dying within the euthanasia debate and the content of the Swiss Constitutional principle of human dignity – with a sideways glance at German and European Law. Guidelines set by Swiss institutions, the American "Patients' Bill of Rights" and recommendations made by the Council of Europe are cited as examples of the codification of patients' rights. The study aims to differentiate the "right to die in dignity". Legalisation of active euthanasia is rejected.

108. Kuhse, H.: 1990, *Menschliches Leben und seine Würde: Fragen des Lebens und des Sterbens*, Series Medizinethische Materialien, Heft 41, Zentrum für Medizinische Ethik, Bochum.

Since terms such as "human dignity" or the "humanity precept" contain more pathos than semantic precision, revolutionary medical developments force us to reveal and defend the values beheld by these formulae. Prolonging life regardless of the circumstances can sometimes amount to a violation of dignity. It is a creature's interests, and not life *per se*, which are morally significant. In addition, we speak of dignity when referring to autonomous beings capable of making decisions. The dignity norm therefore means that we have to accept as binding the decisions reached freely by patients capable of judgment.

109. Auer, A.: 1977, 'Die Unverfügbarkeit des Lebens und das Recht auf einen natürlichen Tod', in A. Auer, H. Menzel and A. Eser, *Zwischen Heilauftrag und Sterbehilfe*, Heymanns, Köln *et al.*, pp. 1–51.

3.4 *Psychiatry*

110. Abelson, R.: 1978, 'Psychotherapy and personal dignity', *Psychoanalysis and Contemporary Thought* **1**, 1978, 203–210.
111. Abelson, R. and Margolis, J.: 1978, 'A further exchange on psychotherapy, personal dignity and persons', *Psychoanalysis and Contemporary Thought* **1**, 1978, 227–236.
112. Borsi, G.M. (ed.): 1989, *Die Würde des Menschen im psychiatrischen Alltag*, Vandenhoeck & Ruprecht, Göttingen.
113. Bruder, J.: 1989, 'Die Würde des verwirrten alten Menschen. Zum Verständnis von Würde', in G. Borsi (ed.), *Die Würde des Menschen im psychiatrischen Alltag*, Vandenhoeck & Ruprecht, Göttingen, pp. 58–66.

Department of Philosophy
University of Bremen
Germany

NOTE

* Translated into English by Sarah L. Kirkby

GEORGE KHUSHF

THE SANCTITY OF LIFE: A LITERATURE REVIEW

In the debates about issues such as abortion, infanticide, euthanasia, suicide, and the death penalty, there are often appeals to the "sanctity of life." It is assumed that this is an ancient and well defined ethical principle on which a sound argument can be based. In order to evaluate this common assumption, the origin, development, and current usage of this phrase needs to be traced. For this a literature search was conducted. Some of the results along with an annotated bibliography are here provided.

For this literature search the following databases were used:

(1) Epic database: 4800 libraries, including the Libraries of Congress in 26 countries, but excluding periodicals; (2) the CD-ROM of Philosopher's Index, which contains bibliographic citations to philosophical journal articles and books from 1940 to the present; (3) the CD-ROM of Religion indexes, which comprise four American Theological Library Association publications that list periodicals from 1949 to the present; (4) Medline, an on line database produced by the National Library of Medicine and indexing data published in clinical medicine and the biomedical literature from 1966 to the present. In each case the search was done for the intersection of "sanctity" and "life", and it included titles, abstracts, and key words. The search was restricted to the English phrase, but not to English language articles and books.

After the search was conducted, many of the articles and books were obtained and the references of these works were checked, in order to identify any citations on the sanctity of life that did not show up in the initial search. This was especially important for the identification of early

293

K. Bayertz (ed.), Sanctity of Life and Human Dignity, 293–310.
© 1996 *Kluwer Academic Publishers. Printed in the Netherlands.*

articles, since the databases for articles extended back no earlier than 1940.

Well over one hundred articles and books on the sanctity of life have been identified in this literature survey, and 143 representative citations are listed in the annotated bibliography. Nearly all citations are after 1948 and all but two are from the twentieth century. This is not because the search focused only upon this time period. Although the indexes used for articles was limited to the latter half of the twentieth century, the Epic database for books includes libraries with extensive holdings from previous centuries. Further, as noted above, an effort was made to check the references of many of the citations that have been identified, in order to trace older sources.

One of the most interesting results of this literature search is that current articles and books on the principle of the sanctity of life do not cite pre-twentieth century uses or discussions of this principle. Much older sources are cited in the discussion of related issues such as abortion, euthanasia, or the ordinary/extraordinary care distinction, and these sources are often used to argue that the sanctity of life is ancient. But the older sources do not explicitly refer to the sanctity of life. They refer only to results that people today seek to develop as a consequence of the principle of the sanctity of life. This absence of pre-twentieth century sources is surprising, since both advocates and detractors designate the sanctity of life as an ancient principle of the Judeo-Christian and Hippocratic traditions. If it is indeed ancient, why are there no references to ancient uses of the principle in the extant literature on the topic, and why were no pre-twentieth century sources identified in the search?

There are several possible reasons for the failure of this search to identify pre-twentieth century references to the sanctity of life: (1) the search was restricted primarily to the English phrase, and older uses may have been in another language; (2) the databases for articles was very limited in time, and the phrase may not have played a conspicuous enough role in the more distant past to have been included in a book title; (3) the principle may have been designated by another name; e.g., one may have spoken of the image of God, or the infinite value, uniqueness or dignity of human life, and these terms were not included in the search; in fact, earlier uses of such alternative phrases are found in the literature, although it is not clear that they serve the same function as the sanctity of life; (4) the principle may have been implicit or taken for granted, and it is only in recent times that conditions have been sufficiently altered to require explicit reflection. One thus cannot conclude from this search that the sanctity of life is not an ancient principle.

There is good reason to believe that the phrase was commonly used well before the earliest citation provided in this survey. In *The Better America Lectures* (1921) Newell Dwight Willis has a chapter titled "The Security of Property as the Logical Inference from the Sanctity of Life," and he argues that "[t]he founders of the [U.S.] Republic . . . saw that civilization was based upon the sanctity of life and security of property" ([102], p. 59). Here the sanctity of life seems to be regarded as an alternative expression for the commandment not to kill, and it is viewed as the fundamental principle on which the commandment not to steal is based. However, no account is given of the meaning of this principle or its justification. It is clear from the text that the author is not being innovative in his use of terminology, and the phrase "sanctity of life" must have been current in usage. Unfortunately, Willis does not provide any citations, which would have assisted in tracing back the principle. Still, it is clear that he is drawing on an older tradition, one that is also manifest in 1915 with Andrew Klarmann's use of the sacredness of life as a basis for rejecting abortion.

Although no conclusions can be reached regarding how long the principle of the sanctity of life has served an important function in ethical thought, one important result does follow from this survey: recent, twentieth century reflection on the principle does not draw directly on older accounts of the sanctity of life (if there are such accounts), at least not explicitly, since no citations to such accounts are provided in the current literature.

With respect to book titles, the earliest use of the phrase "sanctity of life" in its technical sense as an ethical principle occurs in 1957 with *The Sanctity of Life and the Criminal Law*, by Glanville Williams. By "in its technical sense as an ethical principle" I refer to the "sanctity of life" as a principle that is used in ethical argumentation and that applies to all human lives, not just those who are religious. Before 1957 the phrase is indeed used in book titles, but it has a different meaning. Thus, for example, one has Dinsdale Young's *The Sanctity of Life: Secular and Mid-Week Sermons*, which was published in 1932. In this case the focus is on how secular matters of life such as work, meal-time, and family parties can be regarded as sacred. Young's use of the phrase in the title shows that in 1932 "sanctity of life" was not yet restricted to its present technical meaning as an ethical principle. For the broader religious meaning, one can even go back much earlier to Jeremy Taylor's *The Great Exemplar of Sanctity and a Holy Life* (1667), but here the sanctity and holiness of life is not something that characterizes all human lives; it is a quality of a holy or sanctified person, and primarily a quality of Jesus Christ. The meaning of "sanctity of life" in these cases is so different from that of the ethical concept at issue in this review that one cannot view these sources as relevant to this search.

Glanville Williams' book is on issues in reproduction, abortion, suicide and euthanasia, and it draws heavily upon an earlier work by Joseph Fletcher, which addressed many of the same themes. Fletcher's 1954 *Morals and Medicine* was a ground-breaking work, not just for its discussion and critique of the sanctity of life, but also for its role in initiating sustained reflection on bioethical issues. This book by Fletcher incorporates an earlier article of his, "Our Right to Die", published 1951 in *Theology Today* as a response to an essay by John Sutherland Bonnell titled "The Sanctity of Human Life." Bonnell rejects euthanasia, stating without references that "Christianity has never ceased to emphasize the sanctity of human life and the value of the individual, even the humblest and lowliest, including the afflicted in mind and body" ([2], p. 201). In his rejoinder, Fletcher argues that there is an important distinction between "mere life" and personality, and orthodox Christians do not regard life itself as sacred. He further dismisses the sanctity of life doctrine as "a form of vitalism or naturalistic philosophy" ([3], p. 206), and then goes on to argue against Willard Sperry, who called on Albert Schweitzer's "reverence for life" as a basis for rejecting euthanasia. Fletcher suggests that such a notion is Hindu or Buddhist, not Christian.

The essays by Sperry and Schweitzer are the oldest ones to which reference is made in the current literature on the sanctity of life. It is also of interest that Schweitzer sees himself as being somewhat innovative, describing in his autobiography how he "struggled to find the elementary and universal conception of the ethical which [he] had not discovered in any philosophy" and how the phrase "reverence for life" came to him in an inspirational flash (quoted in [1], p. 988). Fletcher's criticism of the sanctity of life as naturalistic and vitalistic, and his attribution of an Eastern religious origin, also indicate that in 1951 the principle may not yet have been as closely associated with the Christian and Western traditions as it is today. However, Fletcher is notoriously unreliable when he attempts to characterize what is or is not traditionally Christian, and his attempt to disassociate the phrase "sanctity of life" from Christian tradition may also have been a part of his broader rhetorical strategy of criticizing that tradition and advancing his particular form of humanism in the name of Christianity. Since the essay by Bonnell and the earlier books by Willis and Klarmann seem to point toward results that contradict those of Fletcher, we cannot come to a definitive conclusion regarding how closely the phrase was identified with a traditional Christian ethic in the first half of the twentieth century.

For representative arguments in favor of the sanctity of life, the reader may consider the references below by Barth, Jakobovitz, John-Stevas, Ramsey, Shils, Callahan, and Boyle. For representative criticisms, one may consider the writings of Fletcher, Williams, Clouser, Singer, and Kuhse. An outstanding bibliography can be found in the book by Kleinig.

ANNOTATED BIBLIOGRAPHY

ARTICLES

1. Sperry, W.: 1948, 'Moral problems in the practice of medicine with analogies drawn from the profession of the ministry', *The New England Journal of Medicine* **239**(26), 985–990.

   The Dean of the Harvard Divinity School looks at issues in medicine. In a section on euthanasia, he argues that the professions of the minister and doctor both presuppose a "feeling for the sacredness of life".

2. Bonnell, J.: 1951, 'Sanctity of human life', *Theology Today* **8** (May–July), 194–201.

   The author criticizes euthanasia because it undermines the belief in the sanctity of life, and asserts that "Christianity has never ceased to emphasize the sanctity of human life and the value of the individual, even the humblest and lowliest".

3. Fletcher, J.: 1951, 'Our right to die', *Theology Today* **8** (May–July), 202–212.

   Distinguishing between "personality" and "mere life", the author argues that "there is no cause, in a rational or Christian outlook, to regard life itself as sacrosanct". He criticizes as "vitalists" those who argue against euthanasia.

4. Daube, D.: 1967, 'Sanctity of life', *Proceedings of the Royal Society of Medicine* **60**(11), 1235–1240.

5. Dunstan, G.R.: 1967, 'Sanctity of life: a professional ethics not enough', *Proceedings of the Royal Society of Medicine* **60**(11), 1240–1241.

6. Medawar, P.: 1967, 'Sanctity of life', *Proceedings of the Royal Society of Medicine* **60**(11), 1242–1243.

7. Callahan, D.: 1968, 'The sanctity of life' (with commentaries by J. Pleasants, J. Gustafson and H. Beecher and a reply by Callahan), in D. Cutler (ed.), *Updating Life and Death: Essays in Ethics and Medicine*, Beacon Press, Boston, pp. 181–250.

Callahan looks to the principle of the sanctity of life as the basis for moral consensus in a pluralistic society, seeking to mediate the religious approach of Paul Ramsey, which focuses on the "alien dignity" of life in God, and the secular approach of Edward Shils, which focuses on a protoreligious experience of the intrinsic value or worth of life.

8. Ten, C.L.: 1969, 'Religious morality and the law', *Australasian Journal of Philosophy* **47**, 169–173.

    Criticizing Basil Mitchell's *Law, Morality and Religion in a Secular Society*, this essay defends the arguments made by Glanville Williams in his *The Sanctity of Life and the Criminal Law*. Religious justifications of law are to be ruled out, and thus cannot play a role in restricting abortion or euthanasia. Religious considerations are relevant, however, in the case of mandatory rules that impose a burden on individuals against their religious convictions.

9. Smith, H.L.: 1971, 'Abortion, death, and the sanctity of life', *Social Science and Medicine* **5**(3), 211–218.

10. Brody, B.: 1973, 'Abortion and the sanctity of human life', *American Philosophical Quarterly* **10**, 133–140.

    The author argues that abortion would not be morally permissible if the fetus has the full rights of a human person.

11. Clouser, K.D.: 1973, ' "The sanctity of life": an analysis of a concept', *Annals of Internal Medicine* **78**(1), 119–125.

    By a careful analysis of the sanctity of life concept, the author concludes that "as it is generally used . . . [it] is only a slogan. . . . There is no place for it in the formal structures of ethics."

12. Pendergast, R.: 1973, 'Sanctity of life', *Annals of Internal Medicine* **78**(6), 979–980.

    A response to Clouser, arguing that the concept of the sanctity of life is an expression of a basic intuition that "human life is somehow related . . . to the mystery that overhangs all finite existence."

13. Maurita, M.: 1973, 'Sanctity of life', *Hospital Progress* **54**(11), 56–59.

14. Fletcher, J.: 1973, 'Medicine and the nature of man', in R. Veatch, W. Gaylin and C. Morgan (eds.), *The Teaching of Medical Ethics*, A Hastings Center Publication, Hastings-on-Hudson, New York, pp. 47–58.

    Distinguishing between human and personal life, and criticizing the idea that human life is sacred, the author develops criteria of personal life such as minimal intelligence, self-awareness, self-control, control of existence, and communication.

15. O'Rourke, K.D.: 1974, 'Rationale and implications of sanctity of life commitment', *Hospital Progress* **55**(2), 57–59.

16. Baier, K.: 1974, 'The sanctity of life', *Journal of Social Philosophy* **5**, 1–6.

    Four interpretations of the sanctity of life are developed: (1) sacred process, where the whole of the life process is sacred; (2) sacred individual, where the natural life span of every individual is sacred; (3) sacred essence, where that which constitutes the uniquely valuable distinguishing feature of humanity is sacred; and (4) sacred essence, where there is a sacred goal to which every rational being is committed. The author looks at what is worth preserving in each of these interpretations.

17. Peden, C.: 1974, 'The "sacred natural process" interpretation', *Journal of Social Philosophy* **5**, 6–8.

As an alternative to Baier's account of the sanctity of life ideal, which is regarded as unrealistic, the author argues that the goal of every person is to lead a life that is optimal, and thus individuals will sometimes advance their own well-being at the expense of others.

18. Jaggar, A.: 1974, 'The sanctity of life as a humanist ideal', *Journal of Social Philosophy* 5, 8–11.

The author shows how the normative conclusions that Kurt Baier seeks to arrive at by means of the concept of the sanctity of life can be reached by other preferable principles. She then claims that Baier's sanctity of life ideal is too narrow, argues for a new and broader interpretation, and shows how this new ideal leads to important but neglected moral considerations.

19. Brill, M.: 1974, 'The sanctity of life', *Jewish Spectator* 39, 60–61.

20. Spring, C.: 1974, 'The ethical and legal implications of genetic screening, counseling, and treatment', *Religious Humanism* 8, 177–182.

Ethical and common law tradition in England and the United States seeks to preserve a delicate balance between the integrity and freedom of the individual and the common good. Looking at the concepts of the quality of life, sanctity of life, and individual freedom and privacy, the author considers the way issues in genetics pose a challenge to the delicate balance.

21. Bokser, B.Z.: 1975, 'Problems in bio-medical ethics: a Jewish perspective', *Judaism: A Quarterly Journal of Jewish Life and Thought* 24 (spring), 134–143.

Evaluations of the legitimacy of biomedical techniques for manipulating the life process should be guided by the respect for the sanctity of life.

22. Farris, S. and James, W.: 1976, 'Reflections on the sanctity of human life', in J. Hinchcliff (ed.), *Religious Dimension*, Rep. Prep.

23. Miles, J.: 1976, 'Jain and Judaeo-Christian respect for life', *Journal of the American Academy of Religion* 44 (September), 453–457.

By a comparison of Jain and Judeo-Christian views, the author argues that the latter tradition subordinates the sanctity of life to the sovereignty of God. The key ethical question then becomes "who speaks for God?" In American democracy it has been the majority, but this is now challenged in the abortion debate.

24. Cobb, W.: 1977, 'Abortion and the sanctity of life', *Encounter* 38 (summer), 273–287.

25. Dixon, S.: 1977, 'On speaking clearly while counseling about life and death', *The Journal of Pastoral Care* 31 (December), 252–263.

The author argues that four different Old Testament understandings of life and sanctity have led to competing positions on the sanctity of life, and Greek metaphysical reflection has influenced Christian understandings of personhood. These metaphysical-theological positions predetermine ethical decisions about matters such as abortion, and the author recommends that counselors be aware of this when working with those who do not share their metaphysics.

26. Singer, P.: 1977, 'Utility and the survival lottery', *Philosophy* 52, 218–222.

Theoretically, the total number deaths could be reduced by sacrificing randomly chosen individuals and using their organs for spare parts. This possibility seems to produce a conflict between utilitarianism and the sanctity of life. The author argues, however, that utilitarianism can account for why such a survival lottery is not justifiable.

27. Koop, C.E.: 1978, 'The sanctity of life', *Journal of the Medical Society of New Jersey* **75**(1), 62–67.

28. McCormick, R.: 1978, 'The quality of life, the sanctity of life', *The Hastings Center Report* **8**, 30.

29. Mills, R.: 1978, 'Family and synagogue: partners in a search for the holy', *Journal of Religion and Health* **17** (April), 130–135.

    The quality of daily living is enhanced by a sacred sense of life, and the family is regarded as the key element in nurturing an awareness of the sanctity of life. The synagogue should provide support and inspiration for the family in this task.

30. Nelson, R.: 1978, 'On life and living: the Semitic insight', *Journal of Medicine and Philosophy* **3**, 129–143.

    The author argues for a Semitic understanding of human life that is unitary in contrast to a Greek dualism, and distinguishes between qualitative dimensions of life (bios, psuche, and zoe) that help in ethical decision-making, and that enable understanding of the quality and sanctity of life.

31. Reich, W.: 1978, 'Quality of life', in W. Reich (ed.), *The Encyclopedia of Bioethics*, vol. 2, pp. 829–840.

    Arguing that both the quality-of-life and sanctity-of-life ethics advance a respect for life, the author distinguishes between them by aligning the former with a consequentialist moral theory and the latter with a deontological one. He then develops the debate over the sanctity of life in terms of the debate over a deontological position.

32. Reines, C.: 1978, 'The self and the other in Rabbinic ethics', in M.M. Kellner (ed.), *Contemporary Jewish Ethics*, Sanhedrin Press, New York, pp. 162–174.

33. Rosner, F.: 1978, 'Use and abuse of heroic measures to prolong dying', *Journal of Religion and Health* **17** (January), 8–18.

    The author argues that "quality of life" embodies a concept of worthiness tied to social status and personal character, and that the use of this in place of "sanctity of life" involves an irreparable harm to society. Instead, humanity should recognize that it does not possess absolute title to human life, and it has the task of preserving, dignifying and hallowing that life.

34. Singer, P.: 1978, 'Value of life', in W. Reich (ed.), *The Encyclopedia of Bioethics*, vol. 2, pp. 822–829.

35. Baier, K.: 1979, 'Technology and the sanctity of life', in K. Goodpaster and K. Sayre (eds.), *Ethics and Problems of the 21st Century*, University of Notre Dame Press, pp. 160–174.

36. Hartt, J.N.: 1979, 'Creation, creativity, and the sanctity of human life', *Journal of Medicine and Philosophy* **4**(4), 418–434.

    The author examines the way in which the principle of the sanctity of life is grounded in Christian doctrines of creation.

37. Adelson, L.: 1980, 'The gun and the sanctity of human life; or the bullet as pathogen', *Pharos* **43**(3), 15–25.

38. Harper, T.D.: 1980, 'State-funded abortions: judicial acquiescence in the sanctity of a physician's medical judgment', *Journal of the Medical Association of Georgia* **69**(4), 313–315.

39. Burke, M.: 1981, 'The sanctity of human life', *Irish Nursing News* (May), 4–6.

40. Kuhse, H.: 1981, 'Extraordinary means and the sanctity of life', *Journal of Medical Ethics* **7**, 74–82.

The essay argues against the viability of the distinction between withdrawing extraordinary vs. ordinary care. Thus the sanctity of life doctrine, which rejects quality of life considerations, either requires that life must be preserved at any cost or it is inconsistent.

41. Barkley, R.: 1982, 'The attitude toward abortion in middle English writings: a note on the history of ideas', *Communio: International Catholic Review* **9** (summer), 176–183.

Attempting to counter the contention that the idea of the sanctity of unborn human life is a modern reaction against liberalism, the author considers the view of abortion in Middle English writings.

42. Fackenheim, E.: 1982, 'The spectrum of resistance during the Holocaust: an essay in description and definition', *Modern Judaism* **2** (May), 113–130.

The Nazi logic of destruction aimed at producing a self-loathing among Jewish victims that would lead to suicide. Resistance in both armed forms and in prayer and song testified to the sanctity of life.

43. Robertson, G.: 1982, 'Dealing with the brain-damaged old – dignity before sanctity', *Journal of Medical Ethics* **8**, 173–177.

Extreme attempts either to preserve or terminate life are regarded as medically, morally and socially unacceptable.

44. Rachels, J.: 1983, 'The sanctity of life', in J. M. Humber and R.F. Almeder (eds.), *Biomedical Ethics Reviews*, Humana Press, Clifton, New Jersey, pp. 29–42.

The author criticizes Eastern and Western approaches to the sanctity of life, and argues for an approach that does justice to the two different senses of "life".

45. Regan, R.: 1983, 'Abortion and the conscience of the nation', *Human Life Review* **11**(2) (reprinted in *Fundamentalist Journal* **3**:1, 19–25).

Commenting on the 1973 U.S. Supreme Court ruling in Roe v. Wade, Regan argues that diminishing the value of the unborn leads to the diminishing of all human life. Society must choose the sanctity of life rather than the quality of life, thus affirming the right to life without which all other rights have no meaning.

46. Sevensky, R.: 1983, 'The religious foundations of health care: a conceptual approach', *Journal of Medical Ethics* **9**, 165–168.

The sanctity of life is regarded as one of nine religious ideas that constitute the ethical foundation and heritage of medical practice and health care.

47. Singer, P.: 1983, 'Sanctity of life or quality of life?', *Pediatrics* **72**(1), 128–129.

Arguing that the acceptance of abortion and of the withholding and withdrawal of life-saving treatment from the terminally ill involve a fundamental movement away from a sanctity of life ethic, the author criticizes the Reagan Administration's attempt to require treatment of severely handicapped infants.

48. Brouwer, M.: 1984, 'Quality of life and the sanctity of his will', *Fundamentalist Journal* **3**(8) (September), 24–25.

49. Buchanan, P.: 1984, 'Life: quality vs. sanctity', *Fundamentalist Journal* **3**(3) (March), 66.

50. Gilmore, A.: 1984, 'The sanctity of life versus quality of life – the continuing debate', *Canadian Medical Association Journal* **130**(2), 180–181.

51. Koop, C.E.: 1984, 'Surgeon General Koop and the fight for the newborn: responsibility of government and Church to promote sanctity of life', *Christianity Today* **28**(5) (March 16), 36–37.

The author discusses federal regulations to protect handicapped infants and suggests that churches can restore a sense of the sanctity of life in society by providing caring alternatives to abortion, infanticide, and euthanasia.

52. Lively, R.: 1984, 'My tears have been my food: on state murder and the sanctity of life', *The Other Side* **155** (August), 11.

53. Moss, T.: 1984, 'The modern politics of laboratory animal use', *Science Technology and Human Values* **9** (Spring), 51–56.

    The author considers the sanctity of life as an underlying concern driving the current political movement towards stronger federal regulation to ensure the welfare of laboratory animals.

54. Anderson, S.: 1985, 'Sanctity of life issues bring a variety of demonstrators to Washington, DC', *Christianity Today* **29**(10) (July 12), 40f.

    An overview of a demonstration by 1300 Christians in which 248 were arrested for civil disobedience. Participants protested against the death penalty, South Africa, abortion, and poverty, all of which were regarded as a violation of the sanctity of life.

55. Hirschfeld, M.J.: 1985, 'Ethics and care for the elderly', *International Journal of Nursing Studies* **22**(4), 319–328.

    The sanctity and quality (or dignity) of life are both unsatisfactory for directing nursing intervention when proxy judgments are necessary. In their place a Jewish ethical perspective is advanced that sees dependence and uncertainty as inevitable.

56. Anonymous: 1985, 'Upholding the sanctity of life' (letter), *Canadian Medical Association Journal* **134**(8), 866, 868.

57. Taylor, S.G.: 1985, 'The effect of quality of life and sanctity of life on clinical decision making', *Aorn Journal* **41**(5), 924, 926, 928.

58. Mayr, H.: 1986, 'Das Prinzip der Heiligkeit des Lebens', *Conceptus* **20**, 39–40.

    After differentiating several concepts of life (human, conscious, non-conscious) and considering the arguments for the positive value of life and against killing, the author provides a minimal formulation of the principle of the sanctity of life.

59. Cahill, L.S.: 1987, 'Sanctity of life, quality of life, and social justice', *Theological Studies* **48** (March), 105–123.

    The ethical issues surrounding new life-prolonging medical technology call for an appropriate balance between two important concerns in Catholic moral theology: (1) the sanctity of life and the responsibility to sustain individuals, and (2) the common good and social justice.

60. Geisler, N.: 1987, 'Sanctity of life', in K. Kantzer (ed.), *Applying the Scriptures*, Zondervan, pp. 139–160.

61. Lower, J.: 1987, 'Teaching self-worth and the sanctity of life', in J. Hoffmeier (ed.), *Abortion: A Christian Understanding and Response*, Baker, pp. 139–148.

62. Mori, M.: 1987, 'The philosophical basis of medical ethics', *Social Science and Medicine* **25**:6, 631–636.

    There are two ethical approaches to medical ethics that are based on two conflicting principles, the sanctity of (human) life and the disposability of mere biological (human) life. The author argues for the latter approach.

63. Norberg, A. and M. Hirschfeld: 1987, 'Feeding of severely demented patients in institutions: interviews with caregivers in Israel', *Journal of Advanced Nursing* **12**(5), 551–557.

A study of 60 health care workers shows that in Israel decisions on forced feeding of senile demented patients is based on a Jewish sanctity of life ethic, which requires that life be preserved.

64. Olson, M.: 1987, 'The sanctity of life: from heresy to hope', *The Other Side* **23** (July-August), 2.
65. Almond, B.: 1988, 'Philosophy, medicine and its technologies', *Journal of Medical Ethics* **14**(4), 173–178.

The author seeks to establish a middle way between a utilitarian, quality of life ethic and a deontological, sanctity of life ethic.

66. Braine, D.: 1988, 'Human life: its secular sacrosanctness', in D. Braine and H. Lesser (eds.), *Ethics, Technology and Medicine*, Avebury, Aldershot, pp. 54–63.

The author seeks to develop a secular basis for why innocence makes a difference to the sacrosanctness of the life in question.

67. Davis, J.: 1988, 'Raping and making love are different concepts: so are killing and voluntary euthanasia', *Journal of Medical Ethics* **14**(3), 148–149.
68. Hostler, J.: 1988, 'The sanctity of life and the sanctity of death', in D. Braine and H. Lesser (eds.), *Ethics, Technology and Medicine*, Avebury, Aldershot, pp. 64–75.

Criticizing the hedonism that assumes the consciousness of death is an evil, the author develops an argument for the sanctity of death that is analogous to arguments for the sanctity of life.

69. Kuhse, H.: 1988, 'Sanctity of life and the role of the nurse', *Australian Nurses Journal* **18**(2), 10–12.
70. Philip, J.: 1988, 'The sanctity of life: the Christian consensus', in I. Brown and N. Cameron (eds.), *Medicine in Crisis: A Christian Response*, Rutherford House Books, pp. 14–25.
71. Singer, P. and Kuhse, H.: 1988, 'Resolving arguments about the sanctity of life: a response to Long', *Journal of Medical Ethics* **14**(4), 198–199.

Thomas Long has argued that there is an irreconcilable metaphysical difference between the sanctity and quality of life ethics, which, in turn, makes it impossible for one side to refute the other. Singer and Kuhse disagree with Long.

72. Sproul, R.: 1988, 'The Christian and the sanctity of life', in J. Boice (ed.), *Transforming our World: A Call to Action*, Multnomah Press, pp. 71–79.
73. Boyle, J.: 1989, 'Sanctity of life and suicide: tensions and developments within common morality', in B. Brody (ed.), *Suicide and Euthanasia*, Kluwer Academic Publishers, Dordrecht, pp. 221–250.
74. Brody, B.: 1989, 'A historical introduction to Jewish casuistry on suicide and euthanasia', in B. Brody (ed.), *Suicide and Euthanasia*, Kluwer Academic Publishers, Dordrecht, pp. 39–75.

The author argues that traditional Jewish casuistry on suicide and euthanasia in not just committed to the sanctity of human life, but seeks to balance several different values of which the great value of human life is just one.

75. Goldworth, A.: 1989, 'The real challenge of "Baby Doe": considering the sanctity and quality of life', *Clinical Pediatrics* **28**(3), 119–122.
76. Singer, P. and Kuhse, H.: 1989, 'The quality/quantity-of-life distinction and its moral importance for nurses', *International Journal of Nursing Studies* **26**(3), 203–212.

Rejecting a sanctity of life ethic, the authors argue that acting in the best interest of patients must involve quality of life judgments.

77. B'edard, J.G.: 1990, 'Geriatric shouting: ethical reflections', *Canadian Nurse* **86**(1), 34–37.

The author considers the debate between a sanctity and quality of life ethic in the context of reflection on ethical issues in geriatric care.

78. Berseth, C.L.: 1990, 'Evaluating the effect of a human values seminar series on ethical attitudes toward resuscitation among pediatric residents', *Mayo Clinic Proceedings* **65**(3), 337–343.

The attitude of residents toward resuscitation of children was surveyed before and after a seminar series on medical ethics. It was found that the resident's willingness to change as a result of the seminar series was correlated with their views on the sanctity of life.

79. Callahan, D.: 1990, 'Current trends in biomedical ethics in the United States of America', *Boletin de la Officina Sanitaria Panamericana* **108**(5–6), 550–555.

The author sets forth the debate over the meaning of the sanctity of life and its opposition to the quality of life as one of the five most important topics in U.S. bioethics.

80. Caralis, P.V.: 1990, 'Withdrawal and withholding of life-supporting food and fluids, one state's struggle', *Journal of the Florida Medical Association* **77**(9), 821–828.

The author argues that the sanctity of life must not be discarded in the rush to acknowledge the quality of life.

81. Dean, H.E.: 1990, 'Political and ethical implications of using quality of life as an outcome measure', *Seminars in Oncology Nursing* **6**(4), 303–308.

This essay evaluates the legitimacy of using quality of life as an outcome measure that plays a role in allocating health care resources. Problems that arise from the opposition between a quality and sanctity of life ethic are considered.

82. Harms, D.L. and Giordano, J.: 1990, 'Ethical issues in high-risk infant care', *Issues in Comprehensive Pediatric Nursing* **13**(1), 1–14.

The authors consider controversy regarding the 'salvaging' of high-risk infants, and regard the sanctity of life as one of many ethical concerns involved in this controversy.

83. Lagaipa, S.J.: 1990, 'Suffer the little children: the ancient practice of infanticide as a modern moral dilemma', *Issues in Comprehensive Pediatric Nursing* **13**(4), 241–251.

The authors challenge nurses to reflect on how the shift from a Judeo-Christian sanctity of life ethic to a new quality of life philosophy will impact their practice.

84. Brewin, T.B.: 1991, 'Sanctity of life', *Medico-Legal Journal* **59**(1), 36–40.

85. Cameron, N.: 1991, 'The seamless dress of Hippocratic medicine', *Ethics and Medicine* **7**(3), 43–50.

The sanctity of life together with philanthropy are understood as constituting the moral fabric of the Hippocratic medical tradition.

86. Kass, L.: 1991, 'Death with dignity and the sanctity of life', in B. Kogan (ed.), *A Time to be Born and a Time to Die: The Ethics of Choice*, Aldine De Gruyter, Hawthorne, NY. R.M. Green responds.

The sanctity of life means that "life is something before which we stand (or should stand) with reverence, awe, and grave respect – because it is beyond us and unfathomable." On this basis the author argues against those who advance euthanasia in the name of "death with dignity".

87. Kuhse, H.: 1991, 'Severely disabled infants: sanctity of life or quality of life?', *Baillieres Clinical Obstetrics and Gynecology* **5**(3), 743–759.

88. Kuuppelomaki, M. and Lauri, S.: 1991, 'Ethical decision making of nurses associated with the feeding of demented patients and terminally ill elderly cancer patients in 7 countries', *Hoitotiede* **3**(4), 146–153.

A seven country international study considered whether nurses working with demented patients and those suffering from terminal cancer would feed a patient who refused to eat. It was found that nurses who fed patients primarily based their decision on the sanctity of life.

89. Cowley, L.T., Young, E. and Raffin, T.A.: 1992, 'Care of the dying: an ethical and historical perspective', *Critical Care Medicine* **20**(10), 1473–1482.

A historical overview of the sanctity and quality of life, arguing that medieval Christianity suppressed the debate by giving priority to the sanctity of life. Ancient Greeks and Romans gave greater weight to the quality of life.

90. Geisler, N.: 1992, 'The natural right', in W.B. Ball (ed.), *In Search of a National Morality: A Manifesto for Evangelicals and Catholics*, Baker Book House, Grand Rapids, and Ignatius Press, San Francisco, pp. 112–128.

The author seeks to develop a natural law justification of the sanctity of life and argues against abortion on this basis.

91. Hornett, S.: 1992, 'The sanctity of life and substituted judgement: the case of Baby J', *Ethics and Medicine* (Summer), 2–5.

A consideration of the ruling of the English Court of Appeal in the case of Baby J, questioning the court's use of quality of life assessment and substituted judgment.

92. May, W.: 1992, 'The sanctity of human life', in W.B. Ball (ed.), *In Search of a National Morality: A Manifesto for Evangelicals and Catholics*, Baker Book House, Grand Rapids, and Ignatius Press, San Francisco, pp. 103–111.

Rooting the sanctity of life in the creation of humanity in the image of God, the author argues against all forms of intentional killing and contraception.

93. Poliwoda, S.: 1992, 'Criteria for determining death in Jewish religion', *Diskussionsforum Medizinische Ethik* (April), v–vii.

In Jewish thought, life ends with the final breath. This understanding is based in part on the sanctity of life.

94. Schlomann, P.: 1992, 'Ethical considerations of aggressive care of very low birth weight infants', *Neonatal Network* **11**(4), 31–36.

The author seeks to move beyond the sanctity versus quality of life debate, and account for the complexities in addressing the aggressive treatment of at-risk infants.

95. Zwierlein, E.: 1992, 'The quality-of-life doctrine and the sanctity-of-life', *Wiener Medizinische Wochenschrift* **142**(23–24), 527–532.

When the "quality of life" is used in place of sanctity of life to define the distinct quality of human life rather than just external circumstances, then it can become a basis for inappropriate discrimination.

96. Weldon, G.: 1993, 'The sanctity of life – and death', *Imprint* **40**(2), 88.
97. Callahan, D. *et al.*: 1994, 'The sanctity of life seduced: a symposium on medical ethics', *First Things* **42** (April), 13–27.

   Callahan argues that technology has seduced and distorted medicine and unless this seduction is resisted we will not be able to resist the movement toward euthanasia and assisted suicide. Others respond to his essay.
98. Strong, C.: 1994, 'What is the "inviolability of persons"?', in R. Blank and A. Bonnicksen (eds.), *Medicine Unbound: The Human Body and the Limits of Medical Intervention*, Columbia University Press, New York.

   The inviolability-of-persons view is regarded as problematic because of the rigidity with which it adheres to rules and fails to consider the plurality of values, but the sanctity of life should be given prima facie status along with other values.

BOOKS OR JOURNAL ISSUES DEVOTED TO THE SANCTITY OF LIFE

99. Taylor, J.: 1667, *The great exemplar of sanctity and holy life according to the Christian institution: described in the history of the life and death of the ever-blessed Jesus Christ . . . : with considerations and discourses upon the several parts of the story, and prayers fitted to the several mysteries, in three parts*, Printed by J. Flesher for Richard Royston, London.
100. 1671, *The Christian's crown of glory, or, Holiness the way to happiness: shewing the necessity of sanctity, or a Holy life, from a serious consideration of the life of the Holy Jesus, who is Christ our sanctification: also a plain discovery of the formalist or hyppocrite: together with the doctrine of justification opened and applied*, Printed for Tho. Passenger, London (a microform reproduction of the original in the Bodleian Library).
101. Klarmann, A.: 1915, *The Crux of Pastoral Medicine: The Perils of Embryonic Man* (5th ed.), F. Pustet, New York.

   Seeking to limit "pastoral medicine" to abortion and fetal care, the author will prohibit direct killing on the grounds that "human life is sacred, that is, that it has a value set upon it by God, who created it for supernatural purpose."
102. Willis, N.D.: 1921, *The Better America Lectures*, The Better America Lecture Service, Inc., New York.

   The sanctity of life is developed as the basis of civilization, which protects human life and property.
103. Young, D.T.: 1932, *The Sanctity of Life: Secular and Mid-week Sermons*, Epworth Press, London.

   The author considers the way in which secular matters of life can be regarded as sacred.
104. Schweitzer, A.: 1946 (1923), *The Philosophy of Civilization, Vol.2: Civilization and Ethics*, C.T. Campion (tr.), A&A Black, London.

   Arguing for an ethically based worldview that prefers existence to nonexistence and thereby affirms the intrinsic value of life, the author develops the reverence for life as the foundation of ethics. From that approach to life arises the impulse to raise existence to its highest level of value, and this process of progress constitutes true civilization.

105. Fletcher, J.: 1954, *Morals and Medicine*, Princeton University Press, Princeton, New Jersey.

A sustained criticism of the sanctity of life based on the contention that "self-consciousness marks the frontier between thou and it". Where such self-consciousness or personal life is absent, the life in question need not be accorded the protection or privileges that are given to persons.

106. Harnack, A.: 1957 (1900), *What is Christianity?*, T. B. Saunders (tr.), Harper & Row, New York.

The fatherhood of God, brotherhood of man, and infinite value of the human soul are seen as the essence of Christianity. The author elaborates on the "supernatural value" of individuals and their status as the eternal in time.

107. Hirschfeld, G.: 1957, *An Essay on Mankind*, Philosophical Library, New York.

In order to counter the fragmentation of humanity into opposing groups, the author seeks to develop a unitary conception of humanity based on the sanctity of human life, the opposition of human life to nonhuman life, and the priority of human interests over all others.

108. Williams, G.: 1957, *The Sanctity of Life and the Criminal Law*, Knopf, New York.

Drawing heavily on the thought of Fletcher, the author addresses issues in contraception, abortion, infanticide, suicide and euthanasia.

109. Jakabovits, I.: 1959 (updated 1975), *Jewish Medical Ethics*, Bloch Publishing Co., New York.

The author develops the Jewish view on the sanctity and infinite value of human life. "Infinity being indivisible, any fraction of life, however limited its expectancy or its health, remains equally infinite in value." As a result euthanasia is illicit.

110. Barth, K.: 1961, *Church Dogmatics*, Vol. III/4, par. 55, T. and T. Clark, Edinburgh, Scotland.

Criticizing Schweitzer's "reverence for life" as naturalistic and vitalistic, the author argues that Christians should value life as God's gift. The reverence for life is then the pause before that mystery or plenitude of human life, which indicates its transcendent origin and calls for a distance and respect. Barth also further develops the "reverence for life" (thou shalt not – the negative) into the "treasuring of life" (affirmation of life – the positive).

111. Daly, C.B.: 1962, *Moral Law and Life; an Examination of the Book: The Sanctity of Life and the Criminal Law*, Clonmore and Reynolds, Dublin.

A strong criticism of Glanville Williams from a Roman Catholic perspective.

112. Jackson, D. MacGilchrest: 1962, *The Sanctity of Life*, Tyndale Press, London.

Based on an address given at a meeting of the Christian Medical Fellowship on March 14, 1962.

113. John-Stevas, N.St.: 1963, *The Right to Life*, Holt, Rinehart and Winston, New York.

On the basis of the Christian view that "life is not at the absolute disposal of the holder but is a gift of God in whose control it lies", the author argues that there is a limit on what one can do to relieve suffering. He considers issues at the beginning of life, end of life, capital punishment and just war theory.

114. Rosenberg, L.: 1966, *Is the Sanctity of Life Absolute?*, Friends Book Centre, London. Reprinted from *The Friends' Quarterly*, v. 15, no. 7, 1966.

    The author defends the sanctity of life and argues against euthanasia, but argues that treatment should be withheld from those who are overmastered by disease.

115. Ramsey, P.: 1967, *The Sanctity of Life: In the First of It*, reprinted from the Dublin Review, Spring 1967.

116. Shils, E. (ed.): 1968, *Life or Death: Ethics and Options*, Reed College, Portland, Oregon.

    The papers included in this volume were presented at a symposium on the sanctity of life held March 11–12, 1966 at Reed College, Portland, Oregon. The essays by Shils and Ramsey are especially influential on later literature on the sanctity of life.

117. Kindregan, C.: 1969, *The Quality of Life: Reflections on the Moral Values of American Law*, The Bruce Publishing Company, Milwaukee.

    The author argues that the law best advances the quality of life when it upholds the sanctity of life.

118. Callahan, D.: 1970, *Abortion: Law, Choice, and Morality*, The Macmillan Co., New York.

    The author incorporates the arguments of his 1968 essay "The Sanctity of Life" into his book as a basis for establishing moral policy on abortion.

119. Kaplan, A.: 1970, *Individuality and the New Society*, University of Washington Press, Seattle.

    This is volume 2 of the Sanctity of Life series, of which the Shils, 1968 book was volume 1. This is a collection of essays from a conference held at Reed College, addressing social conditions and individuality in the United States from 1960 to the time of the book's publication.

120. Ramsey, P.: 1970, *The Patient as Person*, Yale University Press, New Haven and London.

    The author introduces a deontological dimension into medical ethics, countering teleological, consequence-based ethical reflection such as that of Fletcher. The sanctity of life is associated with the deontological constraints. "Only a being who is a sacredness in the social order can withstand complete dominion by 'society' for the sake of engineering civilizational goals."

121. Hill, B.: 1973, *Education and the Endangered Individual: A Critique of Ten Modern Thinkers*, Teachers College Press, New York.

    Education seeks to foster individuality, but the scope of that individuality has recently been narrowed. This book emphasizes the sanctity of the individual as a check against the inroads of society on individuality.

122. Kohl, M.: 1974, *The Morality of Killing: Sanctity of Life, Abortion and Euthanasia*, Peter Owen, London.

123. Brody, B.: 1975, *Abortion and the Sanctity of Human Life: A Philosophical View*, MIT Press, Cambridge, Massachusetts. (Reviewed by Curley, Edmund: 1976, in *The Review of Books in Religion* 5, 9.)

    Arguing that the fetus becomes a human being before birth, the author concludes that only in very rare circumstances is abortion justifiable and that there should be laws against it.

124. Crane, D.: 1975, *The Sanctity of Social Life: Physicians' Treatment of Critically*

Ill Patients, Russell Sage Foundation, New York. (Reviewed by Smith, David: 1976, in *The Hastings Center Report* **6** [June], 31–34.)

125. Glover, J.: 1977, *Causing Death and Saving Lives*, Pelican Books, London.

   The author develops the equivalence of "taking life is intrinsically wrong" and the "sanctity of life", and he argues against the principle. He seeks, however, to maintain a kernel of what is implied by the principle, namely that taking life is normally directly wrong.

126. Grisez, G. and J. Boyle: 1979, *Life and Death with Liberty and Justice: A Contribution to the Euthanasia Debate*, University of Notre Dame Press, Notre Dame.

   The definition of death, refusing treatment, and euthanasia are considered from both jurisprudential and ethical viewpoints, with the former based on liberty and justice and the latter on the sanctity of life. Consequentialist approaches are criticized.

127. Keyserlingk, E.: 1979, *Sanctity of Life or Quality of Life in the Context of Ethics, Medicine and Law: A Study*, Law Reform Commission of Canada, Ottawa, Canada.

   Providing an overview of the sanctity and quality of life principles, the author argues that (1) public policy should protect the sanctity of life and euthanasia should be illegal, but (2) death of persons should be established by a quality of life criterion (brain death), and thus merely biological life need not be protected.

128. Wallis, Jim: 1980, *Abortion: A Convergence of Concern*, Sojourners, vol. 9

129. Saltzman, S.: 1982–83, *The Sanctity of Life in Jewish Law*, Thesis (D.H.L.), Jewish Theological Seminary of America, 1982. Microfiche, 1983, University Microfilms, Ann Arbor, Michigan.

130. Wurzburger, W.: 1983, *The Quality and Sanctity of Life*, Sound Cassette, National Interfaith Coalition on Aging, Athens, GA.

   This is a lecture by Dr. Walter Wurzburger, rabbi and president emeritus of Synagogue Council of America, and he speaks about euthanasia, abortion, suicide, living wills and other questions from a Jewish ethical and religious perspective.

131. Channer, J.H. (ed.): 1985, *Abortion and the Sanctity of Human Life*, Paternoster, Exeter. (Reviewed by Jacob, W.M.: 1986, in *Theology* **89**, 407–408; by Rodd, C.S.: 1986, The Expository Times **97**:10, 290–291; by Biggar, Nigel: 1989, in *Churchman: Journal of Anglican Theology* 103:4, 377–379.)

   A collection of essays evaluating the historical, religious, and philosophical dimensions of the abortion debate and generally arguing for a pro-life position.

132. Sanctity of Life Task Force, The Lutheran Church Missouri synod, Board for Social Ministry Service: 1986, *Sanctity of Life Resource Guide*, Board for Social Ministry Services, The Lutheran Church Missouri Synod, St. Louis, MO.

133. Board of Social Ministry, Lutheran Church, Missouri Synod: 1987, *Social Ministry Planning Manual*, Board of Social Ministry, St. Louis, MO.

   Has a section on the sanctity of life that comes just before another section on health and healing.

134. Fraser, D. (ed.): 1987, *The Sanctity of Life: Conference, June, 1987*, Evangelical Round Table (3rd.) , Eastern College, Eastern Baptist Theological Seminary, St. Davids, PA.

   This overviews the pro-life movement in the United States, together with a discussion of religious aspects of the abortion and suicide debates and just war theory.

135. Kuhse, H.: 1987, *The Sanctity-of-Life Doctrine in Medicine: A Critique*, Oxford University Press, New York/Clarendon. (Reviewed by Kaye,Bruce: 1989 in *St Mark's Review* **137** [Autumn], 38–39.)

Focusing on the distinction between killing and letting die, the author challenges the consistency of those who would simultaneously affirm (1) that all human lives are equally valuable and inviolable, irrespective of quality, and (2) there is a limited duty to preserve life, thus it is appropriate to withhold life-sustaining treatment.

136. Lamb, D.: 1988, *Down the Slippery Slope: Arguing in Applied Ethics*, Croom Helm, London.

Those calling on the sanctity of life in debates on abortion, euthanasia, and genetic engineering often appeal to a slippery slope argument. This book argues that the moral philosopher should not dismiss such argumentation.

137. Mason, J.K.: 1988, *Human Life and Medical Practice*, Edinburgh University Press, Edinburgh.

A discussion of the historical development in the United Kingdom from a sanctity of life to a quality of life stance. There are chapters on terminal illness, euthanasia, suicide, brain damage and death, neonaticide, abortion and embryocide.

138. Baird, R. and Rosenbaum, S. (eds.): 1989, *Euthanasia: The Moral Issues*, Prometheus Books, Buffalo, NY.

In this collection there is an article by C. Everett Koop, 'Sanctity of life versus quality of life', and one by Joseph Fletcher on 'The wrongfulness of euthanasia'.

139. Swaby-Ellis, and Dawn, E.: 1990, *Sanctity of Life: 6 Studies for Individuals and Groups*, InterVarsity Press, Downers Grove, Illinois.

140. Swindoll, C.: 1990, *Sanctity of Life: The Inescapable Issue*, forward by James Dobson, Word Publishing, Dallas, TX.

141. Kleinig, J.: 1991, *Valuing Life*, Princeton University Press, Princeton, New Jersey.

An extensive review and evaluation of the different ways in which life can be valued.

142. Dworkin, R.: 1993, *Life's Dominion: An Argument About Abortion, Euthanasia, and Individual Freedom*, Alfred A. Knopf, New York.

The author argues that the abortion debate does not just rest on whether the fetus is a person; it depends on the implicit assumption that human life has intrinsic, sacred (or inviolable) value. This value is not associational but genetic, and differences over the legitimacy of abortion depend on how one weighs the two modes of creative investment (the natural and human) into that life.

143. Privitera, S. *et al.*: 1993, *The Sanctity of Life: Physical, Cultural and Religious Problems*, Quaderni Di Bioethica E Cultura, Instituto Siciliano di Bioethica.

This is a collection of articles on the meaning of life, especially its mystery and sanctity. There are essays on the Muslim perspective, and on the meaning and sanctity of aging.

Department of Philosophy and Center for Bioethics
University of South Carolina
Columbia, South Carolina
USA

NOTES ON CONTRIBUTORS

Kurt Bayertz is Professor of Philosophy at the University of Münster, Germany.

Dieter Birnbacher is Professor of Philosophy at the University of Dortmund, Germany.

H. Tristram Engelhardt, Jr., Ph.D., M.D., is Professor, Department of Medicine, as well as Community Medicine and Obstetrics and Gynecology, Baylor College of Medicine; also Professor, Department of Philosophy, Rice University; Adjunct Research Fellow, Institute of Religion and Member, Center for Medical Ethics and Health Policy, Houston, Texas.

Björn Haferkamp is Assistant at the Department of Philosophy, University of Bremen, Germany.

Martin Hailer is Research Assistant at the Ecumenical Institute, University of Heidelberg, Germany.

Martin Honecker is Professor of Systematic Theology and Social Ethics (evangelische Fakultät, University of Bonn) and Member of the Academy of Sciences of Northrine-Westphalia, Germany.

Ludger Honnefelder, Dr. phil., is Professor and Director, Department of Philosophy, University of Bonn, Germany, and Director of the Institute for Science and Ethics at Bonn.

James F. Keenan, S.J., S.T.D., is Associate Professor of Moral Theology, Weston School of Theology, 3 Phillips Place, Cambridge, Massachusetts 02138, U.S.A.

George Khushf, Ph.D., is Humanities Director, Center for Bioethics and Assistant Professor, Department of Philosophy, University of South Carolina, Columbia, South Carolina, 29208, USA.

311

K. Bayertz (ed.), Sanctity of Life and Human Dignity, 311–312.

Helga Kuhse, Ph.D., is Director, Centre for Human Bioethics, Monash University, Clayton, Melbourne, Victoria, Australia.

Anton Leist, Ph.D., MA, is Professor of Philosophy at the University of Zurich, Switzerland. He is director of the Arbeits- und Forschungsstelle für Ethik which is part of the Zurich philosophy department.

Wolfgang Lenzen, Dr. phil., is Full Professor of Philosophy at the University of Osnabrück, Germany.

Volker von Loewenich, Prof. Dr. med., Head of the Department of Neonatology, Zentrum der Kinderheilkunde, Frankfurt a.M., Germany.

John C. Moskop, Ph.D., is Professor of Medical Humanities, East Carolina University School of Medicine, Greenville, North Carolina, USA.

Thomas Petermann, Dr. phil., is Deputy Director of the Technology Assessment Office of the German Parliament; Lecturer on Political Science at the Albert-Ludwig-Universität, Freiburg, Germany.

Mary Rawlinson is Professor of Philosophy, State University of New York at Stony Brook, Long Island, USA.

Dietrich Ritschl, Ph.D., D.D., is Professor of Systematic Theology, and Director of the Ecumenical Institute and Internationales Wissenschaftsforum, University of Heidelberg, Germany.

Stephen Wear, Ph.D., is Clinical Ethicist, State University of New York at Buffalo, USA.

Kevin Wm. Wildes, S.J., Assistant Professor of Philosophy, Department of Philosophy and Senior Research Scholar of the Kennedy Institute of Ethics, Georgetown University, Washington, D.C., USA.

INDEX

Aaron, H. 227
abortion xii, 4, 8f, 13f, 33, 50f, 54, 118,
    131ff, 140, 144f, 152, 179, 201,
    203f, 208, 213, 293, 296
abuse 20, 32, 131
action 20–25, 27, 29–32, 34f, 51, 78, 83,
    87, 127, 162, 164f, 166f, 169, 179,
    184, 186, 212
Age of Enlightenment 257
Alexy, R. 102
allocation 213, 223, 226
altruism 252
Amery, J. 96
anthropocentrism 88
anthropology 79, 140, 250f
Antigone 162, 170
Aquinas, Th. 9f, 15, 140
arbitrariness 125, 132, 263
Arendt, H. 83
Aristotle 95, 146f, 150, 157, 179
Arndt, A. 237
autonomy xiii, xiv, 14, 20, 22, 27, 33,
    77, 80, 82, 87, 89, 112, 116, 125,
    129, 133f, 147, 180f, 192–196,
    212, 221, 226f, 241, 245, 247, 258,
    268

Bacon, F. 75, 86
Baird, E. v. 14
Barth, K. 2, 297
Baumgarten, A. G. 181

Bayertz, K. viii
Beauchamp, T. 245f, 249
Benda, E. 108, 127
benefit 147, 155, 202, 211, 226, 235, 238
Bentham, J. 147, 267
Bernardin, J. 13
Betke, K. 237
bioethics xiii, 107f, 114, 179, 195, 201,
    204f, 212, 242, 244–248, 251,
    263f, 265, 272, 285
biomedicine 241, 243f, 251, 254, 257
biotechnology 126
Birnbacher, D. 45
Blackmun 253
Blackstone, W. 241
Bloch, E. 261
blood-relationship 162, 170–173
Boethius 140f
Bonnell, J. S. 204, 296f
Bowes v. Hardwick case 253
Boyle, J. 297
Brandt, R. 197
Bullock, W. 7

Callahan, D. 12, 205, 222
Calvin, J. 96
capacity 86, 163, 183f, 191f, 194f
Cardinal de Lugo 15
Carnap, R. 149
categorical imperative 110f, 116, 207
causation 32

character 161
Childress, J. 245f, 249
Chomsky, N. 192
Christian theology 73, 203f, 222, 250
Cicero 73, 260
Clouser, K. D. 205, 222, 297
Coleman, G. 5
community 20, 167, 229, 264
Condorcet, M. de 76, 86
consciousness 118ff, 131, 141f, 152, 164f, 166, 168f, 170ff
constitution 79, 117
cooperation 78
critical care 202, 213f, 221–227
Cronin, D. 218
Cruzan, N. B. 206, 242, 252ff, 258
Cruzan v. Director of Missouri Department of Health case 253
Cruzan v. Harmon case 242, 252f

Davidson, D. 197
death 21–24, 30f, 33, 78, 119, 148, 170, 258
decision 22f, 34, 133, 150, 227, 264f
Declaration of Euthanasia 4, 19, 25
democracy 262
deontology 245, 264
desire 35, 76, 181, 183–186, 191
determination 156f, 161, 163f, 167
development 155, 163, 225, 236
Descartes, R. 75, 84f, 86
Devlin, P. 241, 267–270
Diamond, E. 7
Diderot, D. 77, 81, 84
Diskurs-Ethik 234
disposition 119f, 180, 185f
DNA analysis 128–130, 132
Donceel, J. 8
Donum vitae 8
Dürig 124
duty 81, 117, 124, 162, 164ff, 170ff, 177, 180ff, 229, 232, 237, 241, 265, 269

Ehrenreich, B. 174
Eibach, U. 45
Einbecker Empfehlung 237

embryo research 108, 117
Engelhardt, H. T. Jr. 70, 146, 153f, 157, 221–227
English, D. 174
environmental ethics 177ff
Enquête Commission "Opportunities and Risks of Gene Technology" 125, 129
entity 150ff, 163
equality 161, 211, 261, 270
Erlangen Baby xvii, 232
essence of man 108
ethical consensus 272
ethical relativism 258
ethics 19, 109, 117, 140, 149, 212
eugenic indication 131
euthanasia xii, xvi, 4, 7, 13, 22–31, 33, 47–50, 54, 174, 201, 204, 245f, 258, 266, 270, 285, 290f, 293, 296
– non-voluntary euthanasia 27, 29f, 35
– voluntary euthanasia 19ff, 26–31, 47–50, 54, 109, 204
evolution theory 85
existence 110f, 165, 235
extracorporal fertilisation 109

faith 4, 107, 250
family 62–65
fate 77ff, 166
father 63–66
feelings 77f, 87, 150, 161, 172
feminism 63f, 170–173
Fletcher, J. 204, 217
forgiveness 162, 166f
Forster, G. 78
Frankena, W. 205, 222, 252
freedom 14, 79, 86, 110f, 127, 132, 134f, 142, 144, 156, 161, 169f, 172, 184, 227, 257, 259, 262, 267–271
Fuchs, J. 5

gene and reproduction technology 107, 123, 285, 287–290, 296
genetic analysis 126, 128, 131ff
genetic code 129
genetic manipulation 107, 270

genetic testing 124, 127, 132f, 135f
German Basic Law (/Grundgesetz) xviii, 14, 79, 81, 111ff, 117, 123f, 126f, 131, 231, 282–285
German Embryonenschutzgesetz 53, 108
Gert, B. 116
Gethmann, C. F. 146
Glendon, M. A. 13f
Glover, J. 147
God xvi, 3, 5, 9–13, 73, 75ff, 95ff, 140, 142, 193ff, 203, 206, 210, 212, 215, 222, 272
Goethe, J. W. 237
goods 33, 110, 112f, 117, 167, 180, 204, 211, 216, 244f
government 134, 171, 215, 225, 269

Habermas, J. 104
Haferkamp, B. ix
happiness 141f
Hare, R. 153, 156
harm 20f, 26f, 29f, 32, 35, 40, 54, 78, 163, 222, 235, 244, 264, 267–270
Hart, H. L. A. 267–270
Hartmann, K. 157
Hastings Center 11, 251f
health care 111, 202, 209, 215, 225, 248, 254
health policy 92
Hegel G. W. F. 161f, 164–174
Heuss, Th. 136, 260
Hippocratic Oath 33, 235
Hobbes, Th. 31, 97, 101, 103, 266
Hoerster, N. 153
holiness xii, xv, xviif, 250
Honnefelder, L. 197, 227, 232
honor 79, 260
Hufeland, Ch. W. 237
human genus 80–83, 88
human nature 78, 87, 181, 194, 212, 223, 260
human species xiv, 114f, 230
humanity 75, 77, 80, 83, 115f, 182, 207, 222, 272
Hume, D. 74, 76, 161, 180, 184–193, 197, 223f

identity 115, 130, 140ff, 148, 150–154, 163, 168, 170
imago-dei-concept xiv, 73f, 96, 98, 250f, 261, 286, 294
individual 20, 81f, 88, 112, 116f, 123ff, 125, 127–130, 133f, 149, 154ff, 170, 204, 209, 212, 214, 231, 233, 245, 248, 257, 260, 269
individuality 78, 108, 142, 154, 205, 232
infanticide 131, 208, 213, 293
infringement 108, 112f, 134, 145
Innocence III. 75
institution 78, 135, 269f
insurance 134f
integrity 113, 117, 130, 163, 205, 271
intention 32f, 34, 112, 164
interests 11, 40ff, 108, 111, 114, 116f, 124f, 129, 132, 135, 146ff, 150, 153, 164, 198, 204, 264f, 270f
intuitions 211f
inviolability xi, 1, 3f, 7, 14, 60, 144, 156, 203, 229, 258f, 262, 264
in-vitro-fertilisation 51–54, 135
Irenaeus 250
Irigaray, L. 171

Jakobowitz, I. 297, 307
Jefferson, Th. 100
John-Stevas, N. St. 297
Jonas, H. 178, 182
Jonsen, A. 246, 249
judgement 108, 161ff, 165f, 173f, 184, 191, 211, 266, 269
jus naturale 210
justice xiii, xv, 97, 129, 147, 149, 183, 192, 221, 226f

Kant, I. xiv, 77, 81, 95, 98f, 103, 110f, 116, 141–146, 153, 161, 177, 180–184, 194f, 197, 207–210, 217, 222, 230f, 245, 261f
Katz, J. 57
Keenan, J. F. , S.J. 107, 222
Keown, J. 33
Klarmann, A. 295, 297
Kleinig, J. 297

Koch, T. 47, 235
Kommers, D. 14
Kripke, S. A. 150
Kuhlmann, A. 232
Kuhse, H. vii, 7f, 13, 91, 229–232

Lamb, D. 46, 117
language 171, 191f, 216
law 4, 19, 35, 79, 109, 117, 143, 162–165,
    170, 203, 213, 215, 241, 259,
    265–271
– constitutional law xivf, 79, 99ff
– moral law 181, 230, 241, 266
– natural law 77, 99, 107f, 210, 241,
    244f, 257
– positive law 15, 144, 266
Leibniz, G. W. 197
Leist, A. 52
legal enforcement of morals 267ff
legal moralism 268
legislation 124, 128, 266, 270
Lewis, C. 148
Lewis, D. T. 203
liberalism 110
liberty 20, 22, 27, 33, 110f, 113, 117, 119,
    134, 192, 211, 245, 247, 262
life 2, 12, 19f, 28, 32, 74, 78, 109, 115,
    117, 123, 131, 150, 161f, 165ff,
    171f, 203f, 206, 213, 215f, 222f,
    226f, 233, 253, 257f, 263f, 271f
– beginning of life 150, 154, 156f, 202
– equality of life 2, 7, 57f
– good life 145, 183, 191, 193, 249
– medical end-of-life decisions 21–24,
    31–34, 36, 150, 242
– protection of life 109, 222, 263
– quality of life 202, 214, 216
– respect for life 11, 205f, 272
Locke, J. 97, 101, 103, 110, 141ff, 148
Lockwood, M. 150
Long, Th. 13
Looman C. W. N. 34
Lyotard, J.-F. 241

MacIntyre, A. 243
Mackie, J. L. 196

MacQuarrie 11
Manetti, G. 73f
Marquis, D. 50
McCormick, R. 7
medical care 254
medical doctor 26, 28, 31f, 34, 92, 131,
    270
medical ethics xii, 92ff, 101, 104, 135f,
    139, 251
Merkel, A. xviii
metaphysics 140, 146, 149, 153, 156,
    161ff, 210, 230
Mill, J. St. 20, 75, 190, 267–270
mind-body-problem 163, 204
Minkowski, H. 149
Missouri Supreme Court 242
Mitchell, B. 269
Moore, G. E. 177ff, 182, 196
Montaigne 161
moral authority 12, 212f, 215, 223, 226,
    247, 249, 257f, 265
moral casuistry 111, 242, 244, 246f, 249
moral decay 268f
moral imperative 166
moral paternalism 268, 270
moral Populism 269
moral theology 1f, 5, 96, 213, 229
morality xi, 20, 77, 103, 116, 156, 167,
    194, 207, 209f, 211f, 230, 234,
    241f, 249, 258f, 265–271, 283

National Commission for the Protection
    of Human Subjects 249
National Socialism 267
nature xiv, 75–88, 108, 115, 123, 140f,
    147, 153, 156, 172, 179, 207, 212,
    247, 259
needs 116, 225
neonatologist 229, 232f, 235
Nietzsche, F. 166
Noonan, J. 8
norms xv, xix, 76, 79, 113, 123, 167ff,
    173, 264f, 268, 272f

object 80, 86ff, 124ff, 129, 132, 152, 178,
    185

objectivity 82
ontology 149f, 153, 178

Palazzini 3
Parfit, D. 140, 148f, 151, 156
Pascal, B. 75
paternalism 147
patient 6f, 11, 20f, 23–36, 59f, 67, 92f, 111, 139, 148f, 202, 214, 216, 225, 232f, 235, 237f, 246f, 252, 257f, 264f, 270
Patzig, G. 156
perfectibility 76f, 80, 86
persecution 110f, 113
person 5, 19, 53, 85, 119, 125, 139–144, 146, 148–157, 161–164, 166, 177, 181f, 183f, 189, 191, 194, 197, 204f, 208, 213f, 215f, 223, 226f, 230, 237f, 247, 252, 262ff, 271
personal identity 141, 150f, 153f, 163, 251
personality 126f, 128f, 130, 143, 205, 230
Petrarca, F. 73
Picco della Mirandola, J. 73, 75, 96, 261
Pijnenborg, L. 34
Pinkard, T. 162, 169, 172f
Pius XII. 214
Plato 95, 150
Ploch, M. xvii
pluralism 109, 130, 243f, 257f, 269, 271f
politics 109, 117, 124, 265f
Polyneices 170
post-modernity 212, 241f, 247, 263
potentiality 115f, 119f, 149, 152, 154f, 208
Pound, R. 91
predisposition 126f, 131, 133
preference 148, 198, 211, 230f, 258
preference utilitarianism 149, 211, 230, 245
prenatal diagnostic 130, 133
privacy 14, 111, 126f, 129, 133, 136
property 115, 118ff, 162, 178, 222, 252
prosperity 211
protection of life xiv, 109

Protestant Reformation 209
public policy 20f, 28, 30, 202, 213
punishment 110, 267f

Quinn, K. 6f
Quinn, W. V. O. 149

Rachels, J. 47
Ramsey, P. 2, 6, 204, 222, 297
rationality 75, 77, 82, 87, 89, 181f, 191, 208–211, 222, 244, 247f, 249
Rawls, J. 97, 104, 147, 177, 180f, 183f, 191f, 195ff, 223, 227
reciprocity 152, 162, 165
Reich, W. 6f
Reiter, J. 54
respect 146, 161f, 215, 222, 226f
responsibility xiii, 12, 14, 30, 87, 141, 143, 145, 150, 156, 161f, 174, 222, 225, 227f, 250, 263, 265
rights 101ff, 109ff, 112f, 115, 119, 127, 130–135, 142, 144, 167, 181, 183, 206, 264f, 271f
 – animal rights 12ff, 223
 – elementary rights 108
 – fundamental rights 231f, 271
 – human rights 79, 81, 111, 113, 117, 144f, 146f, 259f, 262f, 265, 267, 272
 – minimal rights 109, 112, 115, 119f
 – negative/positive right 110
 – right to life 7, 109, 132, 139, 142, 204, 229, 231f, 264
Rousseau, J.-J. 76
Runggaldier, E. 151
Ryder 113

Sallust 3, 10, 15
Sartre, J. P. 197
Schiller, F. 110
Scholastic Theology 251
Schopenhauer, A. 110
Schwartz, W. B. 227
Schweitzer, A. 42, 217
science xix, 39, 81–88
secular state 242, 248f

security 135, 211, 265
self-certainty 164ff, 170
self-determination 77f, 80, 87, 124, 127, 136, 147, 166, 171, 205, 263f
self-interest 141
self-respect 110f, 117, 119
self-unfolding 80ff, 87f
Seneca 95f
sensibility 13, 211
slippery-slope argument 26, 35, 48, 238
Shils, E. 1, 12
Short, C. 203
Simons, P. 151
Singer, P. vii, 8, 13, 43, 49, 54, 105, 114, 147f, 151, 155ff, 194, 197, 217, 229–232
Skinner, B. F. 85f
Skowronek, H. viii
Smith, H. 204
Smuts, J. 104
society 6, 13, 21, 33, 79, 107, 117, 131, 222, 232f, 267–270
solidarity 78, 156, 248, 250
Spaemann, R. 96, 234
species 12, 114f, 142, 147f, 177, 205
speciesism 113, 144, 148
Sperry, W. 217
Spiegelberg, H. 109
state 20, 112f, 128, 162
St. John-Stevas, N. 204
St. Paul 2
Steinfels, M. 13
Stevens 253, 255
Strawson, P. F. 150ff
Suarez, de Vitoria 15
subject 11, 80, 82, 88, 115, 130, 140f, 143f, 150ff, 156, 162, 166ff, 171
subjectivity 73, 77, 80–88, 119, 148, 163, 171, 177
suicide 9f, 20, 22, 24, 33, 293, 296
Supreme Court of Missouri 242, 252, 258
surrogate motherhood 135

Taylor, J. 296
technology xiff, 4, 78, 81ff, 88, 123, 265
teleology 76, 154, 264
Tooley, M. 148
Toulmin, S. 246, 249
treatment 22ff, 25, 29, 31, 33, 36, 110
tyranny 112ff, 170

utilitarianism 41f, 147, 149, 235
Universal Declaration of Human Rights xiv, 79f, 92, 100
universality 144f, 168, 172, 211, 244, 247, 263
universalizability 166, 171, 173

van Delden, J. J. M. 34
van der Maas, P. J. 34
values xvf, 14, 20, 33, 42f, 76, 101, 107f, 114–118, 123ff, 131–134, 139, 148, 162, 177–180, 182, 184, 186, 188, 192–195, 204, 209f, 214, 216, 221ff, 227, 249, 257ff, 261f, 249, 268ff
violation 124, 126, 131f, 133f, 210

Wade, R. v. 14
wants 185ff, 198
Warnock, M. 52f
Warnock Report 150, 157
Weber, L. 7
Weber, M. 112
Weingart, P. viii
welfare 262f
Wellmer, H.K. viii
Wiggins, D. 151f
Wildes, K. Wm. S.J. viii, 257ff
will 11, 77, 135, 197, 203, 207
Williams G. 2, 204
Willis, N. D. 295, 297
Wolfenden-Report 267
world 74, 88, 161, 179

Young, D. 295

23. E.E. Shelp (ed.): *Sexuality and Medicine*. Vol. II: Ethical Viewpoints in Transition. 1987 ISBN 1-55608-013-1; Pb 1-55608-016-6

24. R.C. McMillan, H. Tristram Engelhardt, Jr., and S.F. Spicker (eds.): *Euthanasia and the Newborn*. Conflicts Regarding Saving Lives. 1987 ISBN 90-277-2299-4; Pb 1-55608-039-5

25. S.F. Spicker, S.R. Ingman and I.R. Lawson (eds.): *Ethical Dimensions of Geriatric Care*. Value Conflicts for the 21th Century. 1987 ISBN 1-55608-027-1

26. L. Nordenfelt: *On the Nature of Health*. An Action-Theoretic Approach. 2nd, rev. ed. 1995 ISBN 0-7923-3369-1; Pb 0-7923-3470-1

27. S.F. Spicker, W.B. Bondeson and H. Tristram Engelhardt, Jr. (eds.): *The Contraceptive Ethos*. Reproductive Rights and Responsibilities. 1987 ISBN 1-55608-035-2

28. S.F. Spicker, I. Alon, A. de Vries and H. Tristram Engelhardt, Jr. (eds.): *The Use of Human Beings in Research*. With Special Reference to Clinical Trials. 1988 ISBN 1-55608-043-3

29. N.M.P. King, L.R. Churchill and A.W. Cross (eds.): *The Physician as Captain of the Ship*. A Critical Reappraisal. 1988 ISBN 1-55608-044-1

30. H.-M. Sass and R.U. Massey (eds.): *Health Care Systems*. Moral Conflicts in European and American Public Policy. 1988 ISBN 1-55608-045-X

31. R.M. Zaner (ed.): *Death: Beyond Whole-Brain Criteria*. 1988 ISBN 1-55608-053-0

32. B.A. Brody (ed.): *Moral Theory and Moral Judgments in Medical Ethics*. 1988 ISBN 1-55608-060-3

33. L.M. Kopelman and J.C. Moskop (eds.): *Children and Health Care*. Moral and Social Issues. 1989 ISBN 1-55608-078-6

34. E.D. Pellegrino, J.P. Langan and J. Collins Harvey (eds.): *Catholic Perspectives on Medical Morals*. Foundational Issues. 1989 ISBN 1-55608-083-2

35. B.A. Brody (ed.): *Suicide and Euthanasia*. Historical and Contemporary Themes. 1989 ISBN 0-7923-0106-4

36. H.A.M.J. ten Have, G.K. Kimsma and S.F. Spicker (eds.): *The Growth of Medical Knowledge*. 1990 ISBN 0-7923-0736-4

37. I. Löwy (ed.): *The Polish School of Philosophy of Medicine*. From Tytus Chałubiński (1820–1889) to Ludwik Fleck (1896–1961). 1990 ISBN 0-7923-0958-8

38. T.J. Bole III and W.B. Bondeson: *Rights to Health Care*. 1991 ISBN 0-7923-1137-X

39. M.A.G. Cutter and E.E. Shelp (eds.): *Competency*. A Study of Informal Competency Determinations in Primary Care. 1991 ISBN 0-7923-1304-6

40. J.L. Peset and D. Gracia (eds.): *The Ethics of Diagnosis*. 1992 ISBN 0-7923-1544-8

41. K.W. Wildes, S.J., F. Abel, S.J. and J.C. Harvey (eds.): *Birth, Suffering, and Death.* Catholic Perspectives at the Edges of Life. 1992 [CSiB-1]
ISBN 0-7923-1547-2; Pb 0-7923-2545-1

42. S.K. Toombs: *The Meaning of Illness.* A Phenomenological Account of the Different Perspectives of Physician and Patient. 1992
ISBN 0-7923-1570-7; Pb 0-7923-2443-9

43. D. Leder (ed.): *The Body in Medical Thought and Practice.* 1992
ISBN 0-7923-1657-6

44. C. Delkeskamp-Hayes and M.A.G. Cutter (eds.): *Science, Technology, and the Art of Medicine.* European-American Dialogues. 1993   ISBN 0-7923-1869-2

45. R. Baker, D. Porter and R. Porter (eds.): *The Codification of Medical Morality.* Historical and Philosophical Studies of the Formalization of Western Medical Morality in the 18th and 19th Centuries, Volume One: Medical Ethics and Etiquette in the 18th Century. 1993   ISBN 0-7923-1921-4

46. K. Bayertz (ed.): *The Concept of Moral Consensus.* The Case of Technological Interventions in Human Reproduction. 1994   ISBN 0-7923-2615-6

47. L. Nordenfelt (ed.): *Concepts and Measurement of Quality of Life in Health Care.* 1994 [ESiP-1]   ISBN 0-7923-2824-8

48. R. Baker and M.A. Strosberg (eds.) with the assistance of J. Bynum: *Legislating Medical Ethics.* A Study of the New York State Do-Not-Resuscitate Law. 1995   ISBN 0-7923-2995-3

49. R. Baker (ed.): *The Codification of Medical Morality.* Historical and Philosophical Studies of the Formalization of Western Morality in the 18th and 19th Centuries, Volume Two: Anglo-American Medical Ethics and Medical Jurisprudence in the 19th Century. 1995   ISBN 0-7923-3528-7; Pb 0-7923-3529-5

50. R.A. Carson and C.A. Burns (eds.): *The Philosophy of Medicine and Bioethics.* Reprospective and Critical Appraisal. (forthcoming)   ISBN 0-7923-3545-7

51. K.W. Wildes, S.J. (ed.): *Critical Choices and Critical Care.* Catholic Perspectives on Allocating Resources in Intensive Care Medicine. 1995 [CSiB-2]
ISBN 0-7923-3382-9

52. K. Bayertz (ed.): *Sanctity of Life and Human Dignity.* 1996
ISBN 0-7923-3739-5

KLUWER ACADEMIC PUBLISHERS – DORDRECHT / BOSTON / LONDON